Prescribed

PRESCRIBED

Writing, Filling, Using, and Abusing the Prescription in Modern America

Edited by

Jeremy A. Greene

and

Elizabeth Siegel Watkins

The Johns Hopkins University Press
Baltimore

 Published 2012
Printed in the United States of America on acid-free paper
9 8 7 6 5 4 3 2 1

The Johns Hopkins University Press
2715 North Charles Street
Baltimore, Maryland 21218-4363
www.press.jhu.edu

Library of Congress Cataloging-in-Publication Data

Prescribed : writing, filling, using, and abusing the prescription in modern America / edited by Jeremy A. Greene and Elizabeth Siegel Watkins.
p. ; cm.
Includes bibliographical references and index.
ISBN-13: 978-1-4214-0506-3 (hdbk. : alk. paper)
ISBN-13: 978-1-4214-0507-0 (pbk. : alk. paper)
ISBN-10: 1-4214-0506-7 (hdbk. : alk. paper)
ISBN-10: 1-4214-0507-5 (pbk. : alk. paper)
I. Greene, Jeremy A., 1974– II. Watkins, Elizabeth Siegel.
[DNLM: 1. Prescriptions—history—United States. 2. Government Regulation—history—United States. 3. History, 20th Century—United States. 4. Professional Autonomy—United States. 5. Substance-Related Disorders—history—United States. QV 11 AA1]
LC classification not assigned
615.1'4—dc23 2011029814

A catalog record for this book is available from the British Library.

The Johns Hopkins University Press uses environmentally friendly book materials, including recycled text paper that is composed of at least 30 percent post-consumer waste, whenever possible.

To the memory of Harry M. Marks (1946–2011),
our mentor, colleague, and friend

Contents

Abbreviations

AARP	American Association of Retired Persons (AARP became the complete name of the association in 1999)
AIDS	acquired immune deficiency syndrome
AMA	American Medical Association
AMI	Advocates for Medical Information
ANA	American Nurses' Association
APhA	American Pharmaceutical Association
BCH	Boston City Hospital
BTC	behind-the-counter
CDAPCA	Comprehensive Drug Abuse Prevention and Control Act
CDER	Center for Drug Evaluation and Research (of the FDA)
CRR	Center for Reproductive Rights
DAWN	Drug Abuse Warning Network
DEA	Drug Enforcement Administration
DES	diethylstilbestrol
DESI	Drug Efficacy Study and Implementation
EKG	electrocardiogram
FBN	Federal Bureau of Narcotics
FDA	Food and Drug Administration
FDCA	Federal Food, Drug, and Cosmetic Act
FWHC	Feminist Women's Health Center
GAO	Government Accountability Office
GNP	Gross National Product
HEW	Health, Education, and Welfare, Department of
HIV	human immunodeficiency virus
HMO	Health Maintenance Organization
IMR	Institution for Motivational Research
IPS	interim policy statement

IUD	intrauterine device
IUPAC	International Union of Pure and Applied Chemistry
LSD	lysergic acid diethylamide
NDTI	National Disease and Therapeutic Index
NIDA	National Institute on Drug Abuse
NPA	National Prescription Audit
NPC	National Pharmaceutical Council
NRC	National Research Council
NRTA	National Retired Teachers Association
NWHN	National Women's Health Network
OFP	Office of Family Planning
OTC	over-the-counter
PCS	Prescription Card Services
PDR	*Physicians' Desk Reference*
PMA	Pharmaceutical Manufacturers Association
PPI	patient package insert
PTA	Parent-Teacher Association
sNDA	Supplemental New Drug Application
STI	sexually transmitted infection
UCLA	University of California, Los Angeles
UCSF	University of California, San Francisco
USP	United States Pharmacopeia
WCC	Women's Capital Corporation
WHCS	Women's Health Care Specialist
WHO	World Health Organization

Prescribed

Introduction

The Prescription in Perspective

Jeremy A. Greene and Elizabeth Siegel Watkins

In March 1963, the regional journal *Southwestern Medicine* offered to teach its readers "how to write a prescription in the second half of the twentieth century."[1] For doctors, patients, and pharmacists, the postwar decades had ushered in an era of therapeutic plenty, as novel classes of "wonder drugs" sprang into existence. Popular enthusiasm over these new and modern medicines—from penicillin to antipsychotics to anti-ulcer agents to chemotherapeutic cures for cancer—visibly elevated the status of pharmacotherapy from a relative backwater of therapeutics to a relentlessly modern field that now formed one of the principal dimensions of the therapeutic encounter. As such, noted the article's author, Texan family physician W. E. Lockhart, the modern prescription pad was now as potent a symbol of the modern physician as was the stethoscope, and physicians neglected the prescription at their own peril. "Not rarely," he quipped, "this is the only tangible reward which the patient has in return for the green stuff he gives your secretary."[2]

Although the prescription had been a staple of therapeutic encounters for millennia, the modern prescription of Lockhart's concern stood as a proxy for understanding several key changes in the form and content of medical practice taking place in the mid-twentieth century. An entirely new kind of commodity—the prescription-only nonnarcotic drug—had come into being as a corollary of the 1938 Federal Food, Drug, and Cosmetic Act (FDCA) and was formally incorporated into federal statute with the passage of the Durham-Humphrey Amendment of 1951. At the same time, an exuberant and seemingly endless pipeline of patent-protected, individually branded, and heavily marketed prescription pharmaceuticals was actively reinventing the American pharmaceutical industry, transforming it from a relatively small economic sector into the paradigmatic example of the technologically driven postwar boom.

The American pharmaceutical industry continued its almost unbounded growth into the twenty-first century. Domestic sales of pharmaceuticals increased from $500 million in 1945 to $2 billion in 1963, to $40 billion in 1990, to $200 billion in 2004, to $300 billion in 2009. In the late 1950s, the rate of return for pharmaceutical companies was 20 percent, as compared to 10 percent for all manufacturing. From 1995 to 2002, the pharmaceutical industry was the most profitable sector in the United States, averaging more than 17 percent annual net profit after taxes.[3] In the fifty years from 1959 to 2009, the annual growth rate of drug sales in the United States fell below 5 percent just three times, and two of those were in the Great Recession years of 2007 and 2008.[4]

These extraordinary growth rates and sales figures mirrored annual numbers of prescriptions that spiraled ever upward. At the end of the 1950s, retail pharmacies were filling some 550 million prescriptions each year.[5] That number ballooned to 2.1 billion in 1994 and to 3.6 billion in 2005, enough for twelve prescriptions a year for every man, woman, and child in the country.[6] The number continued to rise, in spite of the recession economy, reaching 3.9 billion in 2009. Although prescription drugs account for only 10 percent of national health care spending, the prescription itself has achieved a high degree of material and symbolic significance in the doctor-patient encounter, as recognized by W. E. Lockhart fifty years ago.

This volume takes the modern prescription—until recently an iconic slip of paper with handwritten instructions, now increasingly an electronic medical record—as its object of inquiry and dates it as a product of the mid to late twentieth century. Our goal is to push our understanding of modern biomedical therapeutics beyond the drugs themselves, to shine a spotlight on various actors and their interactions concerning how these medications have been used. We use the prescription as a shorthand reference for a set of complex relations among the producers, providers, and consumers of medicine, and we shift our focus from the biographies, life cycles, and trajectories of individual drugs and specific classes of drugs to explore the processes by which these medications have moved through the social geography of health care.[7]

Recent years have seen a surge in popular concern over the development, marketing, and utilization of pharmaceuticals by Americans, along with the steady development of critical scholarship in the history, sociology, and anthropology of prescription drugs. Earlier works that portrayed the history of pharmaceuticals as a triumphant story of material progress have led to more critical and scholarly analyses of the social, cultural, political, and economic complexities of pharmaceutical development and utilization. To date, however, none have focused on the history of the prescription itself: as a professional boundary marker, as a gradient of knowledge and power, as a therapeutic performance, as a means to define limits of consumer safety, as a means of tracking the behaviors of physicians, and as a key site for examining the ebb and flow of public trust in the cures we consume and the professional, commercial, and regulatory frameworks that support them.

This volume represents a collaborative vision on the part of ten historians of American medicine whose work has converged to illuminate our understanding of the prescription as an object of mutual interest and urgent relevance in understanding changes in the material, political, economic, cultural, and epistemological dimensions of therapeutic authority that took place over the second half of the twentieth century in the United States. Although we recognize that, for as long as people have claimed to be physicians, they have been issuing "things" we would call prescriptions (notable examples are found in the Ebers papyrus, a record of ancient Egyptian medicine that dates to circa 1550 BC), we have chosen to focus on the prescription in late-twentieth-century America. Prescriptions took on novel features and distinct meanings in the middle of that century, as they became the newly required legal gateway for consumer access to this newly created category of goods called "prescription drugs."

We are interested in how the prescription gained relevance and power at midcentury, as American medicine was itself undergoing a therapeutic revolution. The prescription in the late twentieth century can be understood as an object, as an action, and as a series of processes by which social networks and professional identities of doctors, patients, nurses, and pharmacists have been formed and re-formed over time. Training our lens on the prescription—

instead of on the disease, the drug, the patient, or the physician—allows for subtle explorations of the problems of medical power, the politics of therapeutic authority, and the complex relations between knowledge and practice in recent and contemporary medicine. As the prescription circulated among physicians, nurses, pharmacists, government regulators, pharmaceutical manufacturers, and, of course, patients, it acted as a key boundary object. That is to say, it simultaneously linked these disparate groups and served to mark their different roles and relative power. These linkages themselves proved highly dynamic. For example, while the authority to prescribe in the 1950s and 1960s was tightly linked to the authority of the medical profession, that distinction had eroded by the end of the century as psychologists, nurse practitioners, and physician assistants all gained access to some of the power of the prescription. The prescription as an object itself also began to lose materiality by the early twenty-first century, when new electronic prescribing systems increasingly displaced the prescription pad as a means of designating medication use. Collectively, the chapters in this volume investigate the prescription in its historical contexts over the past sixty years, to explore how its power has been upheld and resisted, supported and critiqued, subverted and transformed by the various stakeholders in the modern biomedical enterprise.

Early Histories of the Prescription

For centuries, the prescription has played a role in defining medical practice, structuring the doctor-patient relationship, and shaping access to therapeutics. Important works in the history of ancient, medieval, and early modern medicine have focused on the prescription as an intentionally inscrutable talisman of the medical guild,[8] as a proprietary token of therapeutic politics between competing practitioners in the medical marketplace,[9] and as a primary source reflecting the evolving roles of doctors and patients within therapeutic encounters.[10] Relatively little historical attention, however, has been paid to the role of the prescription in the twentieth century, a time in which, paradoxically, its roles in the definition of the physician as professional, the patient as consumer, and the limits of state power were uniquely salient.

The omission of the prescription from the historical gaze is itself relatively recent. The history of the prescription was a subject of keen interest to medical and pharmaceutical educators of the late nineteenth and early twentieth centuries. Dr. Otto Wall's *The Prescription: Therapeutically, Pharmaceutically,*

Grammatically, and Historically Considered, first published in 1885 to critical acclaim, went through four revisions before Wall's death in 1922.[11] Wall, the director of revision for the United States Pharmacopeia (the nation's compendium of "official" drugs), portrayed the prescription as a central feature of the art of medicine, a key aspect in the formation of the individual physician's therapeutic style. "Correct prescription writing," said Wall, "is an accomplishment which is to the physician what elegant clothes are to a gentleman, or a handsome frame to a fine painting. If it is not an essential part of his education, it at least displays his other acquirements to best advantage."[12]

For Wall and his contemporaries, to attend to the history of the prescription was to insist on the interpersonal, humane face of an increasingly impersonal scientific medicine; history both humanized the practice of medicine and dignified the profession.[13] In this narrative, the invention of the prescription represented a crucial dividing line between primitive prehistory and ancient Western history coincident with the rise of writing and the organization of complex social forms. When the physician and the priest were combined in one figure, they argued, there was no need for a prescription: one simply prayed for relief from illness. But once spiritual and somatic healers could be distinguished, and somatic healers were segmented into many types, one could speak of the beginnings of a scientific approach to medical knowledge. In discussing the Ebers papyrus, Wall notes that "the Egyptian method of practicing medicine was known and prevailed through all these lands, and . . . prescriptions were written by one class of practitioners and dispensed by others wherever Egyptian science had penetrated."[14]

Classicist approaches to prescription history began their investigations with etymology. The term *prescription* was of Roman origin, traced back to the Latin term *praescriptum,* meaning "written before" (i.e., for the guidance of one who would subsequently prepare a therapeutic substance). The symbol *Rx* could be traced back to the word *receptum* (recipe, or receipt) in that the prescription reflects a form of knowledge received by apothecary or patient to guide a therapeutic formulation—although some dispute existed over whether the Rx symbol itself emerged from the Latin term, or from a modification of the Greek symbol for Jupiter, or from an iconic form of the Egyptian sun-god Horus.[15] Wall and his colleagues referred to Late Antiquity as a "Dark Ages" when prescription knowledge was lost to the West but preserved and refined in the Arabic world, and they identified a subsequent "Renaissance of the prescription" in early modern Europe—indeed, in these narratives the prescription is foregrounded as one of the tools that helped define the modern professions of medicine and pharmacy. To the medical humanist, the history of the prescription was nothing less than the history of Western civilization. Not surprisingly, non-Western prescriptive traditions—for example, the complex system of Chinese medicine or the Ayurvedic prescriptive practices of the Indian subcontinent—were nowhere to be found in these accounts.[16]

For most of their recorded history, prescriptions were highly idiosyncratic and contained closely guarded secrets, as a highly personalized form of communication between physician and pharmacist, crucial to both but belonging to neither.[17] Galen of Pergamon, the great systematizer of classical medicine in the second century AD, was well remembered for extravagant and ornate prescriptions requiring numerous ingredients and several lines of specialized instructions for apothecaries to follow in producing his cures. Well into the early twentieth century, significant legal battles continued to erupt in Europe and North America over who owned the prescription: the physician, the patient, or the pharmacist. Textbooks for pharmacy students emphasized the secrecy of the prescription. In part, the continued use of Latin in prescriptions—when so many other realms of medical description had shifted away from Latinate terms toward the polysyllabic notation of biomedicine—was justified as a means of maintaining the secrecy of those parts of the prescription that might still be the private property of the physician. Scoville's *Art of Compounding* (1914), a central text for generations of American pharmacy students, emphasized the dangers that would befall pharmacists who betrayed the secrecy of the prescription. "Latin affords secrecy," Scoville insisted. "This last reason should be remembered and respected by the pharmacist, and inquiries regarding the composition and nature of a prescription should be answered with caution. It may be unobjectionable to answer frankly such questions when asked by people of high intelligence and with honorable motives but in most cases an evasive reply, which conveys little or no information, is advisable. Prescriptions often betray secrets which should be zealously guarded."[18]

Medical accounts of the prescription in the first decades of the twentieth century considered its ownership "a mooted question," reflecting the notion that the prescription could never properly be parsed into the property of doctor, pharmacist, or patient. Instead, it traced a shared line that connected all three along unstable equilibriums of responsibility, property, and ownership. Bernard Fantus advised medical students in his popular *Text Book on Prescription-*

Writing (1905) that the prescription must always be understood to be in part the property of all three. "Patients are inclined to think that they have paid for the prescription, and that it is therefore their personal property," he noted. "This, however, is a misconception of the very nature of a prescription. The written prescription is simply an order from physician to pharmacist . . . The patient has, of course, a right to do what he pleases with a prescription; he may not have it filled at all; but once the order has been filled by a properly authorized person, it passes into the hands of that individual."[19] This early-twentieth-century account of the differing conceptions of the prescription between physicians and patients (with no mention of the pharmacist's perspective) foreshadowed the contested terrain traversed by the prescription later in the century.

The Prescription in the Early Twentieth Century

Two profound changes contributed to a dramatic shift in the role of the prescription in American social history by the mid-twentieth century. First, the physical form of the prescription underwent a radical revision in terms of the language and phraseology used by physicians to communicate with pharmacists. Second, a series of federal rulemaking and legislative acts greatly expanded the role of the state in regulating the medical marketplace in general and in the sale of drugs in particular.

First, the physical prescription itself. Faced with an archive of prescriptions stretching back over the twentieth century, the historian is immediately struck by changes in the material culture of the scrip—the shift from fountain pen to ballpoint, from high-quality paper to thin sheets stamped with boilerplate forms, often with the name of a drug or pharmaceutical firm preprinted on the blanks. One change, however, is evident above all others: the shift in notation from Latin to English, both in the form of the prescription and in the naming of its contents. In part this shift is linked to the decline of Latin as an international scientific language; in part it is due to the rise of new synthetic chemotherapeutics whose names did not fit as easily within the Latin grammar of the prior materia medica. Indeed, it appears that the prescription was one of the last aspects of medicine to leave Latin behind. By the late nineteenth century, as a working knowledge of Latin became less important in medical training, the Latinate form of the prescription had shifted from an individualized recipe toward a standard formulation. Texts to help accomplish this feat, such as Gerrish's *Prescription Writing Designed for the Use of Medical Students Who Have Never Studied Latin,* lamented that "every teacher of medicine in this country must have been impressed with the fact that a large proportion of the students in our schools have no knowledge of the Latin language . . . In no other class of cases, however, is this deficiency of education so apparent as with regard to the terminology of the pharmacopoeia, especially when it becomes necessary to make the grammatical changes required in giving directions for compounding medicines."[20]

Gerrish's solution was to provide a functional primer in pidgin Latin, apologizing to readers that while his system may produce "little more than a parrot-like command of a few words and expressions," it would be "far better for a man to write a prescription correctly, even in the most automatic way, than to blunder through it disgracefully, as so many habitually do, and thus expose himself to the ridicule of apothecaries' shop-boys."[21] Similar accounts of Latin-illiterate pharmacists-in-training, however, suggest there was likely little need to fear the apothecaries' shop boys. By 1914 Scoville's textbook of pharmacy complained about the paucity of Latin training among pharmacists, noting that most druggists in practice simply memorized stock phrases.[22]

The Latin prescription of the early twentieth century was based upon these standardized phrases structured into a template with three functional domains: the *inscription,* or ingredients, the *directions* to the compounder, and the *signature,* or directions for the patient. For example, the following prescription from Scoville's textbook illustrates what he considered to be a "classical" prescription, whose abbreviated Latin spelled out a prose paragraph of instructions to the compounding pharmacist:

Rx. Morphinae Sulphatis ____ grana dua
Sacchari ____ drachman unam
Misce et divide in chartulas duodecim,
Signatura:—Capiat unam nocte[23]

This is a rather simple prescription for morphine powders. The ingredients are morphine sulfate and sugar. The instruction begins with the Rx symbol itself, which is a verb, *recipie* (take), with its first object *grana* (grains, a specific unit of weight), its second object *drachman* (a drachm, another unit of weight). The directions for compounding stand below the ingredients, and the signature stands indented at the last line. Translated, this scrip forms a prose paragraph with clear instructions to pharmacist and patient alike: "Take two grains of morphine sulfate, one drachm of sugar, mix and divide into twelve powders. Let the patient take one powder at night."

This Latinate structure was already beginning to dissolve in the early twentieth century. Signatures were increasingly written in English to aid in translating the course of therapy to the patient, and some physicians had taken to writing the directions in English as well, acknowledging that most pharmacists—and most physicians—no longer had sufficient knowledge of Latin for compounding instructions to be communicated effectively in this manner. Although the form of the prescription still maintained its structure of inscription, directions, and signature, by the middle of the century, the Latin had almost entirely disappeared. A roughly equivalent prescription would now read:

Rx: morphine tablet, 20mg
24
sig: i tablet q4–6h prn pain

The tablets of morphine specified above were bought premade from the manufacturer, their dosage prespecified. The instruction "#24" means literally that a pharmacist should count twenty-four of these pills and put them in a bottle. The signature to the patient is a mix of English and vestigial Latin abbreviation—"sig" for signatura, "i" (roman numeral) for one tablet, "q4–6h" (every 4–6 hours), "prn" (as needed) for pain relief.

Two key transformations had conspired to squeeze Latin almost completely out of the prescription. First, Latin as an international language of biological, chemical, and medical science had itself been jettisoned in favor of a new language of synthetic chemicals and biopharmaceuticals: the language of International Union of Pure and Applied Chemistry (IUPAC) chemical names and new nonproprietary nomenclature, overseen by international bodies such as the League of Nations and the World Health Organization (WHO).[24] Indeed, even before WHO held its first assembly in 1948, an interim committee had met to address the problem of harmonizing the global materia medica as each country shifted from Latin toward new and frequently discordant scientific vernaculars. At the same time, the act of compounding, previously thought of as the principal domain of the pharmacist and the rationale for pharmacists' extended education in the science of pharmacology, had declined as the pharmaceutical industry had shifted its product lines increasingly toward prefabricated "specialty" remedies. These prepackaged pills, capsules, and solutions were sold by the manufacturer ready-made for consumption by patients: all the pharmacist had to do was to stock them, count them, and put them into a bottle for the patient. With the rise of the specialty drug, the percentage of prescriptions that required compounding declined from 75 percent in the 1930s to 25 percent in the 1950s to just 1 percent in the 1970s.[25] As the job of the retail pharmacist shifted from that of interpreter and coconspirator to that of pill counter, the form of the prescription shifted from a set of explicit instructions for an individually tailored remedy to a permission slip in a uniquely mediated form of mass consumption.

Turning now to the role of the state in the medical marketplace, we find that scholarly interest in the politics of the prescription in recent American history has centered on interpretations of how the state restricted the access of individuals to therapeutic compounds.[26] The prescription drug took on this unique political and economic valence in the second half of the twentieth century, after the 1951 Durham-Humphrey Amendment to the Federal Food, Drug, and Cosmetic Act wrote into federal law a distinction between drugs that could be sold directly to consumers (over the counter) and drugs that could be dispensed only with the prescription of a physician, a dentist, or a veterinarian. This legislation delineated new terrain for the prescription in relation to the pharmaceutical industry and its markets, the state and its regulatory responsibilities, the health professions and their respective boundaries, and the patient and his agency as a consumer.

The Pure Food and Drugs Act of 1906, which created the Food and Drug Administration (FDA) as a consumer protection agency, assumed that decisions regarding "self-medication"—the choice to consume drugs—were ultimately the patient's prerogative. The role of the FDA was not initially to restrict which drugs could be sold but to ensure that drugs—most of which also led a double life as poisons—were not sold with incorrect labels that made misleading claims. This regulation of claims and labels presupposed a rational consumer who could make decisions for herself once proper information was provided. In 1914, however, the passage of the Harrison Narcotic Act established that one category of addictive drugs—narcotics—was so dangerous to individuals and to society at large that sales needed to be restricted through closely guarded professional channels (i.e., by prescription only, and even then with close supervision). But for all other drug categories, consumers still had the authority to decide which drugs to take, with or without the counsel of a physician.

In October 1937, more than one hundred Americans, many of them children, died after consuming Elixir Sulfanilamide, a new formulation of sulfa drug suspended in ethylene glycol—the toxic component found in antifreeze. Largely in response to this tragedy, the U.S. Congress passed the Federal Food, Drug, and Cosmetic Act (FDCA), which gave the FDA broad powers to ensure the safety of drug products. Given that all drugs might potentially be poisonous, it would be impossible to distinguish between acceptable and unacceptable levels of drug safety without careful consideration of the context in which a given drug would be used. A side effect that made a drug unsuitable for treatment of mild allergies, for example, might be fully tolerable in the treatment of a fatally infectious disease. Section 502 (f) of the 1938 law determined that a drug could not be safely marketed

> unless its labeling [bore] (1) adequate directions for use; and (2) such adequate warnings against use in those pathological conditions or by children where its use may be dangerous to health, or against unsafe dosage or methods or duration of administration or application, in such manner and form, as [were] necessary for the protection of users: *Provided*, that where any requirement of clause (1) of this paragraph, as applied to any or drug or device, [was] not necessary for the protection of public health, the Secretary [would] promulgate regulations exempting such drug or device from such requirement.[27]

The phrase "not necessary for the protection of public health," contained the loophole that allowed prescription-only drugs to come into existence. If a drug was designated to be dispensed only under the prescription of a physician, a dentist, or a veterinarian, the precise burden of the safety label was "not necessary for the protection of public health." By declaring their products to be available only by prescription, manufacturers could avoid paperwork, regulatory delay, and liability.

When lobbying for the bill that ultimately became the 1938 FDCA, W. G. Campbell, then commissioner of the FDA, insisted that the law had not been designed to create a category of prescription-only drugs. "There is no issue," he stated, "from the standpoint of the enforcement of the Pure Food and Drugs Act about self-medication. This bill does not contemplate its prevention at all."[28] And yet in the decade that followed, the FDA promulgated regulations by which manufacturers could declare which of their drugs could be labeled for consumers to self-medicate and which should be restricted to sale by prescription only. It is worth pointing out here that pharmacists, not manufacturers, bore the legal brunt as these actions were enforced. For more than a decade, in the confusing marketplace that ensued, the same drug could be produced by different manufacturers with some designating it prescription-only and others designating it over-the-counter. If a pharmacist erred in dispensing the prescription-only version, he would ultimately be liable to prosecution. Challenges to the FDA's power to prosecute pharmacists rose to the level of the Supreme Court, which upheld the FDA's power in a wide-reaching 1948 decision, *U.S. v. Sullivan*, ultimately leading to the passage of the Durham-Humphrey Amendment of 1951.

The radical nature of the Durham-Humphrey Amendment in transforming the prescription into a barrier against free access to a set of commodities was first explored by the economist Peter Temin. Temin's 1979 article "Origin of Compulsory Prescriptions"—followed quickly by his 1980 monograph *Taking Your Medicines*—explained the FDA's restriction of certain drugs to prescription-only status in terms of the expansive tendency of state regulatory functions toward what he (and other classical economists) saw as the increasingly inappropriate restriction of free markets. Temin's article and book stood for years as the definitive history of drug regulation at the FDA, until historian of medicine Harry Marks, in a 1995 article, criticized Temin for producing an overly simple and convenient narrative of what had been a more complex political reality. Marks suggested that any historical account of the FDA that focused on its expansive powers must also acknowledge its *limitations:* chiefly, that the main province of the FDA's regulatory reach in the realm of drugs was limited to labeling. In other words, the FDA's powers centered on (and were largely restricted to) its ability to determine what kinds of claims companies could make about their products and what kind of instructions were adequate for safe use. The prescription-only category was a recognition of the shortcomings of the label, which were criticized not only by the regulators, but also by the pharmaceutical manufacturers themselves. A letter from Winthrop Chemical Company to the FDA noted in 1938: "The warning that Phenobarbital is contraindicated in large doses in nephritic subjects is ineffectual and meaningless to a lay consumer who does not know that he is suffering from nephritis (even if the labeling uses the synonym 'Bright's Disease'). Furthermore, a layman lacks the knowledge and experience to determine what quantity constitutes a large dose or an excessive amount. In some cases a layman may not recognize the presence of untoward effects specified in the warning until considerable harm has resulted from continued use of the drug."[29] Trying to determine safety in regard to a label required a sense of how, and by whom, the safety of a drug could most properly be understood. What Temin derided as an expansion of state authority was interpreted by Marks as an explicit attempt by government regulators (the FDA officials) to accommodate at once consumers' needs, manufacturers' potential liability, and physicians' autonomy.

In Marks's retelling, the formalization of the prescription drug market was hammered out in a series of conflicts between the passage of the 1938 act and the clarification of the 1951 amendment, which needed to be understood not—as Temin had suggested—as an autonomous expansion of the state, but rather in the context of a politics of deregulation that characterized the immediate postwar era. President Harry Truman's attempt to augment the regulatory system of the New Deal in the wake of the wartime economy precipitated a backlash among pharmaceutical manufacturers (along with much of the American industrial sector) against government regulation of business.[30] Within this broad critique of the expanding role of the state in economic affairs, the prolonged wrangling of the FDA and drug manufacturers over the precise wording of their labels had itself become the symbol of state overreach. Exempting a drug label from such wrangling by restricting the drug to a physician's prescription was seen as a corrective to the FDA's undue influence—in other words, as a corrective to the excesses of state involvement. By the time of the 1951 amendment, drug firms were opposed not to the existence of the "prescription-only" category but merely to the purview of the FDA to determine which drugs fell into that category. Manufacturers wanted to continue with what had by then become the status quo, that is, each firm's determining which of its products should be sold by prescription alone.

In the continuing scholarly debate about the role of the state in limiting consumer access to the class of prescription-only medications, historians such as John Swann and political scientists such as Daniel Carpenter have pointed out that the FDA is often inappropriately treated as a monolithic entity, whose actions can be reduced to the statements of its commissioner.[31] But the FDA is more accurately understood as a living, adaptive institution, which took initiative in redefining the prescription through a cumulative series of rule-making decisions about the precise meanings of technical terms, for example, *safety*, *efficacy*, *labeling*, and *indications*. The creation and subsequent expansion of the category of prescription drugs was achieved neither by executive fiat nor against the interests of the growing pharmaceutical industry, but rather by the complex give-and-take of negotiations among many parties outside and inside the FDA.[32] These midcentury debates over the limits of regulation reflected new fault lines in the political economy of professional turf battles, battles that were in part fueled by the expansive fervor of the postwar drug boom and in part driven by shifting cultural expectations about medicines, the pharmaceutical industry, the physician, and the state. At their center was the modern American prescription.

The Modern American Prescription

Amid the shifting cultural expectations of the postwar era, the prescription served as a tool for the fashioning of professional and lay identities around the consumption of pharmaceutical products. Although on the surface of things, the Durham-Humphrey Amendment appeared to clarify a specific part for each actor—the physician ordered the drug by writing the prescription, the pharmacist translated the doctor's written order into a product by filling the prescription, and the patient (with or without the help of an insurance company) purchased the prescribed medication and consumed it—this ostensible chain of command belied a much less stable situation. As the chapters in this volume show, the late twentieth century was marked by several significant challenges to the therapeutic authority of physicians and a proliferation of "alternate" spaces in which the narrowing function of the prescription could be inverted or repurposed. Although the prescription initially stood as a marker of the physician's dominion over therapeutics, it swiftly became a political site for the contestation of that authority.[33]

At stake in contests over the prescription was the definition of therapeutic rationality. Medicine's exclusive claim to the prescription pad was based on a restrictive understanding of the forms of knowledge necessary to safely wield powerful legend drugs: that is, a knowledge of pharmacology and pathophysiology combined with practical clinical experience. Yet in the years following the passage of the Durham-Humphrey Amendment, consumers, regulators, and other providers began to question the purported rationality of the physician and the sufficiency of these limited forms of knowledge in guaranteeing consumer safety. As doctors continued to indiscriminately prescribe wonder drugs such as the broad-spectrum antibiotic Chloromycetin (chloramphenicol) in the 1950s, 1960s, and 1970s, in spite of mounting evidence of fatal side effects, reform factions within the medical profession also derided the "irrational" use of drugs by physicians.[34] As both Scott H. Podolsky (chapter 2) and Jeremy A. Greene (chapter 10) demonstrate in this volume, these therapeutic reformers within the medical profession compiled databases of prescription data as evidence in their arguments to persuade their colleagues to change their prescribing habits from "irrational" to "rational" forms, with varying results.

While the medical profession aimed to regulate itself with no outside intervention, legislators, judges, and government regulators saw otherwise. Efforts to rein in physicians' authority and autonomy were undertaken by members of all three branches of government. From the executive branch, directives issued by agencies such as the Food and Drug Administration and the Drug Enforcement Administration delimited physicians' prescription practices, although, as Elizabeth Siegel Watkins shows in chapter 4, these regulations could be reversed as a result of an executive order. Attitudes and actions toward prescriptions, drugs, and medicine in general shifted according to the political leanings of the party in control of the White House.

Legislators in both the House of Representatives and the Senate repeatedly turned their attention to prescription drugs. In 1958 Representative John Blatnik and the House Subcommittee on Legal and Marketing Affairs of the Committee on Government Operations held a four-day hearing titled "False and Misleading Advertising of Prescription Tranquilizing Drugs." From 1959 to 1961, Senator Estes Kefauver and his Anti-Trust and Monopoly Subcommittee of the Committee on the Judiciary held a series of hearings initially focused on unfair drug pricing, and in 1967 Senator Gaylord Nelson and the Subcommittee on Monopoly of the Committee on Small Business began what eventually turned into ten years' worth of hearings on "Competitive Problems in the Drug Industry." These inquiries, and the many more that followed over the next three decades, targeted the pharmaceutical industry, but physicians also attracted the attention of investigators for their role as prescribers of drugs.

The judicial branch weighed in as well on the matter of physician autonomy and authority in prescribing drugs. Individual physicians, such as those whose stories are told by David Herzberg in chapter 9 and Marcia L. Meldrum in chapter 8, and nonphysicians, such as the feminist activists in chapter 6, by Judith A. Houck, could find themselves brought to court for questionable prescribing practices. Other legal cases took up more general issues of who did and did not have the right to prescribe drugs. Viewed against the backdrop of court cases, legislative bills, and FDA mandates from the first half of the twentieth century, as described by Nicolas Rasmussen in chapter 1, the Durham-Humphrey Amendment of 1951 appears less as a watershed in the political history of the prescription and more as yet another instance of the modern trend toward government intervention in and oversight of the prescription and its players. The medical profession and the pharmaceutical industry were united in their critiques of the government's attempts to "prescribe medicine" to physicians.

The decade of the 1970s stands out in particular as a crucible for legislative, regulatory, interprofessional, and lay challenges to the status quo of physicians' monopoly over the prescription. A few examples from the studies in this volume serve to illustrate. The Comprehensive Drug Abuse Prevention and Control Act of 1970 restricted doctors from unlimited prescribing of controlled substances such as narcotics, stimulants, depressants, hallucinogens, and anabolic steroids. The Joint Commission on Prescription Drug Use was authorized by Congress to figure out how to establish a surveillance system for prescription drugs and, by implication, physicians' prescribing habits. Forty states passed laws that allowed pharmacists to substitute generic equivalents for the brand-name drugs prescribed by physicians. The FDA mandated patient package inserts for oral contraceptives and estrogen products and considered implementing the same requirement for all prescription drugs. Nurse practitioners and physician assistants began to assert what they saw as their rights to prescribe drugs for patients, and some states formally recognized those rights. Self-help clinics were opening across the country, offering an alternative to the doctor's office.

The boundaries between various groups of health care professionals and between professionals and nonprofessionals with respect to their roles, rights, and responsibilities regarding the prescription were never clearly demarcated. These blurred borders caused friction, which erupted at times into outright conflict. In their chapters in this volume, Dominique A. Tobbell (chapter 3) and Julie A. Fairman (chapter 5) describe in close detail the nature of such conflicts in the roles of pharmacists and nurse practitioners, respectively, in the clinical contestation over the prescription. These interprofessional disputes were paralleled by lay attempts to gain some measure of involvement in the prescription process. Starting in the 1970s, consumers and their advocates argued for greater access to the information encoded in the written prescription, as demonstrated by Watkins in her study of the proposed patient package insert, and for greater access to some of the drugs themselves, as shown here by Heather Munro Prescott in her investigation of the debate over moving oral contraceptives from prescription-only to over-the-counter status (chapter 7). Perhaps the most significant challenge to physician authority in the late twentieth century was mounted by feminist health activists who set up lay-controlled clinics for the provision of well-woman care, as chronicled here by Houck (chapter 6). The prescription can be considered not only as a site for cultural negotiations of identities and political contestations of authority; it also functions as a critical commodity in the economies of knowledge and material resources within American health care. As Herzberg argues in chapter 9, prescribing itself could become a nakedly commercial act. The marketplace of drugs as restricted commodities generated a secondary marketplace for prescriptions, in which pseudopatients would pay for the privilege of access to the recreational drug of their choice. The multiple secondary markets of prescriptions and prescribing feature prominently in several of the chapters in this volume, often in unexpected ways. Tobbell and Watkins explain how mandates to carry stocks of brand-name drugs and patient package inserts, respectively, were seen by pharmacists as creating an undue financial burden that motivated them to lobby for generic drug substitution privileges and against the FDA mandate for written information for drug consumers. The cost differential between birth control pills sold by prescription only or over the counter loomed large in the debates investigated by Prescott, and competition for patients (and their dollars) played an important role in the contest, described by Fairman, between physicians and nurse practitioners over the right to prescribe. Greene examines the value of aggregated prescription data as a commodity that itself could be bought and sold by pharmaceutical marketers to fine-tune their marketing practices and to build their own data empires in an emerging information economy. Collectively, these authors use the prescription as a lens to focus on the historical development of some of the most pressing economic concerns presented by our current system of health care.

As an object that sits at the boundary between licit and illicit substances, the prescription plays a key role in mapping the borderlands of American consumer culture. Rasmussen describes how social fears about barbiturate use in the 1940s led to stricter federal controls over potentially addictive drugs in the 1950s, but he points out that the prescription-only status of these pharmaceuticals failed to diminish the continued abuse of and addiction to these drugs. By the 1970s, as Herzberg tells us, ten of the top fifteen most abused drugs were prescription medications, and they accounted for two-thirds of all drug abuse in America. By 2009, 7 percent of Americans reported that they had taken a prescription drug for a nonmedical reason in the previous month, more than the percentage reporting use of cocaine, heroin, hallucinogens, and inhalants combined.[35] That same year, 2.2 million Americans used a prescription pain reliever for an unprescribed reason for the first time, almost as many as those who first experimented with marijuana (2.4 million).[36] Meldrum describes the complex relationships that have developed between patients who need narcotic regimens as legitimate medical treatment for chronic pain and physicians who attempt to prescribe them, amid lingering suspicions of abuse.

While certain classes of prescription drugs frequently could traverse the porous boundaries between legal and illegal use, this traffic typically flowed in one direction. That is, several medications that had been approved by the FDA for the treatment of medical conditions later found nonmedical uses, but so-called street drugs did not pass from illegality into the approved pharmacopoeia. (Perhaps the only exception is the highly contested medical use of marijuana. Although sixteen states have legalized medical marijuana, cannabis, along with heroin and LSD, is still classified by the federal government as a Schedule I drug.) Once prescription drugs—from barbiturates in the 1950s to Valium and Quaalude in the 1970s to Vicodin and OxyContin in the 1990s—began to be abused, they occupied a space that was simultaneously licit and illicit. As Prescott describes in chapter 7, many other medications have traversed an analogous boundary from prescription-only to over-the-counter status. This switch signifies an important point in the biography of a drug, often coinciding with the expiration of a product patent; the relinquishing of the prescription here marks a moment when the sites of consumer decision making, regulatory oversight, and the understanding of the drug as a branded commodity change simultaneously.

Finally, like pharmaceuticals themselves, prescriptions can be considered to be informed materials—objects that circulate meaningfully through lay and professional networks largely because of the forms of knowledge claims they can make.[37] Several of the chapters in this book interrogate the prescription as an object of encoded knowledge and privileged information. While a universal sense of shared medical knowledge—about diseases, therapeutics, and drug indications and contraindications—is required for the prescription to circulate meaningfully, each prescription also implies a specific or local form of medical knowledge about specific drugs for specific patients in specific circumstances. The epidemiology of evidence-based medicine jostles for position with physicians' empirical experience and individual judgment. From the 1950s through the 1970s, as Podolsky and Tobbell show, physicians claimed that only they had the knowledge to decide which drug to prescribe; they resisted both antibiotic control programs and pharmacists' requests for substitution privileges. In the 1970s and again in the 1990s, as Watkins and Prescott demonstrate, consumers and their advocates rejected the notion of the prescription as a privileged document and demanded access to the information contained therein. For Greene, the information encoded within a prescription takes on new meaning when multiplied by the thousands in aggregated data sets and taken as an object of research in its own right.

The Plan of the Book

Each chapter in this volume highlights a distinct facet of the role of the prescription in modern American life. Together, the ten chapters reflect a broad range of therapeutic and diagnostic areas, from the heroic treatment of acute infectious disease, to the management of chronic asymptomatic conditions, to psychoactive drugs, to contraception. Individually, they use the circulation of prescriptions and the traces they leave behind to reveal insights into the cultural history of the linked enterprises of medicine and prescription pharmaceuticals in the United States. Collectively, they illuminate four common themes. First, they explore how the prescription acts as a boundary marker to delineate professional identities (e.g., physician, pharmacist, nurse practitioner) as well as lay identities (e.g., patient, advocate, pill fiend, drug seeker). Second, they demonstrate how prescriptions act as political sites for both the consolidation and the expansion of medical authority and for popular agitation against restrictive and hierarchical systems of medical care. Third, they investigate the commercial nature of the prescription enterprise and its impact upon the practice of medicine and the evolution of the American medical consumer; and fourth, they analyze the forms of knowledge-making required to produce prescriptions and the means by which that information is both encrypted and decoded in the circulation of prescriptions as material objects.

The first set of chapters explores government efforts to regulate the medical profession and the pharmaceutical industry by legislating or mandating oversight of prescription practices. In chapter 1, Rasmussen employs the concept of "moral panic" to explain the conflation of concerns about drug addiction and dangerous pharmaceuticals, specifically barbiturates, that rose to a crescendo in the 1940s and again in the 1960s and to which the "prescription-only" category was codified as a partial solution. He interprets both the passage of the Durham-Humphrey Amendment to the Federal Food, Drug, and Cosmetic Act in 1951 and the Comprehensive Drug Abuse Prevention and Control Act in 1970 as federally legislated efforts to mitigate drug harm by regulating prescriptions and prescribers. Looking inside the prescribers' domain, Podolsky addresses the persistent problem of antibiotic overuse in chapter 2. Focusing on the years from the 1950s through the 1970s, he describes how the medical profession successfully resisted efforts from without and within to restrict (or "rationalize") antibiotic prescribing, by supporting voluntary education programs as a means of preserving professional autonomy in the face of threatened regulation. In chapter 3, Tobbell investigates legislative challenges and counterchallenges at both the federal and state levels over the physician's right to designate brand-name versus generic drugs in the 1950s, 1960s, and 1970s. Her analysis brings another set of actors to the stage, namely pharmacists, who argued on both economic and professional grounds for the right to make decisions about the substitution of drug products in filling prescriptions.

The next four chapters address challenges to physicians' prescriptive power from lay groups and other health care professionals. Chapters 4 to 6 focus on the turbulent decade of the 1970s, while chapter 7 looks at a more recent episode in which restricted access to prescription drugs once again came into question. Watkins revisits the history of the patient package insert in the 1970s in chapter 4; she, like Tobbell in the previous chapter, takes pharmacists as the focus of her attention and asks how they reacted to consumer-generated proposals to expand access to drug information. In chapter 5, Fairman details the efforts of nurse practitioners to build on the de facto prescribing practices of nurses—practices that dated to the late nineteenth century—by earning the de jure right to prescribe in the late twentieth century. By making explicit rules around these previously tacit arrangements, this measure forced nurse practitioners to confront physicians head on about their reluctance to share prescriptive power. In chapter 6, Houck describes the challenges faced by feminists who sought to circumvent doctors in the provision of well-woman care at their lay-run health clinics in the 1970s. These lay providers had to negotiate the terrain of diagnosis and prescription without running afoul of the law. Similar debates over how best to ensure women's access to knowledge about and control over their bodies surfaced again a generation later, as Prescott describes in chapter 7, using two case studies dealing with the role of prescriptions as barriers to access to birth control products. First, she reviews the mid-1990s debate over whether oral contraceptives should be made available without a prescription and the arguments made for and against giving consumers direct access to drugs. Second, she explores how political controversy over emergency contraception (Plan B) led to the formalization of a third category, "behind-the-counter" pharmaceutical products. This move has forced pharmacists back into the role of contraceptive gatekeeper, recalling earlier eras when contraceptive devices such as condoms could be purchased only with the consent of the pharmacist.

The issue of prescription drug addiction introduced by Rasmussen in chapter 2 is taken up again in chapters 8 and 9. These two chapters investigate the gray areas between the licit and illicit prescription and the use of pharmaceuticals; in so doing, they introduce new ways of thinking about the roles of some consumers (are they drug seekers or drug abusers?) and the roles of some doctors (are they gatekeepers or drug pushers?). In these liminal spaces, consumers move away from being passive recipients of doctors' orders, and doctors abandon the Hippocratic imperative to "first do no harm." Meldrum explores how the use and prescription of pain medications has forced both patients and physicians to struggle with issues of identity and stigma. Herzberg, in his study of Quaalude scrip mills in the 1970s, shows how physicians and patients colluded in prescription drug abuse, and he expands the analysis of government regulation of prescription practices to include the drug enforcement system.

In the final chapter, Greene invites us to think more broadly about the social dynamics of the prescription after it has been written by the physician (or other prescriber), filled by the pharmacist, and consumed (or discarded) by the patient. He explains how four distinct groups—pharmaceutical marketers, therapeutic reformers, epidemiologists, and sociologists—have eagerly mined the wealth of information contained in prescription data, starting in the 1950s. Although each attended to its own agenda of understanding and shaping trends in clinical medicine, the possibility arose in the 1970s for a collaborative effort to develop an integrated public database of therapeutic surveillance, an ambitious project that, like many other grand health-system schemes of the 1970s, foundered and collapsed in the early 1980s.

This volume aims to move forward our understanding of the prescription by looking back at its many roles in American culture over the past three-quarters of a century. Who gets to prescribe, what gets prescribed, and how prescriptions and the products they call for are received in society—these aspects of the modern prescription and prescribing have varied over time and according to whose perspective is taken. With our collective lens trained on the prescription, we aim to illuminate the political, cultural, and economic matrix within which various groups of actors have been engaged in the production, regulation, marketing, sale, consumption, and understanding of pharmaceutical products. In exploring these networks of prescribing activity, the chapters in this volume locate the prescription at the heart of the complex social world of American medicine.

CHAPTER 1

Goofball Panic

Barbiturates, "Dangerous" and Addictive Drugs, and the Regulation of Medicine in Postwar America

Nicolas Rasmussen

In late February 1945, with the European battles of World War II nearing an end, a medicine problem vaulted to the forefront of national attention in the United States. As a prominent piece in the widely read *Saturday Evening Post* explained, the good folk of America were silently being polluted by a demon drug:

> We would be outraged at the suggestion that we are becoming a nation of drug fiends. We do not smoke opium nor do we tattoo ourselves with hypodermic needles. Nevertheless there are among us, as long ago as 1939, enough users of sleeping pills—"barbiturates" to the doctor, "goof balls" or "red devils" or "yellow jackets" to the addict—to account for the sale of 2,200,000 doses a day, and their use was rapidly increasing. Today, the figures show that . . . we are using almost three times as much.[1]

The piece's identification of the reader with the opium fiend and the needle junkie, those most stigmatized figures, aimed to shock. It went on to cite further shocking figures and facts on the ballooning national appetite for barbiturates, modern pharmaceuticals popularly used as sleeping pills. Their menace was threefold. First there were accidental overdoses and suicides, which had grabbed headlines in late 1944 with a rash of celebrity deaths, including radio evangelist Aimee Semple McPherson and Hollywood starlet Lupe Velez. These sad events had brought to light expert opinion that the ubiquitous drugs not only made it too easy to kill oneself but also actually produced suicidal acts by making the user more depressed or forgetful of pills already taken. Second, there was a major risk of harmful addiction to barbiturates even when used as prescribed, such that one could not live without them, suffered personality change, and tended to have deadly accidents under the influence. Third, they were a menace to public order, because the pills, aptly called "goofballs" for the disinhibited behavior they could produce, had grown popular for recreational use among wayward youth, "underworld characters," and the other undesirable elements vaguely figured above as "the addict."[2]

If this *Saturday Evening Post* feature flagged the onset of a national panic, the presence of the surgeon general's endorsement on its title page marked it as no random journalistic excess. The goofball panic lasted for years, and as we shall see, it involved struggles among the drug industry, its federal regulators, the medical profession, public health advocates, law enforcement, and the self-appointed defenders of public morality. It not only left its mark in the 1951 Durham-Humphrey legislation that codified the category of "prescription only" pharmaceuticals but still reverberates in today's drug control regime.[3]

Moral Panic and Drug Addiction in America

In this chapter I treat the public discourse around barbiturates from the mid-1940s to the late 1950s in terms of a "moral panic." Well known in criminology, moral panics are peculiarly dramatic popular reactions against some social threat. New or previously insignificant phenomena suddenly loom large as extraordinary menaces to right thinking and conduct, demanding action and, in the process, making transgressors into stereotyped, deviant "folk devils." Since its birth around 1970, the moral panic concept has fruitfully been invoked to explain episodes of legislation or heightened enforcement, or social "control waves" more generally. Departing from the typical use of the term for any disproportional popular reaction, I follow recent calls to return the concept to its anthropological and psychoanalytic roots by taking "panic" more literally as a dissociative phenomenon spurred by a perceived menace to moral and symbolic economies. Thus a true "moral panic" is a vigorous, morally charged social reaction involving issues, some of them unconscious, that are much broader than any explicit threat. Focusing on the emotional tone and semiotics of the reaction, this definition does not depend on how "real" the explicit threat may be.[4] On this understanding, the historian may analyze modern panics in just the same way as seventeenth-century witch hunts or manias, putting the question of proportionality to one side (where it might be better addressed by other, more quantitative disciplines such as, in the case of drugs, epidemiology). And in the American context, no social problem has proved more repetitively productive of moral panic throughout the past century than the issue of drug addiction.

As many historians have discussed, between 1900 and the 1920s the term *addiction* ceased to refer to an inconvenient, commonplace medical condition and became a state of rare moral decadence and irretrievable, possibly contagious criminality. These fevered imaginings have been traced to Puritan values of self-control and virtuous labor, in the context of a nation enduring disruptively rapid urbanization, industrialization, and immigration of non-Anglo-Saxon cultural aliens. (While a shift like this took place in many Western countries, nowhere was it stronger than in the United States.) At the same time as the addict's image mutated from a harmless if unmanly victim—of medicine, typically—to a deeply deviant, often racialized felon beyond the social pale, the official definition of addicting drugs narrowed sharply to include only the "narcotics": cocaine, morphine, and derivatives of morphine such as heroin and, later, marijuana. First came federal legislation against smoking opium and efforts to remove the "narcotics" from patent medicines. Then, beginning with the Harrison Narcotic Act of 1914 and subsequent legal decisions, users of the stigmatizing drugs were dealt with by increasingly harsh federal law enforcement. Scientific opinion rallied to a newly narrow concept of addiction. Pharmacologists defined the condition as a physiological compulsion manifest in tolerance and withdrawal, rooted in the nervous system's permanent adaptation to drugs. Psychologists and psychiatrists doubled the addict's spoiled identity with (in Erving Goffman's terms) a character blemish to match the physical abomination that pharmacologists had located in his nerves: an inborn psychopathic cowardice and laziness in the face of challenge that drew him to narcotic escapism and then made him incapable of resisting abstinence pains and cravings.[5]

Medicine retreated. Physicians who still wished to prescribe "narcotics" were required to have special registration and keep special records, subjecting them to police surveillance. Allowed medical uses were sharply restricted, as Marcia L. Meldrum discusses in chapter 8. The Harrison Narcotic Act created the category of prescription-only drugs, a regrettable federal intrusion into physician discretion for American medicine's leaders, but one fitting contemporary reform efforts within the profession.[6] Policing what historians call the "classic era of narcotic control," from 1930 the entrepreneurial commissioner of the Federal Bureau of Narcotics (FBN), Harry Anslinger, built a national law enforcement empire on the fear of drug addiction by stoking the anxieties of self-appointed defenders of morality such as the YMCA, the Parent-Teacher Association (PTA), and the American Legion.[7]

What I argue in this chapter is that another moral panic linking "dangerous pharmaceuticals" and "drug addiction" emerged around barbiturates in the mid-1940s. Public health professionals cultivated and attempted to harness that panic, but the medical profession allied itself with the pharmaceutical industry to limit the threatened wave of external controls that might result from the perceived crisis. The 1951 Durham-Humphrey prescription-only law remained as an institutional residue. Further unfinished business from the 1940s was pursued for another two decades by the Food and Drug Administration and other entities concerned with public health, and it was finally resolved in 1970, on the heels of another moral panic linking "dangerous pharmaceuticals" and "addiction" in the 1960s. That resolution ultimately brought further federal restrictions on prescribers as well as manufacturers, in the name of controlling drug abuse. Thus the distinctively extreme American anxiety over drug addiction has left lasting marks on physician prescribing and the pharmaceuticals industry, outcomes that escape analysis in strict terms either of professional and economic interests or bureaucratic rationality.

Barbiturate Use: From Problem to Panic

By the time the United States entered World War II in late 1941, the barbiturate drugs were widely viewed as problematic but were not yet the subject of panic. This family of synthetic sedatives had become popular during the 1920s for insomnia; the pills largely replaced old opiate and chloral-based sleeping drafts and were also used for daytime calming or sedation in the manner of bromides. By the 1930s more than a dozen different barbiturate compounds were in common use, and annual consumption exceeded around 1 billion 1-grain sedating doses (a figure that averaged the standard doses of the most common compounds) nationally. Newspaper stories of sleeping pill suicides and recreational misuse became increasingly commonplace. Presumably reflecting public opinion, by early 1940 some twenty-six states had adopted laws requiring prescriptions for barbiturate sales, and several also for amphetamines, newer pharmaceuticals that were popular recreationally among the dangerous classes in much the same way as barbiturate "goofballs." Of these twenty-six state laws restricting barbiturates, ten banned prescription refills or limited refills to prescriptions specified as refillable, and a dozen demanded that special records of barbiturate prescriptions be kept by pharmacists or prescribers or both, for inspection by health and law enforcement officers. In a 1939–40 study of "promiscuous" barbiturate use, an American Medical Association (AMA) commission not only supported prescription-only regulations—because the drugs endangered users, contributed to crime and traffic accidents, and produced an "addiction" most "vicious"—but even suggested that "restrictions on prescribing" might be needed too.[8] As we will see, the rank and file of the American medical profession were far from ready to accept this diagnosis by the experts delegated to investigate the problem, let alone their radical recommendation to restrict physician discretion.

The barbiturates had already fallen under federal regulation too. When it came into existence in 1906, the agency that became the Food and Drug Administration (FDA) was charged with ensuring that drugs were accurately labeled, particularly to protect consumers from ingredients like cocaine and morphine in the patent medicines that then dominated the pharmaceuticals market. As the medical profession became more devoted to science, it abandoned "narcotic" remedies and the traditional pharmacopoeia more generally and turned to laboratory-sourced remedies from a reconfigured pharmaceutical trade. New and powerful hormones, vitamins, and synthetic drugs came to market in an increasing stream through the 1920s and 1930s.[9] Agitation to strengthen the agency's power to protect consumers from dangerous synthetic chemicals (in cosmetics and food as well as drugs) gradually mounted in a New Deal environment fostering the nascent consumer's movement, until a pharmaceutical disaster finally generated enough political impetus to overcome industry resistance. A solvent in Massengil's Elixir Sulfanilamide, a product containing one of the breakthrough antibacterial sulfa drugs, killed more than one hundred people in late 1937, and the Federal Food, Drug, and

Cosmetic Act of 1938 gave the FDA legislative authority to require safety testing before the marketing of new drugs. The act also specified that barbiturates should be labeled with a warning that they were "habit forming." The agency combined its new power over safety with its older power over labeling to issue regulations requiring from January 1940 that certain drugs, whether new or old, be labeled "To be used only by or on the prescription of a physician" if unsafe for self-medication. While the choice to apply this prescription-only labeling to pre-1938 drugs technically remained with manufacturers, and some companies overused such labeling on common household remedies to avoid including detailed instructions, the FDA from the start pressured drug firms to ensure that the "potent but dangerous" sulfa drugs, the new stimulant amphetamines, and the "habit-forming" barbiturate sedatives were labeled prescription-only.[10]

The participation of the United States in World War II produced dramatic socioeconomic changes that heightened anxieties about drugs and addiction. Millions of young men were inducted into military service and concentrated in massive camps, with profound effects on nearby municipalities. There was a government-funded boom in war industries, abruptly ending the depression and drawing millions to factory jobs far from their original homes. There was also unprecedented social mobility, with war industry opening well-paying skilled jobs to women and unfavored racial and ethnic groups. As populations underwent major shifts throughout the country, existing social relations and mores were greatly strained. Government concerns about controlling venereal disease escalated in many localities into a virtual witch hunt against not just prostitutes and teenage runaway "Victory girls" but any woman straying from traditional gender roles. "The promiscuous girl, the khaki-wacky and the girl who has become unbalanced by wartime wages and freedom" demanded repressive action—less by military authorities and police than at the hands of volunteer defenders of public decency like the businessman just quoted. Thousands of young women were arrested "on suspicion" of solicitation, for transgressions as minor as ordering a drink or waiting alone for a train and were jailed or confined in treatment centers. Repression fell most heavily on women from racially marked and economically disadvantaged backgrounds, as recent work by Marilyn Hegarty and John Parascandola has described.[11] A related worry about wild youth, their fathers serving overseas and their mothers working long shifts at factories, fueled a wave of public and political agitation over "juvenile delinquency." "Promiscuous" use of barbiturate "goofballs" was a part of this same picture, since they were popular recreationally with soldiers on leave, their female companions, and younger people generally. They were also associated with the exuberant urban Mexican American and African American youth counterculture, targeted with violent outrage by white servicemen in the mid-1943 "zoot suit riots."[12] In the context of this social unrest and hypervigilance to defend the male-dominated, white Protestant social order, the wartime tripling of the consumption of pills associated with loosening of behavioral constraints excited both reasonable worries for the public health and also more emotive reactions (see fig. 1.1).

Escalating Controls on the Panic Wave

By 1944, aggressive enforcement against unprescribed barbiturate use was well under way by public health authorities in many states and cities. In the Los Angeles area, California authorities were investigating hundreds of pharmacies violating the existing code and were prosecuting dozens of druggists, based on tracing barbiturate misuse cases passed along from law enforcement and the military. The investigators were accumulating evidence not only of drugged-and-disorderly servicemen, but also of barbiturates' corrosive social effects. For example, an absent sailor's wife who became addicted to Nembutal (pentobarbital) from a single perpetually refilled prescription "so grossly neglected her baby girl that neighbors had to take care of and feed the baby"; a woman munitions plant inspector was found passed out on the street after a night on the town, her pockets full of barbiturates prescribed to her roommate; and a "sixteen year old Pachuca gang member" became "jagged up" on goofballs with three friends, stole a late-model Oldsmobile, committed a couple of armed robberies, and then fell into a squabble that ended in his own fatal shooting. Such cases show that the pills' perceived threat to decency and the public order (understood in the typical gendered and racialized ways), not just to health, was generating much of the barbiturate issue's urgency even among public health officers. From mid-1944 California health officials began pushing for tighter state regulations involving licensing of barbiturate and amphetamine dispensers and strict inventory control.[13] In New York City, similarly flooded with war workers, servicemen, and other transient young people, public health officials urged that the climbing death rate from and illegal trafficking in barbiturates demanded laws much stricter than the local sanitary code, which already required prescriptions. Soon the health department

Figure 1.1. The full range of "goofball" peril in the 1940s popular imagination. Counterclockwise from upper right: suicide, street crime, urban vice zones, reckless driving, sexual license or prostitution, and addiction or madness. Source: *American Druggist*, September 1945.

cracked down on nonprescription sleeping pill sales and announced stricter new city codes banning refills without physician instructions. Refill restrictions were also multiplying among state laws during the war years.[14]

The FDA tested its power to enforce its own new prescription-only regulations at the retail level in ways shaped by this wartime social and legal context. In 1944 the agency brought action against a Georgia pharmacist named Sullivan for selling small packets of sulfa drugs to military personnel without either prescriptions or labels indicating that one was needed. The agency similarly went after a Waco, Texas, druggist named Fadal who was selling large quantities of barbiturates to war workers and "vagrants," again without either prescription or prescription labeling. Like a long series of other barbiturate-peddling pharmacists prosecuted by the FDA, Fadal pled guilty, but the Sullivan case went all the way to the Supreme Court, ultimately establishing a surer legal basis for the FDA's prescription-only regulatory powers under the 1938 act (as further discussed below).[15] As enforcement targets for its limited investigatory and legal resources, the FDA had chosen drugs whose misuse resonated with popular wartime anxieties. The sulfa drugs were being used by servicemen for venereal disease self-treatment, upon consultation with pharmacists rather than doctors. Barbiturates, as we have seen, were associated with delinquency, licentious behavior, and overdose deaths. Both were "scientific" pharmaceuticals of recent (but pre-1938) vintage and indisputably hazardous (even if the Elixir Sulfanilamide deaths were not actually due to the sulfa component). Given the wider implications of addiction and infectious disease, both drug classes presented problems of consumer protection and public health that were blended inseparably. In all of this the FDA's behavior fit an activist agenda to expand into the latter domain based on statutory authority mainly over the former, as some critics have suggested.[16]

It was in this setting of escalating regulatory action against barbiturates both by public health officials and the FDA, combined with an overcharged atmosphere of morally tinged anxiety over wartime social change, that the Saturday Evening Post ran its "Sleeping Pills Aren't Candy" feature. As noted above, the piece likened the barbiturates to opium and heroin, and the nation to an "addict" or a junkie, that most spoiled and polluted of identities. After dwelling tragically on overdose deaths of celebrities and untold thousands of ordinary victims and somewhat sensationalistically on nonmedical "goofball" use, it pointed the finger of blame mainly at unscrupulous pharmacists, and incidentally also at overprescribing doctors. It went on to discuss policy, suggesting that even restricting barbiturates to prescription-only sale nationally would not suffice, even though Surgeon General Thomas Parran, whose photograph accompanied his endorsement on the article's first page, was quoted in favor of prescription-only status as a solution. Instead, said the article, "What many officials as well as some doctors and druggists would like to see are laws for the sale of barbiturates similar to those for the distribution of narcotics—that is with rigid recording of every grain," perhaps with criminal penalties for possession without prescription. Below Parran—hinting at the identity of the "officials" who wanted Harrison Narcotic Act–style restrictions extended to barbiturates—deputy FDA commissioner C. W. Crawford assured the public that the agency was doing what it could, within its jurisdiction, to control the sleeping pills.[17]

Activist public health and regulatory officials campaigning against barbiturates could not have been displeased with the national outcry of 1945. Newspapers echoed their sentiments that something had to be done about the drugs, particularly "indiscriminate" barbiturate use. Pillars of society like PTA leadership, women's clubs, and church organizations demanded an end to the circulation, especially among youths, of the "addicting" pills that had become "America's Opium." Even commentators with qualms about a new "prohibition" were worried that "the physical and mental vitality of the American people [could] be sapped in much the same manner as that of the Chinese with their opium" if barbiturate overuse continued.[18] If the just-quoted items are not enough to indicate a moral panic, in December 1947 the Hearst newspapers ran a high-profile item in their national Sunday supplement *The American Weekly* entitled "More Victims of the DEVIL'S CAPSULES" (fig. 1.2). Illustrated with the Grim Reaper's hand snatching two children playing doctor, the story discussed accidental barbiturate poisonings and overdoses but also condemned greedy druggists who sold the pills to dubious seekers of "cheap thrills." Quoting the New York City health commissioner and drawing heavily on anecdotes from New York, the piece applauded the city's new "rigid controls" on the diabolical drugs, as well as a bill in Congress to do more federally. This bill to curb barbiturates, on the grounds that they caused suicide, insanity, addiction, and sexual transgression (for instance, slipped into girls' drinks at parties), was sponsored by Representative Edith Nourse Rogers (R-MA) and drafted in consultation with the FDA. It would have formally extended the Harrison Narcotic Act to cover the pills, with physician licensing and strict record-keeping just as with opiates—and everything else that came with it,

By Booton Herndon

More Victims of the DEVIL'S CAPSULES

QUIET breathing came from the darkened bedroom. Two persons were sleeping peacefully too peacefully, for one of them, unknowingly, was bleeding to death.

The breathing became slower, and slower, and finally stopped altogether. Thus Mrs. Helen Proctor met her death, not knowing, not caring, under the influence of sleeping pills. It was coincidence that a tiny incision made by a small piece of glass on which she had stepped had opened an artery.

But it was no coincidence that the woman, lying there bleeding to death in the same room with her husband, did not awaken.

The dangerous pellets, taken to induce slumber, saw to that.

Dr. Amos O. Squire, medical examiner of Westchester County, New York, said that sometime after taking the sedative, Mrs. Proctor, who lived with her husband in the town of Rye, had apparently gotten up and gone to the kitchen. She stepped on a sliver of glass from a broken water tumbler, but didn't seem to realize it. Mrs. Proctor returned to bed to die, still under the influence of the drug.

Mark up another victim to the sleeping pill. In the city of New York alone in 1946, according to Dr. Israel Weinstein, former Commissioner of the Department of Health, 77 persons lost their lives in accidental deaths due to sleeping pills. Many others narrowly pulled through. One woman had an experience which will haunt her the rest of her life.

She was a cultured visitor to New York from California, in the city to seek rare books. After having a few drinks in her hotel room, she took several pills which she had received on prescription, in order to sleep.

As the combined action of the alcohol and the barbiturate began to work, a knock came on the door. She opened it. That's all she recalls.

She awoke in a hospital room, uninjured in any way, but very depressed. What had happened after she admitted the stranger? She will never know. The only clue was that a typewriter, radio and several other articles were missing.

Alcohol and barbiturates, the active ingredient in sleeping pills, do not mix, medical authorities say. Each heightens the effect of the other. A person not affected by the customary dose of sedatives might, on taking that amount after just a small quantity of alcohol, be dangerously affected by the combination.

A Boston couple, returning home after an all-night New Year's Eve party, were well under the weather from celebrating. The husband complained of a headache; the wife staggered into the bathroom for some aspirin; got sleeping pills instead.

He, too, was too drunk to notice the difference. He downed three or four of the pills, then, shortly thereafter, took another dose. His heart couldn't stand both alcohol and barbiturates. He slumped over dead.

Persons looking for cheap thrills have found that sleeping pills taken in combination with beer or whiskey cause a feeling of irresponsible exhilaration, followed by stupor. They call this highball a "Geronimo."

By authority of a new law, the city of New York recently placed rigid control over the prescription and sale of barbiturates. Representative Edith Nourse Rogers of Massachusetts has introduced a bill in Congress which would place these "devil's capsules" under national control.

Dr. G. B. Lal, noted science writer, says, "You've to consider the cupidity of some druggists who deliberately violate the purpose of the law, to get a high price, by selling the sleeping pills in hundreds. They are driving many a person mad, all over America. Most druggists in this country are honest and public-spirited, but there are also the thieving druggists darkening the reputation of their calling."

In Los Angeles, 14 persons died as a result of accidental self-dosage with sleeping pills during 1946. In Chicago, it was estimated that a majority of the 28 accidental deaths due to poisoning during 1946 were caused by barbiturates. In Boston 16 persons died as a result of overdoses of sleeping pills. Dr. William J. Brickley, medical examiner, lists one particularly pathetic case.

A young mother left a bottle of sleeping pills on a low shelf in the bathroom. Her four-year-old daughter found them, and started playing doctor. She gave several pills to her brother, one year younger than she.

The youngster devoured the pills. The mother came home to find her little girl in frantic tears, her little boy dead.

Aimee Semple McPherson, the famous California evangelist, was accustomed to taking barbiturates to quiet her nerves after the strain of a revival service. One night in Oakland she retired to her hotel room.

They found her later, dying. Sleeping pills were found scattered in her bed. No one knows exactly what happened, but those familiar with the effects of barbiturates surmised that she took her usual dose, then, forgetting it in the hypnotic state resulting from the drug, took more.

In New York, the health department files reveal that a physician, suffering from stomach cramps, took three capsules to enable him to sleep. The doctor had to be hospitalized as a result of his own self-dosage.

A professor at the University of Chicago was found unconscious in his bed. His wife called a rescue squad, but an hour and a half of desperate measures failed to save his life. He had been prescribed sleeping pills by his doctor, his wife said, and used them to combat insomnia.

A New York business man, worried over finances, was given ten pills on prescription of his doctor. He took one, but noted no immediate relief. He swallowed another, and another, and still another, until all ten were gone. He sank into a deep coma, but quick action saved his life.

Awaking, he was amazed to find himself in a hospital.

"They weren't marked poison," he protested, "and the doctor gave them to me. I thought it was all right."

Illustrated By LEE CONREY

The Grim Hand of Death Had the Next Move in the Innocent Game of Playing Doctor—When the Little Girl Fed Several Sleeping Pills to Her Brother.

14 December 28, 1947 THE AMERICAN WEEKLY

Figure 1.2. The pharmaceutical as folk devil: barbiturates snatching the innocent at the height of the goofball panic. Source: *Milwaukee Sentinel,* December 28, 1947.

like harsh treatment of noncompliant druggists and doctors and imprisonment of users without prescriptions. This federal measure was debated in Washington for the next four years, discouraged by Anslinger, who was uninterested in extending his reach so far into the domain of medicine, but publicly supported by the FDA. Meanwhile, druggists and other opposing interests rallied, in the hope of "heading off" the pending federal controls with less restrictive state controls.[19]

Industry and Professional Reaction: Damage Control

The pharmacists and trade organizations acknowledged that some change would be required. A September 1945 feature in the *American Druggist* warned that the wave of barbiturate deaths and media attention portended "a wholesale black eye and strangling government regulation" for the entire pharmaceuticals business. It assured the readers that while recent media coverage might be "sensationalized," the FDA, state and local public health authorities, and police departments, as well as "church groups, and other organizations concerned with the moral welfare," were all intent on curbing barbiturates. Strict federal control of the barbiturates as "ordinary dope" was in the offing. While noting the risks of accidental suicide and habit formation, this article stressed the third main danger surrounding barbiturates, their nonmedical use among " 'zoot suiters,' teen-agers, and certain of the poorer elements of the country." And as for who supplied the drugs for nonmedical use, the piece offered the example of "the notorious Fu Manchu, a Negro addict and peddler," said to be the ringleader of a major Los Angeles gang dealing barbiturates stolen from pharmacies and warehouses. Apart from demonizing *nonmedical* barbiturate use by association with a remarkable triple stigma (at once "Negro," junkie, and Asiatic villain), the net effect was to demarcate and diminish the problems associated with medical overuse and to represent the sins of even a lax pharmacist as trivial in comparison with a professionally deviant underworld. This emphasis fit the core argument of the piece: the main offenders were not the druggists but organized crime, sometimes abetted by a few unscrupulous manufacturers and distributors.[20]

In New York the druggists' association and the state Pharmaceutical Council (representing manufacturers and distributors) stalled imposition of the new city health code limiting barbiturate refills for nearly two years, and when the rules were implemented in July 1947, the council brought legal challenge.[21] That year the National Drug Trade Council promoted a "model" state law, restricting sales to prescriptions and explicitly refillable prescriptions, making possession of barbiturates without prescription illegal, and providing for criminal penalties for wholesale supply to anyone other than doctors and pharmacists. There was no limit on physician-approved refilling, nor was there Harrison-style inventory tracking that might enable enforcement. In some states such measures were vigorously pushed by pharmacists fearing strict federal controls like the Nourse Rogers bill or harsher state laws, born of what they saw as a popular barbiturate "hysteria."[22]

The medical profession's response, if the AMA can be taken as its voice, was now similar to that of the pharmacists. "The 'goof ball' panic is on!" began a June 1945 piece in the association's popular organ *Hygeia*. It recounted the successful investigation of widespread barbiturate abuse in Waco, Texas, noting the area's social destabilization by an influx of war industries and workers. FDA action against Fadal's pharmacy dried up this single "uncontrolled source of supply," handily solving that city's barbiturate problem, the piece asserted. Thus the FDA's prescription-only regulation of pharmacists was all that was required, with state enforcement help, to eliminate the "barbiturate menace." The barbiturates themselves were described as "one of the great discoveries of modern drug and medical science" when "used properly under the supervision of a physician." While leading drug addiction scientists had no hesitation about classifying the barbiturates as "addictive" (see below), *Hygeia* skirted the issue by avoiding the term and admitting only that the drugs can produce "habitual reliance for sleep." The drive for stricter controls among state public health officials went unmentioned. Instead, *Hygeia* attributed the impetus for barbiturate control to "regulatory men," and the piece closed with the suggestion that "according to some FDA men the best answer would be a federal law putting the drugs under the jurisdiction of the Narcotics Bureau" (as soon proposed by Nourse Rogers).[23] That the AMA's response was deflationary fits with professional self-interest, as does the implication that behind the "panic" stood the federal government's ambitions to extend regulation to prescribing. (As an aside, the dismissive discourse of "panic" and "hysteria" here strikingly suggests how close to actors' categories lies the moral panic concept.) Both the dismissive tone and the antigovernment posture may signal the AMA leadership's drift into alliance with the drug industry that characterized the 1950s, in contrast to the skepticism and vigilance toward the industry that characterized organized medicine in the interwar era. While it supported restriction of barbiturates and the FDA's other "dangerous drugs" to physician prescription, when the Nourse Rogers bill was introduced, the AMA strongly opposed its prescribing surveillance as "continuation of a trend to confer more and more jurisdiction of the Federal Government over the practice of medicine."[24]

The drug industry's response was similarly predictable in terms of economic interest and was as intransigent as public opinion could tolerate. Responding to the early 1945 flurry of interest spurred by the *Saturday Evening Post* came a piece that October in the middlebrow national monthly *American Mercury*. The author was Theodore Klumpp, a respected former AMA and FDA official and now president of Winthrop pharmaceuticals, a Bayer subsidiary to which Klumpp had been appointed due to government intervention during the war. (Winthrop's two best-selling products were then the barbiturate Luminal [phenobarbital] and the opioid painkiller Demerol [meperidine or pethidine], both very vulnerable to regulation.) After lingering philosophically on insomnia as an inevitable consequence of "modern living" and recommending warm milk, Klumpp's piece described barbiturate sleeping pills as the best medical option, and a sound one given an argument based on proportionality: with billions of doses consumed in 1940, a death toll only in the hundreds testified to their reliability and safety. Echoing the standard moral-psychiatric stigma associated with drug addiction at the time, Klumpp noted that the risk of barbiturate habit formation existed only in individuals lacking "normal self-control," no special abuse liability, and no particular need for regulation.[25]

As the barbiturate panic grew too intense to quash or evade, industry efforts focused efforts on limiting new controls, and particularly on avoiding inventory surveillance. The American Pharmaceutical Association, representing professional pharmacists and some druggists, endorsed the above-mentioned National Drug Trade Council's "model state law," slightly stricter than the similar measure they themselves had proposed in October 1945 (which would have allowed telephone prescription renewals). But even Anslinger criticized the model law for making prescription refills too easy and for failing to define "the responsibility of manufacturers and wholesalers to keep the drugs within legitimate channels." The retail drug industry certainly had reason to worry about the business impact of barbiturate restrictions, because the pills were among drugstores' biggest sellers. In 1948, when the New York Pharmaceutical Council brought suit against the city's moderate record-keeping and refill controls on barbiturates and increased enforcement, one of its grounds was "severe financial loss" to druggists. In response, New York City public health officials expressed satisfaction with the druggists' plight.[26]

Dangerous Drugs: The Wider Issue of Federal Authority, 1948–1951

For the federal government's powers to regulate pharmaceuticals, 1948 was a watershed year. As noted above, in the Sullivan decision, the Supreme Court upheld the FDA's authority to prosecute pharmacists for dispensing "dangerous" drugs without prescription. Thus, the FDA was legally if not legislatively empowered to enforce prescription-only classification on pre-1938 drugs. The Sullivan decision seems to have emboldened the agency in further imposing the standards of scientific medicine on the drug trade, particularly regarding prescription refills. At a 1948 national druggist's convention, FDA commissioner Paul Dunbar struck fear into the hearts of pharmacists in a speech where he likened a prescription to a check written by the doctor, which, once cashed, lost validity. Refilling prescriptions without written specification, and filling or refilling prescriptions issued only orally by doctors, would henceforth be treated as dispensing without prescription.[27] While the agency had continued prosecuting druggists who dispensed barbiturates without prescription after the initial victories against Fadal and Sullivan, in 1948 the FDA followed with a new round of aggressive intervention against pharmacies refilling valid barbiturate prescriptions without written physician instructions. The agency also incensed manufacturers and some doctors by barring certain hormone preparations on the grounds that without demonstrable effectiveness they could not be used safely. Ineffective hormones and unprescribed antibiotics notwithstanding, after Sullivan the barbiturates became the FDA's flagship enforcement priority—with encouragement from President Truman, one inspector recalled later. And according to FDA historian John Swann, from about this point through the entire 1950s, more of the FDA's enforcement efforts targeted barbiturates and (increasingly) amphetamines than all other pharmaceuticals combined.[28]

In 1950 another shift occurred to favor stricter controls, blocking the prevarications of drug trade representatives who avoided the loaded term *addiction* and spoke of barbiturates as merely "habit forming." Use of the habituation-addiction distinction to defend barbiturates had always rested on the shakiest of scientific grounds, since the definitive animal physiology assays for addiction of the 1930s showed that barbiturates and alcohol were both truly "addictive" in essentially the same way as morphine.[29] After the war, Harris Isbell of the Public Health Service's Lexington, Kentucky, narcotics research unit confirmed this in human studies, finding that habitual barbiturate use even produced a potentially fatal withdrawal syndrome. Making headlines nationally in 1950 and attracting congressional attention, Isbell announced, "Addiction to barbiturates is far more dangerous than is addiction to morphine," and called it an increasingly serious national problem that demanded action.[30]

Scientific proof of barbiturate addictiveness took on additional political significance in early and mid 1951. A fresh storm of anxiety over drug addiction arose, fueled by alarms from the Federal Bureau of Narcotics about new heroin supplies flowing from postwar Europe. The situation was exploited by politicians, such as Representative Hale Boggs (D-LA), who were discovering the advantages of declaring "war on narcotics." While Boggs pushed his namesake bill to impose mandatory drug-offense sentences in the House, the Senate Committee on Crime held a highly publicized (and even televised) hearing on the new "narcotics scourge" spreading "moral leprosy" among American youth. In both houses FBN officials testified amply to the magnitude of the renewed "dope" menace. Advocates of prescription drug control, like Nourse Rogers and Lexington hospital director Victor Vogel, attempted to link the now-aging panic over barbiturates to the fresh narcotics alarm, and sleeping pills were named as a big part of the problem in discussions of both "narcotics" addiction and that recurrent popular concern of the era, juvenile delinquency. In 1951 Nourse Rogers again floated her initiative to roll barbiturates into the strengthened narcotics-control legislation with FDA support—and initially in alliance with Boggs. But due to Anslinger's resistance, the Boggs bill passed without covering barbiturates (indeed, this exclusion might have been necessary for the bill's passage, since its extremely repressive measures would have produced a political liability not unlike Prohibition, as Anslinger astutely observed, if applied to such popular drugs).[31]

In the context of the hue and cry about addictive narcotics, *something* would certainly have to be done about "America's opium"—the now undeniably addictive, popular barbiturates denounced by public health officials and defenders of morality. The 1951 Durham-Humphrey amendment, which established the modern category of prescription-only drugs, was presented as the answer. This legislation was initiated because pharmacists sought relief from the heavy toll of prosecutions and convictions by FDA, which (if these were proportional to investigative effort) must have stemmed chiefly from barbiturate sales without prescription. They wanted oral and particularly telephone prescriptions and refills to be authorized. And they wanted an end to the situation of uncertainty created because the same compound might be labeled prescription by one manufacturer but labeled by another for direct sale with instructions for use (never an issue with barbiturates after the FDA prescription-only regulations of 1940). Durham-Humphrey was a compromise brokered by druggists (specifically the National Association of Retail Druggists) with the FDA, in which the druggists could get their certainty and their permission to honor telephone prescribing and the agency could gain legislative authority to do more of what it had already been doing administratively, that is, determining what drugs must be sold only on prescription, based on safety *and efficacy*. The bill as drafted specifically named barbiturates and other habit-forming drugs as prescription-only, and that provision was never challenged. But the drug manufacturers were unwilling to cede to FDA any statutory authority over efficacy, and through aggressive lobbying and congressional airings of concern about "socialized medicine"—in the context of Truman's national health insurance initiative and the early cold war—the drug industry managed to have the efficacy language struck. The legislation as passed empowered the FDA to remove drugs from the prescription-only category if it judged them safe but technically left the decision whether to place drugs in the prescription category with manufacturers. This outcome was, as historian Harry Marks has argued, mainly a loss for the FDA's effort to make pharmaceutical usage more rational and "effective" (although the agency continued to judge safety in the light of efficacy).[32]

Durham-Humphrey was certainly a poor outcome for advocates of stricter barbiturate control. The bill imposed no special regulation on barbiturates other than that they were listed as prescription-only and that telephone refills had to be reduced promptly to writing. Nourse Rogers continued to push her bill for barbiturate supply-chain tracking, licensing, and monitoring of dispensing and prescribing, but with opposition from the AMA and the pharmaceutical industry, and without support from the FBN, it stalled and foundered after the nearly simultaneous passage of the Boggs and Durham-Humphrey acts. The AMA Committee on Legislation explained why the medical profession should oppose Nourse Rogers: "The recently enacted Durham-Humphrey amendment to the Federal Food, Drug, and Cosmetic Act should provide adequate safeguards against the illegitimate distribution and sale of barbiturates that move in interstate commerce" by requiring prescriptions for sale and physician consent for refills. Furthermore,

> legislation providing a federal licensing or registration system of control over the prescribing or dispensing of barbiturates by physicians is unnecessary and would set a dangerous precedent. The Federal Security Agency, through the Food and Drug Administration, might well attempt to extend this type of control to cover all drugs having any significant potentiality for harm. By this means, the Administrator of the Federal Security Agency [Oscar Ewing, the outspoken advocate of national health insurance] could ultimately attain the power to dictate in large measure the practice of medicine in this country.[33]

Thus, in the intensifying cold war atmosphere, fear of socialism trumped fear of addiction, and the FDA was unable to go further in limiting the use of "dangerous drugs" to scientifically approved indications and quantities. Barbiturates—the drugs whose medical use most obviously demanded restriction—had not provided enough of an opening wedge for the federal regulators to expand control over drugmakers or prescribers, no instrument adequate to challenge the growing intimacy of these partners in medicine. The newly codified prescription, imposed by government as a barrier between manufacturer and patient, became invested as a symbol of medicine's autonomous authority and fortified by opponents of regulation as the proper boundary of government's reach. Medically authorized drug use was thus defined as unproblematic and only nonmedical drug use problematic. Federal authority to regulate prescribed medication remained limited to "narcotics." As a consequence, further government efforts at barbiturate control focused even more on the addictive properties and especially the narcotics-like social implications of the drugs.

Thrill Pills versus Medicine in the (Long) 1950s

Although the FDA may have abandoned its ambitions to regulate prescribing after Durham-Humphrey, the agency did not retire from expanding control over dangerous medicines—the leading example of which remained barbiturates. Throughout the 1950s, the FDA continued to push for stronger controls over manufacturers and at least surveillance of prescribers of barbiturates (and, increasingly, amphetamines) in alliance with the defenders of law and order and public decency who had supported the Nourse Rogers bill.[34] Time and again throughout the decade, FDA chief George Larrick stood before Congress together with public health advocates—and often opposed by Anslinger—with efforts to conjure outrage at the social harms of these oft "misused" pharmaceuticals. For the new Senate Subcommittee on Juvenile Delinquency in 1953, Larrick described "thrill seeking exploits of teenagers who got their kicks from Benzedrine [amphetamine] at 'goof ball' parties that turned into sex orgies." Before the special committee on narcotics control laws, sponsored by Senator Price Daniel (D-TX), Larrick related in 1956 how barbiturate "addiction produces a general dissolution of character." He asserted, "We know of men who have held responsible positions but gradually became derelicts through the use of these drugs . . . [Women] no longer take an interest in the home or children, get dirty and slovenly; steal money and sell furniture to get the drug." In 1958 Larrick found a steady ally in Senator Thomas Dodd (D-CT), who used the juvenile delinquency subcommittee to probe the youthful "wave of narcotic addiction," crime, and sexual deviance spurred by "thrill pills"—Larrick's catchphrase unifying the social menace of drugs with the opposing pharmacology: amphetamines and barbiturates. Larrick's testimony was always replete with morality tales, featuring respectable citizens who went "berserk under the influence" and tending carefully to scapegoat weak and stigmatized groups for misuse, such as homosexual "sex deviates preying on juveniles" by selling them pills, rather than naming pharmaceutical makers as villains.[35]

By emphasizing the "narcotic"-like hazards of barbiturates, Larrick essentially sought to create a category of morally and pharmacologically dangerous medicines. Situated between ordinary prescription drugs and "narcotics," they would be controlled but not forbidden, their supply closely monitored and restricted to prevent "diversion" to nonprescribed or inappropriate use. But despite Larrick's best efforts to hitch the public health wagon to the horse of "narcotics" addiction outrage so as to regulate these widely prescribed drugs more closely (without intruding on medical prerogative), and despite sympathetic media coverage with headlines linking barbiturates and amphetamines to "dope" and delinquency, nothing of any consequence was done to limit their consumption between 1951 and 1965.[36] Resistance by the drug industry and organized medicine proved too strong for the FDA, and no regulatory space was opened between benign prescription medicines and inherently bad, addictive "narcotics."

Reasons for this long delay in federal regulation of abused prescription drugs are as numerous and varied as the reasons for its end in 1965. They are too broad to discuss in any detail here, but a few may be identified. For one, the credibility of the medical profession and the pharmaceutical industry were both at a high-water mark in the 1950s, but their allied power to resist declined in the wake of the Kefauver Senate hearings exposing the industry's corrupt practices for manipulating physicians and medical knowledge, and especially after the thalidomide birth-defect catastrophe of 1962. Another reason for the long delay is that, despite a mid-1950s resurgence of delinquency anxieties, competing moral panics overshadowed the worries associated with drugs for most of the 1950s—foremost among which was Communist infiltration, beginning with the Rosenberg "atom spy" trial in 1951, which brought McCarthyist witch hunts on its coat-tails.[37] And as a reason for the eventual passage of the Drug Abuse Control Amendments to FDA's enabling legislation in 1965, one must not neglect the "drug counterfeiting" provisions in those amendments that, their questionable relevance to public health aside, acted as a potent sweetener for the larger pharmaceutical firms. But the legislative delay between 1951 and 1965 cannot be rationalized as an appropriate response to a disappearance and then resurgence of barbiturate oversupply: annual U.S. production of barbiturates hit a peak of 6.3 billion one-grain doses in 1947, declined only to 4.8 billion in 1951 (as wartime export markets dried up), and then gradually climbed back to about 6 billion by 1960—some thirty-three doses per year for every man, woman, and child. Furthermore, only half of national barbiturate consumption was attributable to legitimate medical use in 1951, a proportion that must have further declined as the "minor tranquilizers" partially replaced them during the later 1950s.[38] (On the other hand, it is true that the amphetamine supply exploded dramatically in the interval, annual production rising from around the billion-dose mark in 1950 to an FDA-estimated 8 billion doses in the early 1960s—no more than half of which could be accounted for by medical use. However, until the late 1960s, barbiturate abuse loomed much larger than amphetamine problems in popular and political discourse.)[39]

For whatever set of reasons, by the mid-1960s political and social attitudes around drugs had changed enough that a gap finally opened between prescription medicines and the forbidden, addictive "narcotics." Filling that gap, the Drug Abuse Control Amendments passed under President Lyndon Johnson in 1965. Larrick gained what he long had sought: tracking requirements for manufacturers and wholesalers of barbiturates and amphetamines, criminal penalties for possession without prescription, and law enforcement powers for the FDA. However, despite empowering the agency to fight "diversion" from legitimate channels, the new law did not appreciably reduce the quantities of abused pharmaceuticals consumed in the United States (and hence could not have much reduced the public health harms). Pharmaceutical business channels were evidently part of the problem; for instance, some 90 percent of all amphetamines in street traffic were products of U.S. pharmaceutical firms even in the midst of a late-1960s bathtub methamphetamine epidemic.[40]

Resolution: Regulating the New Domain between "Narcotics" and Medicines, 1965–1971

The way in which the prescription drug overuse problem was reconceived and addressed reflects the complex social and cultural changes associated with the late 1960s and the dramatic escape of the issue from the narrow set of discourses and institutions associated with drug addiction and pharmaceutical regulation since the 1940s. Anslinger retired in 1962, and Larrick followed in 1965. The influence of the FDA, federal law enforcement, organized medicine, and even the drug industry was diluted by a politically resurgent consumer movement and a popular culture more sympathetic to nonmedical drug use and more suspicious of repressive approaches to deviance. In the later 1960s, a drug-related moral panic of traditional repressive flavor did emerge, for instance among the suburban Illinois mothers' organization Grassroots, dedicated to saving middle-class children from parties serving food "treated with LSD" as well as marijuana and heroin. Reactionary panic probably focused most on LSD (never construed as addictive) and its Pied Piper, Timothy Leary.[41] But this time the voices of worried parents were answered by police and school guidance counselors who warned that the prescription "pill popping" example they set for their children put them in a hypocritical position. Thus authorities began echoing consumer advocate and public health critiques of pharmaceutical industry "pill-pushing." The political center and left, from Robert F. Kennedy Democrats to counterculture drug users' activist groups like California's DoItNow, further blurred the old boundaries between addictive "narcotics" and "mother's little helpers"—the bestselling prescription tranquilizers and diet pills and sleeping pills, now said to be equally addictive and potentially deadly. Attributed to far more substances than just the traditional "narcotics," addiction became proportionally more common, less stigmatizing, and less definitive of drugs demanding criminalization.[42]

At the beginning of the 1970s, a new generation of public health researchers went still further than the case-study approach of their predecessors, quantifying the harms of "overprescribed" pharmaceuticals with the same, but freshly refined, instruments that measured illicit drug use. The first systematic household drug-use surveys demonstrated that prescription drugs were indeed widely used nontherapeutically, as much by those holding prescriptions as by those without. Thus, mainstream media reported, the housewives and businessmen of mainstream America were indulging in a prescription-fueled national pill party.[43] Even Vice President Spiro Agnew seemed to agree in 1970 that the United States was on a "collective national [drug] trip" in which "legal drug use" and "hard drugs" like marijuana and heroin were together exacting related, terrible social costs. With such approximate agreement across the political spectrum that pharmaceuticals and addiction were part of the same problem, the stage was set for the Nixon administration's effort to demonstrate its toughness on drugs and simultaneously to include public health and law enforcement in the same "big tent" combined approach to limiting drug demand and harms as well as supply: the Comprehensive Drug Abuse Prevention and Control Act (CDAPCA) of 1970.[44]

This is not the place to analyze the intricate politics of that important act and its evolving realization in increasingly repressive drug control measures after the 1970s. Suffice it to say that once the Department of Justice was empowered to take an approach much less deferential to medicine and the pharmaceutical industry than the FDA ever was, law enforcement quickly dealt with the prescription pill-popping problem. Under the Controlled Substances Act component of CDAPCA (and the related UN Convention on Psychotropics), the predecessor of today's Drug Enforcement Administration (DEA) in 1971 moved amphetamines to the newly created Schedule II on its own administrative authority, disregarding congressional reluctance to do the same legislatively. The second-strictest level of control, Schedule II required prescriptions on special forms that enabled easy tracking and prevented refills, just as Nourse Rogers and other antibarbiturate forces had long been urging, and allowed the agency (in consultation with the FDA) to set quotas limiting production to actual prescription demand. Faced with oversight of prescribing, American doctors in 1972 were only prepared to justify one-tenth as much of the amphetamines as they had prescribed in 1969. DEA followed the same procedure with barbiturates, beginning with the 1973 rescheduling of secobarbital to Schedule II, and by the end of the decade, the abuse of both classes of drugs had become much less important public health problems (even if "prescription mills" continued cropping up, especially for newer drugs; see chapter 9, by David Herzberg). Unfortunate though it may be that public health had to join forces with law enforcement, and moreover that these increasingly strange bedfellows had to work together in policing the medical profession's prescribing to contain the social harms of psychotropic drugs, this is the clear lesson of the barbiturate problem's quick resolution in the 1970s, after the FDA's thirty years of failed efforts.[45] So long as manufacturers can hide behind the medical discretion of the prescribers they have learned to influence so effectively, and users are willing to pay for prescriptions of dubious justification, the prescribing hand can never be free again.

CHAPTER 2

Pharmacological Restraints

Antibiotic Prescribing and the Limits of Physician Autonomy

Scott H. Podolsky

Antibiotic overuse, especially in relation to antibiotic resistance and a perceived failure of antibiotic stewardship, has long served as one of the most visible indictments of therapeutic rationality.[1] Pharmaceutical "education" and patient pressure are acknowledged as important contributors to this process; yet the physician with a prescription pad (or phone or, increasingly, keyboard), at the moment of decision, remains the chief object of scrutiny. Through this lens, the antibiotic-prescribing physician serves to focus concerns regarding therapeutic autonomy itself, from the proper restraints delimiting it to the necessary systems supporting it. More historically, such tension has been defined amid the classic dichotomy between regulation and education.

While such tensions appear today in the context of a seemingly shrinking "antibiotic pipeline" and the potential opportunities offered by computer order entry,[2] they are not new. Rather, the contours of this discussion were defined during a critical period from the early 1960s through the late 1970s. They must be understood against the backdrop of a nearly continuous series of congressional hearings regarding the marketing practices of the pharmaceutical industry; the impact of the removal from the market of the fixed-dose combination antibiotics through the Drug Efficacy Study and Implementation (DESI) process, beginning in the late 1960s; increasing sophistication in data gathering and data analysis regarding physicians' prescribing habits; concerns regarding national health expenditures raised in relation to Medicare and Medicaid; and increasing apprehension regarding antibiotic resistance and its implications.

In the 1950s leading academic infectious-disease-based reformers focused more upon antibiotic marketing than individual "misuse" and contributed to the passage of the Kefauver-Harris Drug Amendments of 1962 and the DESI process, whereby inefficacious drugs could be, respectively, denied entry to or removed from the market. From the 1960s through the 1970s, the attention of reformers shifted from the marketing of apparently inefficacious drugs to the apparently "irrational" prescribing of efficacious ones. As studied by clinical pharmacologists, pharmacists, epidemiologists, and governmental agencies, findings of irrational prescribing were promulgated both in the medical literature and at governmental hearings on the pharmaceutical industry. Yet the limits to the delimitation of physicians' prescribing autonomy with respect to antibiotic wonder drugs had been reached with the passage of the 1962 drug amendments and DESI. And as academic reformers, governmental agencies, and national medical organizations debated regulatory versus educational remedies to the problem, solutions were defined around local, voluntary restrictive programs and, more often, educational initiatives. These programs and initiatives remain with us today, as does the antibiotic overuse problem, reflective of the continuing and interrelated epistemological and social difficulties in delimiting physician prescribing autonomy.

Antibiotic Reformers and the Regulatory Backdrop

To properly situate the increasingly sustained attention to antibiotic prescribing in the 1960s, one must first appreciate the general tenor of the 1950s infectious-disease-based reform activity from which it emerged. As Edward Kass and Katherine Murphey Hayes have related in their history of the Infectious Diseases Society of America, by the time the antibiotics came on the scene, infectious diseases were already on the decline, no longer the common experience of the typical general practitioner. A proto-specialty of infectious disease experts could thus serve as the repository of clinical knowledge about such receding diseases as typhoid and tetanus, and at the same time they could study the emerging classes of antibiotics produced by the pharmaceutical industry and advise clinicians regarding the right drugs for the right bugs in daily hospital or outpatient practice.[3]

Scattered warnings regarding antimicrobial misuse and overuse had dated to the days of the sulfa drugs and penicillin.[4] The advent of such heavily marketed, individually branded, and patented "broad-spectrum" antibiotics as Lederle's Aureomycin (chlortetracycline), Parke-Davis's Chloromycetin (chloramphenicol), and Pfizer's Terramycin (oxytetracycline) in the late 1940s and early 1950s resulted in a growth industry for those who sought to deliver talks (often at state medical society meetings) or write papers drawing attention to the possibility for widespread misuse by clinicians who seemed even less inclined to render specific diagnoses.[5] Cautionary, if anecdotal, tales regarding antibiotic-wielding clinicians' failure to properly treat such increasingly camouflaged nosological zebras as malaria, carcinomatosis, or even tuberculosis, or to overtreat such horses as the common cold, appeared frequently in medical journals throughout the 1950s.[6]

Yet, whether owing to their limited ability to quantify physicians' prescribing activity or because of the increasingly brazen nature of antibiotic advertising amid the advent of the wonder drug revolution, the nation's leading academic infectious-disease experts of the 1950s turned their attention more to the pharmaceutical marketing of what they perceived as poorly conceived drugs than to the misuse of appropriate ones. In particular, such leading clinical investigators as Harvard's Maxwell Finland, the University of Illinois's Harry Dowling, and Ernest Jawetz of the University of California, San Francisco (UCSF) focused upon the marketing of "fixed-dose combinations" of antibiotics (set dosages of more than one antibiotic in a single pill), promoted from the mid-1950s onward in the name of "synergy" (the more-than-additive antimicrobial capacity of antibiotics prescribed in tandem) and broad microbiological coverage. Finland, Dowling, and Jawetz not only challenged the generalizability of "synergy" to most clinical scenarios but also drew attention to such potential concerns as the conflation of side effects, the inadequate dosing of one or more components for a given case, and the potential for drug *antagonism* rather than synergism. More fundamentally, they regarded such combinations as convenient devices by which pharmaceutical companies could link their established antibiotics to less-efficacious emerging antibiotics to render newly patentable entities.[7] To them, such medicines and their marketing represented portents of a future in which bluster would supersede substance, bringing down the entire and increasingly interrelated edifice of medicine and the pharmaceutical industry.[8]

Perceiving an apparent regulatory vacuum in which neither the Food and Drug Administration (FDA) nor any other national agency was empowered to adjudicate the therapeutic efficacy of newly introduced drugs, these reformers and their colleagues, throughout the latter half of the 1950s, unleashed a coordinated series of articles criticizing such remedies. In particular, they attempted to counter the case-series-based "testimonials" upon which such drugs were justified with calls for "controlled clinical studies" that would place the remedies under appropriate scrutiny. Their activities ultimately intersected with the concerns of reformers at the FDA and in Congress, contributing to the eventual passage of the Kefauver-Harris amendments of 1962, which mandated proof of a new drug's efficacy prior to approval. By 1973, as a result of the consequent DESI process, approximately six hundred apparently inefficacious *existing* drug products were removed from the market. DESI was epitomized by the successful removal of the fixed-dose combination antibiotics (most notably Upjohn's Panalba), despite being challenged all the way to the U.S Supreme Court by autonomy-citing segments of the medical profession and the pharmaceutical industry.[9]

Antibiotics and "Irrational" Prescribing

Antibiotic reformers could thus congratulate themselves upon the evaluation of such apparently inappropriate drugs and their removal from the market, concluding an important era in reformer activity.[10] The Infectious Diseases Society of America itself was formed the year after the passage of the Kefauver-Harris amendments; Max Finland was its first president and Harry Dowling its third.[11] Yet, increasing attention had been gathering under the radar throughout the 1950s with respect to the apparently "irrational" prescribing of *appropriate* antibiotics, and such attention was framed by shifting notions of "prophylactic" antibiotics and diagnostic specificity. In 1941 a Virginia practitioner could articulate the ethos of prophylactic antimicrobial administration (as well as unfettered clinical judgment) to his colleagues at the Virginia Medical Association:

> And now I am starting in 1941 to use sulfathiazole and sulfapyridine prophylactically. And why not? It has not been proven to work that way! Not scientific, you say! Remember we are front line soldiers; when we see the enemy we do not have to wait for orders from headquarters through a long line of red tape. We must go for him, without waiting for the attack! Again, it seems to me, that is common sense medicine. What do we fear in grippe or a bad cold? Pneumonia. What do we fear in whooping cough and other contagious diseases, or post-operative? Pneumonia. If pneumonia develops, we have a remedy of proven value. Why wait? Can you tell when pneumonia is going to develop? If it does develop, you would use sulfathiazole or sulfapyridine with confidence. Then why not get the jump on those tough, little bacteria? Kill them before they get a foothold. Why wait for the attack? Bomb their channel ports! Wipe out their bases of supply! Prevent their starting out in the blood stream; meet force with force![12]

Just nine years later, on the December 7 anniversary of another bombing, Maxwell Finland publicly enunciated a counteroffensive at the New York Academy of Medicine, focusing upon the widespread and untested prophylactic use of antibiotics—from their use in preventing colds from turning into pneumonias to their use in preventing elective surgical procedures from turning into postoperative sepsis. Wondering aloud whether such usage might be contributing to increasing staphylococcal antibiotic resistance in particular, Finland posited, "A major portion of the antibiotics actually administered to individuals may be serving little or no definite purpose in terms of curative effect of active infection."[13] That same year, UCSF's Ernest Jawetz focused upon diagnostic specificity in publicly criticizing the use of "shotgun" antimicrobial therapy in febrile illness, insisting that such specificity and the use of the laboratory were essential for the application of "rational therapy."[14]

Throughout the early 1950s, as an uncoordinated series of articles on "the use and misuse of antibiotics" proliferated,[15] a more general rationale for apparent overuse, especially in the outpatient setting, was certainly acknowledged by would-be reformers. Wendell Hall reported at the "Symposium on the Abuses of New Therapeutic Agents" at the annual meeting of the Minnesota State Medical Association in May of 1952:

> We are living in an age in which both physicians and the general public have come to expect miracles of medicine and surgery, leaving little to the powers of Mother Nature. Rare is the physician who has not been urged to prescribe

> all manner of antibiotics for patients with fever or other equivocal evidence of infection. The pace of living and of medical practice is such that all too often the first question is not "What is the cause?" but rather "Should I take sulfa or penicillin?" There can be little doubt that a considerable part of this attitude is the result of the barrage of publicity concerning the new "wonder drugs" loosed upon the layman and physician by popular "science" writers and pharmaceutical companies. Many a physician is loath to treat a respiratory infection without giving antibiotics lest he be subjected to misinformed relatives of the patient.[16]

Allen Hussar, speaking before the second annual national antibiotic conference, held in Washington, DC, two years later, considered such an apparent rationale to include, in addition to the expectations of patient and family, the general safety of the antibiotics, the "urgency" of treating febrile patients before receiving the results of diagnostic testing, and the cost or even unavailability of laboratory evaluation in the first place.[17]

Nevertheless, while such a rationale might be understood, it did not prevent self-defined reformers from increasingly labeling what they deemed inappropriate prescribing as "irrational" behavior. Ernest Jawetz, in his talk "Patient, Doctor, Drug, and Bug" at the 1957 national antibiotic conference, put this most starkly in concluding, "Is it asking too much that in a few areas man behave as a rational being?"[18] The 1950s had witnessed increasing attention paid to the nature and degree of influence of pharmaceutical marketing upon physician prescribing,[19] and by the end of the decade, Jawetz and C. Henry Kempe, in their examination of the problem, echoed Finland and Dowling in attributing a large amount of the blame to pharmaceutical advertising. The pediatrician Kempe concluded, "A rational approach to the selection of the proper antimicrobial agent in treatment of children with bacterial disease requires some basic information and a considerable degree of sales resistance to undue or premature advertising claims."[20] Kempe and his article figured prominently in John Lear's 1959 *Saturday Review* piece "Taking the Miracle out of the Miracle Drugs," which itself played an important role in widening the scope of the Kefauver hearings and their attention to pharmaceutical marketing.[21]

However, as the government began to turn its attention toward such marketing, a lineage of researchers was beginning to turn their attention away from the marketers to an actual enumeration—and evaluation—of antibiotic prescribing habits. In the outpatient setting, the first study took serendipitous advantage of the isolated army-depot community in Igloo, South Dakota, where medical care was available to all for the same affordable price and medical records were freely available to army hospital researchers. William Nolen and Donald Dille found that from January 1, 1952, through December 31, 1956, of the 763 residents of the community, 703 had received antibiotics and that 52.5 percent of such treatments were "not indicated."[22] Over the ensuing decade, as both the pharmaceutical industry and academicians continued to study physicians' prescribing behavior, they moved beyond such natural experiments to prospective studies. Foregrounding the outpatient antibiotic studies were the evolving twin depictions of the common cold as the nosological litmus test for inappropriate prescribing and of the use of chloramphenicol (definitively established by the 1960s as a cause of aplastic anemia) in this particular setting as an even stronger such indicator.[23] Researchers working with clinical pharmacologist Louis Lasagna during his final years at the Johns Hopkins School of Medicine could by the late 1960s report on marketing research data showing that between October 1967 and September 1968, of 1,128 patient visits for the common cold, 60 percent resulted in antibiotic prescriptions.[24] And using their own computerized prescription-recording system in 1968, "monitoring 85% of the prescriptions dispensed in a defined geographic area," Lasagna's group reported not only that antibiotics remained "by far the most commonly dispensed" class of medications, but that chloramphenicol, still frequently prescribed, was "a leading drug in the country" that was mostly prescribed by "general practitioners whom one would not expect to be treating life-threatening infections on an outpatient basis."[25]

Despite such efforts and emerging technologies, it was still easier to capture hospital-based records, perhaps inspired by the nascent effort to promote "therapy audits."[26] In April 1966, Hobart Reimann related the wide variability in antibiotic prescription rates among twelve hospitals in the Philadelphia area,[27] while by 1967 the National Communicable Disease Center (renamed the Center for Disease Control in 1970) reported in its study of seven community hospitals not only that "considerable variability in antibiotic usage [existed] . . . within and among the seven hospitals," but that less than 30 percent of the patients studied had evidence of infections when first administered antibiotics.[28] By 1968 the federally convened Task Force on Prescription Drugs—envisioned by Lyndon B. Johnson as "a comprehensive study of the problems of including the cost of prescription drugs under Medicare"—formally employed the terminology "rational" versus "irrational" drug prescribing. It defined "rational" drug prescribing as selecting "the right drug for the right patient, in the right amounts at the right times," and "irrational" prescribing as "any significant deviation" from such a norm.[29] This was a key moment, as Jeremy A. Greene has described in chapter 10 of this volume, of increasing attention in the Medicare and Medicaid era to the "rational" and "irrational" use of drugs writ large. Antibiotics served, as they had since the 1940s, as a leading edge of this discussion, as qualitative notions of "abuse" and "misuse" generally yielded to more quantified demonstrations of "irrational" prescribing. By 1972 two pharmacists, Andrew Roberts and James Visconti, in their study of hospital antibiotic use titled "The Rational and Irrational Use of Systemic Antimicrobial Drugs," judged only 12.9 percent of all antibiotics administered at a five-hundred-bed community hospital as "rational," with 65.6 percent of the administrations deemed "irrational" and 21.5 percent deemed "questionable."[30]

Thus, in both outpatient and inpatient settings, critiques of "prophylactic" antibiotics—whether to forestall bacterial complications of upper respiratory infections or to forestall surgical infections—were coupled to critiques of apparently insufficient diagnostic testing for offending microbes themselves, producing a general call for a more "rational" use of antibiotics. By the early 1970s, such concerns were likewise fueled by a shift in the "risk" side of the general antibiotic "risk-benefit" equation. Still concerned with individual adverse effects (epitomized by chloramphenicol) and cost (epitomized by the more than doubling in hospital antibiotic costs between 1962 and 1971, after Medicare and Medicaid had been passed),[31] reformers increasingly turned their attention to potential systemic "ecological" consequences, a "pollution" manifested by the increasing number of serious "gram-negative" infections (caused by microbes generally considered normal colonic inhabitants), as well as by the wider emergence of antibiotic resistance itself.[32] Ironically, even though they were grounded in such "rational" concerns, calls for antibiotic delimitation outpaced the production of (what would be admittedly difficult) controlled clinical studies of antibiotic utility in many such "prophylactic" situations.[33] Yet, despite such limitations, by 1972 the reformist position received national exposure through a series of governmental hearings and their fallout. Eighteen years before, at the 1954 national antibiotic symposium, Allen Hussar had asked, "Should we continue to deal with this problem by an occasional speech or paper condemning the erroneous use of antibiotics, or, should a more aggressive effort be made to start a crusade for the rational use of antibiotics?"[34] By the early 1970s, such a crusade was launched in earnest.

Remonstration and Resistance

As with so much discourse around "rational" therapeutics in the late 1960s and early 1970s, a driving force for the crusade seeking the holy grail of rational antibiotic use was Gaylord Nelson's protracted hearings on the pharmaceutical industry.[35] After his examination of chloramphenicol and the fixed-dose combination antibiotics in the late 1960s, Senator Nelson turned his attention to "cough and cold remedies" and the "misuse of antibiotics" in early December of 1972. The tone of the hearings was set the first day, devoted to over-the-counter remedies, by the following exchange between Dr. Sol Katz, of Georgetown University Hospital, and Nelson:

> DR. KATZ: I get kind of distressed at industry saying . . . "Well, we are giving the antihistamine to counteract the stimulating effect of the sympathomimetic," when there is no data to show that the oral sympathomimetics are any good in the first place . . . So if we take out the antihistamine needed to counteract this sympathomimetic and we do not need the sympathomimetic, we do not have a cold preparation. We have no need for one.
>
> SENATOR NELSON: So you need the—
>
> DR. KATZ: You need chicken soup.[36]

By December 7, twenty-two years to the day since Finland first presented his counterattack at the New York Academy of Medicine, Nelson focused exclusively upon antibiotics. He called as his first witness the director of the FDA's Bureau of Drugs, Henry Simmons, who related increasing FDA interest in the problem of apparent antibiotic overuse by presenting a letter from FDA commissioner Charles C. Edwards, which highlighted such "national concern." Edwards had written, reflecting the shift in emphasis of the would-be reformers:

> We at the FDA are seriously concerned about the increasingly massive use of antibiotics in this country. In the past year 2,400,000 Kilograms, or the equivalent of approximately 10 billion doses of antibiotics, were produced in this country and certified for use . . . Antibiotics are being massively used when not

> indicated . . . One result of such use is iatrogenic disease, in the form of unnecessary adverse reactions and the development of super infections. Another is unjustifiable cost. Perhaps most important, antibiotic misuse carries a threat to the future efficacy of the drugs themselves since bacterial resistance can be and is being caused by such use.[37]

Simmons was followed by Harry Dowling, by then emeritus professor, veteran reformer, and seasoned governmental witness, who incorporated the FDA data to assert that while the FDA certified enough antibiotics to provide fifty doses (two full courses) of antibiotics to each American annually, the average citizen likely required an antibiotic only once every five to ten years.[38] The following day, Hopkins's Paul Stolley found himself with a national forum at which to present his published data on the usage of antibiotics to treat the common cold and the persistent overuse of chloramphenicol.[39] Simmons's and Stolley's research and testimonies formed the basis for their later commentary entitled "This Is Medical Progress?" in the *Journal of the American Medical Association.* There they cited the 30 percent increase in national antibiotic utilization between 1967 and 1971—against a 5 percent increase in the U.S. population over the same period—to conclude: "Have we reached the point where the enormous use of antibiotics is producing as much harm as good?"[40]

Dating back to the FDA's proposed removal of the fixed-dose combination antibiotics on the apparent basis of the input of academic "experts" instead of prescribing physicians' views,[41] academicians appearing as governmental witnesses had gone to great lengths to deny their ivory tower removal from the exigencies of daily practice. William Hewitt, founder and head of the UCLA division of infectious diseases, reported at the Nelson hearings on the fixed-dose combination antibiotics in 1969:

> I would like to emphasize that I am not an ivory tower basic scientist . . . I have had a practice of my own for 20 years and even to the present rely upon this type of activity for one-third of the income I derive from professional activities . . . [The men on his particular expert subpanel evaluating the merits of fixed-dose combination antibiotics] were not sitting in libraries writing textbooks and giving lectures to medical students but rather were daily seeing sick patients and caring for their medical and emotional needs. All of us participate liberally in local and national societies, the membership of which consists largely of "general physicians" concerned with both general and specialized medical problems.[42]

Yet by the early 1970s, even as leading reformer (and former Maxwell Finland fellow) Calvin Kunin was expressing his "sympathy" for "the man on the firing line of practice who has to face this barrage" of antibiotic requests from patients and their families, he was apt to be seen as condescending for his analysis of physicians' use of "drugs of fear," a practice that he believed could be obviated if the physician was, among other things, "willing and able to perform and properly interpret a few simple tests, such as Gram stains of exudates and appropriate cultures."[43]

In this setting, and extending in a direct line from previous battles over the removal of the fixed-dose combination antibiotics,[44] backlash was quick to emerge from physicians, not only against being told they were behaving irrationally but also against the consequent implicit threat of the further centralized restriction of antibiotic usage. In January 1974, Nea D'Amelio, associate editor of *Medical Times,* reported on the results of a questionnaire sent to ten thousand family physicians to determine "what the 'accused' had to say about the allegations" of antibiotic overuse; 5,331 physicians took the time to respond. When queried whether they felt "physicians [were] over-prescribing antibiotics," 55.5 percent agreed, but 44.5 percent denied the charge; and when asked whether they agreed "that the average person [did] not require antibiotics more than once in every five or ten years," 89.3 percent responded in the negative.[45]

The prose responses of the offended physicians clustered around several interrelated themes. Most fundamental were simple calls for a reappraisal of the overall cost-benefit analysis of the widespread use of antibiotics. Two years before, in *The Pills in Your Life,* Michael Halberstam (a cardiologist and brother of historian David Halberstam) had set forth the counterargument:

> My most heretical belief about antibiotic therapy, one which cannot be proven but which can be inferred from statistics, is that, given the choice between the purist approach and the admitted overprescription of antibiotics, the nation's health is vastly safer with the latter . . . Another reason why overliberal use of antibiotics is to be preferred to overconservative use (neither is ideal, of course) is that of pure chance. Just as some patients are unlucky enough to be subjected to serious reactions in illnesses they don't really have, so others are fortunate enough to be given antibiotics for completely wrong reasons—and to be cured of an illness the doctor never suspected.[46]

Halberstam was pilloried by Calvin Kunin both in print and in congressional testimony,[47] but similar sentiments were echoed in the *Medical Times* piece. One clinician asked: "Since the massive use of antibiotics for 'fever,' where are the mastoids that you saw—30 or 40 a year, the 40-day hospitalizations for pneumonia, the tubes draining empyema, the rampant fascial dissecting subcutaneous infections? Yes, we *are* 'over-prescribing' antibiotics, and it is good that we are. My God, let's *not* go back to the 'good old days!' "[48]

A more vehement form of backlash released the town-gown frustrations expressed previously with respect to the apparently "ivory tower"-influenced removal of the fixed-dose combination antibiotics:

> How many patients should you watch die from lack of prescribing before you give antibiotics? Those idiots in Washington should try the practice of medicine for awhile, instead of doing it from a test tube.
>
> Are we making more trouble than we are clearing up or preventing? I think *not*, in the absence of some trustworthy studies to the contrary, despite what a few power-hungry physician-bureaucrats, or forgetful, retired professors of medicine may say.
>
> I question whether the honorable emeritus professor of medicine has seen a patient in person from or in his ivory tower for some time. However, in 22 years of vigorous general practice, I haven't had one incidence of super-infection.[49]

Still more deeply, such sentiments were grounded in a general fear of further government-controlled restriction of physician autonomy. One physician from Texas declared, "I'll tell you the only thing I think is being overprescribed and that is a hell of a big over-dose of government being rammed down the esophagus of the medical profession."[50]

Such interrelated themes were united in a letter to the editor of the *Journal of the American Medical Association* written in response to Simmons and Stolley's questioning of "medical progress." Howard Seidenstein, a general physician from New Rochelle, New York, began with a general attack on Orwellian ivory tower pronouncements and perceived amnesia regarding the preantibiotic era:

> It was with a feeling of horror that I read the article by Simmons and Stolley and some of the subsequent comments. I hastened to check the date and was

> only a bit reassured when I saw 1974 and not 1984 . . . It might be that countless thousands of community physicians are as wrong as a $3 bill about antibiotics, but it is just as possible from where I sit that dozens of ivory-tower investigators may be wrong too. My forty years of medical school and general practice span the entire era . . . I recall full well the 40% to 50% mortality of ruptured appendices with generalized peritonitis, the 20% to 25% mortality of pneumococcal pneumonia, and the 100% mortality of streptococcal meningitis . . . I can still hear the howls of incompletely anesthetized young patients during myringotomies that barely reduced the terribly high incidence of "mastoids" and conduction deafness and even cavernous sinus thrombosis secondary to "red ears" . . . Following the "indiscriminate" use of sulfas and antibiotics in my practice, I have not had one single case of mastoiditis develop in 35 years. And although I've had a few close calls with anaphylactic shock following parenterally administered penicillin, I've not yet lost a patient.[51]

Seidenstein's concluding paragraph was most indicative of the prospects facing antibiotic reformers: "More iniquitous than any possible abuse (even of chloramphenicol [chloromycetin]!) are the abuses of power suggested by Simmons and Stolley and backed by Kunin." As related by Seidenstein himself, the action items suggested by the reformers included not the further restriction of offending drugs or prescribers, but only that hospitals should form guidelines and annual reviews of, and feedback regarding, antibiotic prescribing.[52] As such, the sensitized resistance to the proposed measures apparently reflected a deep chord of resentment held by a significant proportion of the medical community against further centralized encroachment—a resentment clearly noted by would-be reformers.

The Limits to Delimitation

The actual measures brought forth in the early 1970s to improve antibiotic prescribing reveal the limits to delimitation itself, especially as perceived by the reformers. More restrictive measures were in fact proposed and debated, especially by Senator Nelson and particularly with respect to chloramphenicol. At the 1967 hearings on chloramphenicol, Nelson had first suggested the possibility of permitting the usage of the drug "only with the written approval of a consultant except in emergencies or special circumstances";[53] and Mark Lepper of the University of Illinois (and a former Harry Dowling fellow) had suggested restricting its usage to hospitals alone.[54] Four years later, Nelson and Phillip Lee—at the time, chancellor of the University of California, San Francisco, and former assistant secretary for health of the Department of Health, Education, and Welfare (as well as chairman of the previously cited Task Force on Prescription Drugs)—lamented the failure of the FDA to regulate chloramphenicol. Lee again proposed to forbid its use in the outpatient setting and instead to restrict it "to hospital use after consultation," even if the patient in question did not require hospitalization, so as to "see this thing brought under rational control."[55] Prescribing general practitioners would then have to subordinate their authority to specialists, despite the potential impact upon the "physician-patient relationship."[56] Lee concluded that what had been thought of "as a problem between the individual physician and the individual patient" needed to be reconceptualized "as a public health problem" so as to overcome the "tacit conspiracy between physician and patient" in the "very act of writing a prescription."[57]

In Lee's framework, given the population-wide impact of inappropriate prescribing, physicians had to be further regulated, "just as Congress had to regulate the auto industry."[58] Yet despite Nelson's ability to persuade several other witnesses to concede to his and Lee's restrictive position during the 1972 hearings,[59] no broad reformist or political will was ever brought forth to qualify the use of any class of antibiotics. Rather, such "inconvenient" restrictive practices were limited to the inpatient setting, and only in local contexts. As far back as the 1950s, with the spread of staphylococcal resistance, scattered hospital staffs in the United States who were appreciative of the apparent selectionist relationship between the use of particular antimicrobials and the development of resistant strains decided to "restrict" the usage of one antibiotic or another so as to save such antibiotics in "reserve" for otherwise resistant strains.[60] Such measures—taken in venues ranging from small community hospitals to large general ones—were typically dependent upon some type of guideline that could be overridden through interaction with a predetermined consultative service. By the early 1970s, John McGowan of Boston City Hospital (BCH) and Maxwell Finland could formally report on the "effects of requiring justification," noting that at BCH between 1965 and 1972, the policy resulted in nearly tenfold reductions in the usage of ampicillin and chloramphenicol during their tenure on the "restricted" list, with implications noted with respect to adverse effects, costs, and antibiotic resistance.[61] Despite the scattered uptake of such measures, they remained local in nature throughout the 1970s. In the late 1950s, when hospitals were first setting up infection control programs in the wake of increasing staphylococcal disease, such programs had been marked by their heterogeneity.[62] A similar pattern appears to have persisted when such more "restricted" components were adopted, with individual hospitals serving as testing grounds for particular approaches.[63] Indeed, supporters were cognizant that they "would be accepted at very few hospitals."[64]

Instead, reformers publicly advocated *educational* measures, devised to support and enhance the autonomously prescribing physician.[65] C. Henry Kempe had concluded in 1958: "Happily, medicine remains a highly individualized vocation. What is more fitting, therefore, than that each of us should work out his own therapeutic philosophy? It is our hope that this philosophy will be at once enlightened and responsible."[66] By the time of the 1972 hearings, Simmons's intended role for the FDA was not increased regulation but rather "for FDA to improve its communications with the practicing medical community so that practitioners can better utilize available data on the safety and efficacy of drugs."[67]

In 1973 and 1974, Senator Edward Kennedy, chairman of the Senate Subcommittee on Health, once again placed antibiotic usage before Congress and the public. At the hearings, James Visconti, by this time the director of the Drug Information Center at the Ohio State University Hospital, updated his previous data to note that in a more recent hospital study, 73.6 percent of all antibiotic use was deemed irrational or questionable, with only 7.2 percent of the prophylactic antibiotics considered rational. Yet Visconti recommended not restriction but rather "the development of more rational methods by which professionals are provided with and use drug information."[68] Advocating for computerized systems for recording prescribing patterns and providing the basis for enhanced utilization review and feedback, Visconti also advanced the role of the clinical inpatient pharmacist to provide "antibiotic information . . . to assist in the rational prescribing of antibiotics." As Elizabeth Siegel Watkins describes in chapter 4, Visconti's efforts epitomized the goals of the "clinical pharmacy" movement to elevate the role of the pharmacist vis-à-vis the medical profession. Visconti was quick to conclude, however, that by bringing in "other professionals who [could] help [the physician], not making decisions for him, obviously not prescribing the drugs for him," he intended such assistance as supportive, rather than restrictive.[69]

Finally, as advanced by Calvin Kunin, if a "war on antibiotics" was to be effective, such individually promising educational approaches had to be comprehensive and coordinated.[70] Fundamental to this depiction, such approaches had to compete with the antibiotic-promoting "educational" efforts of the pharmaceutical industry. It was already clear to the reformers at the 1972 hearings that they were outgunned in this respect.[71] At the 1974 hearings, when Senator Kennedy called to witness the president of Smith, Kline, and French regarding that company's antibiotic "detailing" efforts—which recommended, for instance, the push for antibiotics during the presumed summer off-season on the basis that "air conditioning plus the change of season can make upper respiratory infections tough to beat without antibiotic therapy"—the magnitude of the challenge became still more apparent.[72] To Kunin and other reformers, the antibiotic situation was a "symptom" of the larger "default of post-graduate education to the [pharmaceutical] industry," extending all the way to industrial support of formal continuing medical education itself and necessitating a reconfiguration of the financing and control of postgraduate medical education writ large.[73] With respect to antibiotics, the situation mandated a multilateral countereffort, beginning with educating the public and extending to the teaching of clinical pharmacology in medical school; the provision of guidelines, instruction, and feedback to outpatient and inpatient clinicians alike; and the support of such activities with investment in laboratory facilities.

This was indeed a fecund era for mobilizing around such educational approaches. Jay Sanford's "Guide to Antimicrobial Therapy," now in its fortieth annual edition, dates to 1970.[74] Kunin himself led efforts to produce some of the first guidelines for antibiotic use, initially prepared for the Veterans Administration in an attempt both to standardize notions of rational prescribing and to stimulate further discussion around such issues.[75] The Alliance for the Prudent Usage of Antibiotics—"dedicated to promoting proper antibiotic use and curbing antibiotic resistance worldwide" and today boasting members in more than one hundred countries—was formed in 1981.[76] This era also witnessed the emergence of both academic "counter-detailing" and wider public concern with the apparent industry takeover of formal continuing medical education.[77] Yet the limits to educational measures were apparent even to contemporary investigators,[78] and most of the reformist concerns of the early 1970s remain with us today. Examining the telling manner by which key existing national entities confronted the issue during this formative era foregrounds the ultimate limits to such an educational approach as achieved in practice.

Institutional and Professional Inertia

At the 1972 congressional hearings on the "misuse of antibiotics," Senator Nelson asked the FDA's Henry Simmons about the possibility of more restrictive federal control over such potentially dangerous drugs as chloramphenicol, and Simmons responded that, instead, he would be "speaking to something . . . that I think is a step in the direction that might finally bring us to some effective means of controlling this and other agents in the public interest."[79] What Simmons actually had in mind was a "National Task Force on the Clinical Use of Antibiotics," to be convened by the FDA.[80] Such a group did come to be formed, but control over organizing and convening the meeting establishing it shifted to the AMA, which, in addition to inviting representatives from numerous professional medical and pharmacy associations, also invited representatives from private practice and the pharmaceutical industry.[81] Tellingly, during a full day of discussion and debate at the meeting of the Task Force in March 1973, while Henry Simmons would ask "if limited approvals [by the FDA for antibiotics] might be justified," one of the first "testable" premises that emerged from the meeting was that "physicians do not want the major ruling on antibiotic usage to come from Washington."[82]

By the end of the meeting, participants, focusing upon in-hospital antibiotic use, put forth a motion for hospital antibiotic control entailing that

1. Each hospital should form a committee to monitor antibiotic usage.
2. The committee should develop individualized guidelines for appropriate antibiotic usage both for treatment and prophylaxis. These guidelines should be approved by the Executive Committee of the medical staff.
3. The reports of the Antibiotic Committee would be distributed internally to the medical staff and the Executive Committee.
4. There would be an annual review of antibiotic usage by an outside consultant who would submit written recommendations to the Executive Committee.[83]

However, when even this nonrestrictive motion was voted upon, a "split decision" ensued.[84] Sixteen years previously, when the nation's hospitals confronted increasing staphylococcal infection and antibiotic resistance, a conference convened by the AMA had recommended "that every hospital establish a responsible officer or committee charged with the investigation and control of infections within that hospital and with the institution and practices designed to prevent such infections," thus supplying an important stimulus to the infection control movement in the United States.[85] Yet the 1957 conference members' summary recommendations had focused solely upon "the establishment and enforcement of adequate directions for asepsis and isolation in the hospital," avoiding any mention of restricting prescribing habits.[86] And when Simmons and Stolley published the 1973 recommendations in their 1974 *Journal of the American Medical Association* piece as the output of "an expert committee" that "agreed that there appears to be 'an inappropriate use of antibiotics and a massive overuse,' "[87] six members of the AMA's Department of Drugs (including its chairman, John Ballin) penned a pointed counter-response. Ballin and his colleagues qualified the entire tone of the discussion, pointing to the "possible" overuse of antibiotics, while taking pains to "reaffirm that the AMA Department of Drugs [was] not adopting the position in this commentary that antibiotics [were] not misused any more than it [was] accepting the premise that there [was] massive abuse."[88] Instead, referring to an actual dearth of data on the appropriateness—and consequences—of antibiotic usage in the United States, the AMA representatives called for ongoing data collection and the generation of "suggested guidelines," so as to "obviate any need for restrictive regulations for this class of drugs."[89]

Over the ensuing several years, as the home for those who would generate such guidelines shifted from the AMA to the American College of Physicians, tensions persisted between the desire to formulate guidelines and guidance for antibiotic control programs and the ongoing recognition of uncertainty with respect to the efficacy and consequences of proposed restrictive and educational measures.[90] By 1978, when the National Institute of Allergy and Infectious Diseases convened a two-day symposium titled "The Impact of Infections on Medical Care in the United States," Merle Sande (then at the University of Virginia) noted that controlled clinical studies were "urgently needed" to assess the impact of such measures upon reducing the "emergence of antibiotic resistance" in particular.[91]

Over the next several *decades*, however, as antibiotic resistance became regarded as an international "crisis,"[92] both the broad dream of antibiotic control and more delimited answers to such "urgent" questions remained wanting. In the outpatient setting, apparently irrational antibiotic prescribing for colds and upper respiratory tract infections persisted,[93] finally leading to a "call for action" by the Centers for Disease Control, predicated upon educating the public and practitioners, while supporting clinicians with antibiotic-resistance surveillance data.[94] In the inpatient setting, despite a wealth of collected data, hospital epidemiologists and infectious disease specialists conceded by 1997 that "those who have devoted their careers to the prevention and control of antimicrobial resistance must find the results of their efforts disappointing."[95] John McGowan, who had pioneered the study of "antibiotic stewardship programs" decades before they were labeled as such, could by the early 1990s still lament the need for funding to permit the proper multicenter evaluation of such measures.[96] By 2007, while the data remained admittedly imperfect, McGowan and his colleagues at the Infectious Diseases Society of America and the Society for Healthcare Epidemiology of America could at least formulate a series of "guidelines for developing an institutional program to enhance antimicrobial stewardship," highlighting those measures best supported by evidence; at the same time they identified those measures most in need of further assessment.[97] Yet despite such progress, and despite the apparently widespread belief among leaders of both academic medical centers and community hospitals that "good antibiotic stewardship programs are essential," only 29 percent of the community hospitals surveyed between 2001 and 2003 had instituted an antibiotic approval process.[98] Three decades after the heralded proposal of such measures, stated obstacles to their implementation continued to include both physicians' resentment of "limitations on their autonomy to prescribe" and displeasure "in policing the prescribing of fellow physicians."[99]

Conclusion

As this volume makes clear, while the assumed integrity of the patient-physician relationship has served as a general bulwark against attempts to delimit physicians' prescribing autonomy in twentieth-century America, such a history is more contingent and heterogeneous than it may appear at first glance. During the 1970s, when physicians were ceding control over generic and what came to be termed "controlled substance" prescribing (as Dominique A. Tobbell and David Herzberg describe in chapters 3 and 9), they resisted attempts, in the wake of the removal of the fixed-dose combination antibiotics, to further delimit control over antibiotic prescribing.

With respect to the antibiotic control programs themselves, on the one hand, controlled studies of the efficacy of many of the individual and aggregated components of such programs remain lacking, perpetuating the iterative series of calls for further such studies (and the funding to conduct them). On the other hand, in a setting of relative equipoise, it is clear that in the United States the burden of proof, since the 1970s, has been placed—explicitly or implicitly—upon those who would delimit the autonomy of individual prescribers of antibiotics. Contemporary analysts may draw attention to the historical inertia of decades of patient demands for (and receipt of) antibiotics,[100] yet equal attention should be paid to the historical inertia of the assertion of professional autonomy with respect to antibiotics, an assertion that first crystallized in the decade after DESI and Panalba and has subsequently served as an important barometer of prescribing autonomy writ large.

Historian of medicine Harry Marks, addressing "the science and politics of adverse drug reactions" in the United States, has recently drawn attention to the degree to which the U.S. regulatory system has been predicated upon this preservation of physician autonomy.[101] While Marks draws attention, in this respect, to the increasing entwinement of the concept of medical autonomy with the notion of therapeutic innovation rather than clinical judgment, the antibiotic story points to the persisting attachment of notions of individual clinical judgment to our overall conception of physician autonomy today. This does not inevitably lead to a call for delimiting such autonomy. Rather, as Marks has stated, it serves as a call to further examine its origins and evolution, as a prelude to an examination of both its benefits and its costs.

CHAPTER 3

"Eroding the Physician's Control of Therapy"

The Postwar Politics of the Prescription

Dominique A. Tobbell

In the September 1966 issue of *Physician's Management,* Merck, Sharp and Dohme's director of professional education, Elliot J. Margolis, posed this question: "Will the physician choose the particular brand for his patients or will he leave it to someone else to do so? Or, more broadly, will he retain the right to choose or not choose?"[1] For the brand-name pharmaceutical firm Margolis represented, and the physicians he wrote to, the answer was obvious: the physician should retain complete autonomy over prescription practice, including the decision of whether to dispense a brand-name or a generic drug. For pharmacists, generic drug manufacturers, government officials, and congressional reformers, however, the answer was far less clear-cut.

The question of who has the authority to prescribe a prescription drug, and what form that authority should take, has been a source of continued political, economic, and social contestation since Congress passed the 1938 Federal Food, Drug, and Cosmetic Act. The act and the 1951 Durham-Humphrey Amendment to it designated the category of prescription drug and granted physicians, veterinarians, and dentists sole autonomy over the prescription.[2] Throughout the postwar era, however, various professional groups and congressional reformers sought to circumscribe the physicians' domain over the prescription in the hope of preserving their own professional interests and reducing the economic costs of health care.

Over the course of these decades, there was a series of three distinct yet overlapping attacks on physicians' prescribing autonomy. During the early 1950s, retail pharmacists frustrated by the expense of maintaining full inventories of the rapidly expanding list of prescription drugs regularly substituted cheaper generic name drugs for the brand-name drugs prescribed by physicians. Although substitution was illegal, pharmacists saw it as a necessary economic strategy that would allow them to reduce the amount of inventory they needed to stock, thus reducing their costs. Ultimately, retail pharmacists hoped to create enough pressure on state and federal legislatures to force the legalization of brand substitution. These efforts met with staunch opposition from the drug industry, which argued that the pharmacists were attempting to usurp the physicians' authority to prescribe. The result was a shoring-up of the physician's autonomy as each state passed antisubstitution laws confirming the illegality of substitution.

As Scott H. Podolsky describes in chapter 2, physicians' prescribing practices became a source of political contest during the 1960s and 1970s, as first Senator Estes Kefauver (D-TN), and later Senators Gaylord Nelson (D-WI) and Edward Kennedy (D-MA), introduced legislation that sought to make it mandatory for physicians to prescribe drugs by their generic name rather than their brand name. Although in reality mandatory generic prescribing in no way impeded the physician's authority to prescribe a specific drug, the American Medical Association (AMA) and the drug industry cast this effort at legislative reform as the government's incursion upon the physician's autonomy and a crucial step toward socialized medicine, an outcome the AMA and drug industry warned that every American should fear.

While the efforts of substituting pharmacists had failed in the 1950s, organized pharmacy groups renewed calls for the legalization of substitution in the 1970s. Coinciding with the reform efforts of Nelson and Kennedy, the movement to legalize generic substitution found greater political traction than it had two decades earlier. In the changed regulatory and economic context of the 1970s, the protests of the AMA and the drug industry failed to halt the overturning of state substitution laws in the majority of states by the late 1970s.

In this chapter I examine the prescription as a site of government (both state and federal) intervention into medical practice and as site of contestation over the limits of the physician's autonomy and the expertise and authority of the pharmacist. By contextualizing the contests over prescription practice in the second half of the twentieth century, I provide not only a political history of the prescription but also a broad understanding of the contested nature of clinical practice and health care politics in the postwar era.

Consolidating Hierarchies in the Antisubstitution Debates of the 1950s

The passage of the Federal Food, Drug, and Cosmetic Act in 1938 and the enactment of the Durham-Humphrey Amendment thirteen years later laid the economic and social foundations of the modern pharmaceutical enterprise. In chapter 1, Nicolas Rasmussen describes the "moral panic" surrounding Americans' use of barbiturates that led to passage of the Durham-Humphrey Amendment and details the effects that legislation had on Americans' utilization of narcotic drugs. For our purposes, by distinguishing between those drugs safe enough to be sold over the counter without physician supervision and those that could be sold only under the authority of a physician's prescription, the legislation institutionalized the professional boundaries of the modern pharmaceutical enterprise. Physicians claimed autonomy in their control over the prescription, while pharmacists—in subordination to physicians—accepted responsibility for dispensing the physician's prescription as written. Through the mid-1950s, however, a growing number of pharmacists attempted for economic reasons to shift those professional boundaries. The resulting political struggles, rather than reorienting the professional and economic structure of the pharmaceutical market, served instead to consolidate the professional hierarchies that characterized the modern pharmaceutical enterprise and solidified the *brand-name* prescription drug as its economic foundation.[3]

In the early 1950s, retail pharmacists were struggling to meet the demands placed on them by the ever-expanding market of prescription drugs. They were frustrated, in particular, with the expense of stocking every new product the drug industry introduced to the market, especially those drugs that were rarely prescribed. For many, the expense of maintaining a full inventory of new drugs was proving too great. In an effort to reduce the amount of inventory they needed, some pharmacists were switching cheaper, generic drugs that they held in stock for the more expensive brand-name drugs prescribed by physicians. Although substitution was illegal—forty states had antisubstitution laws or regulations on the books by the 1950s—pharmacists hoped to create enough pressure on state and federal legislatures to force the legalization of brand substitution.[4] They were joined in their efforts by small pharmaceutical firms that, lacking research capacities, manufactured generic versions of brand drugs. For these small firms without the capital to develop sufficient marketing capacities, substitution represented a way to compete in the pharmaceutical marketplace. In February 1952, for example, a group of small drug manufacturers helped to bring a bill before the Michigan state legislature that would permit the substitution by pharmacists of one drug product for another prescribed by a physician.[5]

Retail pharmacists railed, in particular, against the increasing number of drugs released onto the market that were not actually "new products" with unique therapeutic action. They complained that the abundance of so-called me-too drugs (drugs with similar chemical properties and more-or-less identical therapeutic function) resulted in excessive costs to the pharmacist who was obligated to stock all varieties of me-too drugs on the off chance a physician would write a prescription for any one them. Because many of these drugs were rarely prescribed, pharmacists were often left with large inventories that they could not sell. In response, several pharmacists joined with physicians who were "confused and troubled" by the increasing number of me-too drugs and organized an antiduplication drive. Through this campaign, pharmacists sought, first, to restrict the introduction of further me-too drugs and, second, to persuade physicians to write prescriptions for generic drugs rather than brand-name drugs, thus allowing the pharmacists to dispense, in each case, whichever one of the duplicate drugs they stocked.[6]

The pharmacists' substitution and antiduplication campaigns were intrinsically linked reactions to the economic structure of the pharmaceutical industry. Indeed, even as pharmacy leaders in the American Pharmaceutical Association, the National Association of Boards of Pharmacy, the American College of Apothecaries, and the National Conference on State Pharmaceutical Associations condemned substitution, they perceived it to be a symptom of the drug industry's economic practices, specifically their propensity to develop drugs with chemical structures and therapeutic effects nearly identical to those of drugs already on the market. In 1952, for example, the California Pharmaceutical Association's president, John A. Foley, condemned substitution as an "insidious" practice, while calling on manufacturers to "desist from the practice of lifting each other's formulas and ideas." Foley believed that in doing so, "the props would be pulled from under" the substituting pharmacists.[7] In turn, brand-name manufacturers accused substituting pharmacists of threatening the economic foundations and the legal system of drug development. As the trade journal *FDC Reports* warned, "by plugging the use of generic names on [prescriptions]," pharmacists were verging on "destroying the value of pharmaceutical trademarks."[8]

During the spring of 1952, several firms went on the offensive and instituted antisubstitution campaigns aimed at curbing the practice. These included taking legal action against pharmacists guilty of substitution (on the grounds that substituting pharmacists were engaging in unfair competition) and against the small generic-name manufacturers whose imitation drugs were being substituted for brand-name drugs. Leading the way was Abbott Laboratories, which secured a series of court injunctions against New York City pharmacists guilty of substituting other products for prescriptions written for the company's drugs. Theodore Klumpp, president of Winthrop-Stearns, the research arm of Sterling Drug, arguing before an audience of representatives from brand-name manufacturers, admitted, "It is not good business to sue our customers," but he said that because of the "growing conviction that soft words" had failed to stop "this nefarious practice," it was "time to 'get tough with the chiselers.'" In spite of these efforts, a 1953 mail survey of brand-name manufacturers found that 53 percent of responding firms believed that "substitution of their products [was] already widespread," with some firms reporting substitution rates for their prescription products as high as 25 percent in some cases.[9]

In getting "tough," brand-name manufacturers sought to reaffirm the economic—and legal—foundation of the pharmaceutical enterprise and also reassert the appropriate professional relations between physicians, pharmacists, and the prescription. To this end, in December 1953, a group of executives from the leading drug firms organized an industry-wide professional relations program that they hoped would promote "the highest professional standards and ethics in the manufacture, distribution, and dispensing of prescription medication." Although the new organization, the National Pharmaceutical Council (NPC), would have other functions, its main objective was to unite drug manufacturers and pharmacists and squash the practice of substitution and pharmacists' antiduplication drive. In 1954 the NPC's first president, Theodore Klumpp, called on "all those in all branches" of the pharmaceutical industry who shared the purpose of the group to work together to bring their "combined influence to bear against those practices" that, he said, "are undermining the ethical principles of fair competition and fair dealing upon which our industry must rest."[10]

As part of the NPC's antisubstitution drive, executive officers of the NPC traveled throughout the United States "carrying the 'gospel' of prescription brand-names" to state pharmacy boards, state pharmacy associations, and pharmacy students.[11] In 1954 Klumpp delivered such a speech to the Drug, Chemical, and Allied Trades Section of the New York Board of Trade; in it he accused pharmacists who substituted of "cast[ing] a heavy shadow over the entire pharmaceutical family." The substituting pharmacist was guilty of "practicing medicine without a license." After all, added Klumpp, "the pharmaceutical industry—all segments of it—is built on certain principles which have stood the test of time. The key principle is that the physician is boss. It is he who prescribes. It is the pharmacist who compounds the prescription. The pharmacist is professionally, morally, and legally bound to fill that prescription precisely as the doctor wrote it." By overstepping these professional, moral, and legal bounds, the substituting pharmacist not only put the "very foundations" of the pharmaceutical industry at risk but, more importantly, caused grave danger to the patient. Klumpp warned that the substituting pharmacist's "first victim" was the patient, who "at best, got a product of uncertain quality." That pharmacist's "other victims" were "the reputable retail pharmacists in the community," whom their customers might accuse "of overcharging for prescriptions honestly compounded."[12]

In this way, the NPC's campaign sought to preserve the economic foundations of the pharmaceutical enterprise, based as they were on the sanctity of the brand name, by highlighting the professional boundaries the practice of substitution transgressed. In particular, the NPC made clear the dangers inherent in the process by which substituting pharmacists usurped the appropriate professional relationships between physician, pharmacist, and patient. Only physicians had the expertise and authority to write prescriptions for their patients, and it was the duty and professional responsibility of the pharmacist to dispense to the patient the prescription precisely as written by the physician. To challenge the physician's autonomy over the prescription, maintained the NPC, was to put the patient's life in jeopardy.

At the same time, the NPC worked to educate physicians on the dangers of substitution in the hope that they would add pressure to state pharmacy boards to prosecute cases of substitution. In 1957 the NPC published and distributed to physicians a twenty-eight-page booklet that highlighted twenty-four reasons for the importance of prescription brand names to the physician. By explaining that generic drugs were often not therapeutically equivalent to their brand-name counterparts owing to differences, for example, in the drugs' disintegration time, purity, solubility, particle size, quantity of active ingredient, melting point, surface tension, viscosity, and caloric values, the booklet sought to encourage the physician "to (1) specify brand-names on his prescriptions and (2) insist that the pharmacist supply the brand ordered."[13]

Despite the threat substitution posed to their autonomy, physicians were conspicuously silent in the substitution debates of the 1950s. The majority of American physicians were, at that time, preoccupied with the mounting calls for national health insurance as the public grew increasingly agitated with the rising costs of medical care. Inasmuch as pharmacists' substitution practices related directly to the costs of medical care, few physicians, policymakers, or commentators in the 1950s drew the political connection. Thus, the battle over substitution generated little public attention, taking place instead behind the closed doors of state pharmacy boards, in state courtrooms, in the meeting halls of professional pharmacy groups, and in state legislatures.

The NPC's strategy of turning nonsubstituting pharmacists against their substituting colleagues proved successful. In 1956 the new NPC president, Robert Hardt, noted that whereas several years earlier only four or five state pharmacy boards could be counted on to take action against substituting pharmacists, by 1956 the ratio of state boards committed to taking action against pharmacists guilty of substitution had been reversed, with the majority now supporting antisubstitution measures. Furthermore, the rate of substitution had fallen from 14.7 percent in 1953 to 4.3 percent in 1956. Yet for Hardt perhaps the clearest sign of the program's success was the conversion of substitution into a "bad word" in pharmacy circles. A year later, an *American Druggist* survey of brand-name manufacturers found the rate of substitution to have fallen even further to 3.7 percent.[14] And by 1959, forty-four states had put antisubstitution laws—to be enforced by state pharmacy boards—on the books.[15]

By the end of the decade, then, the professional hierarchies within the health care team had been consolidated around the prescription: brand-name manufacturers developed new and innovative prescription drugs; physicians prescribed those brand-name drugs with autonomy; and pharmacists dispensed to patients the physician's prescriptions exactly as written. In the final weeks of the decade, however, amid mounting concerns about rising health care costs, the economic foundations of the pharmaceutical enterprise came under attack. And by the early 1960s the prescription had emerged as a target for health care reform and a site of government intervention.

Prescription Reform in the Kefauver Era

During the late 1950s, the public had become even more concerned about rising health care costs in general and the costs of prescription drugs in particular. Whereas Americans had spent approximately $4 billion (4% of the GNP, or $29.6 per capita) for health and medical care in 1940, by 1955 that figure had more than quadrupled to $17.7 billion (or 4.4% of the GNP and $105 per capita), and by 1960 health care expenditures totaled $26.9 billion (5.3% of the GNP and $146 per capita). Between 1950 and 1960, price increases accounted, on average, for 50 percent of the total increase in health care expenditures.[16]

Despite the stability of prescription drug prices relative to other health care and consumer goods throughout the 1950s, at the end of the decade government and public concern about health care costs focused on prescription drugs.[17] In May 1958, for example, the Citizens' Committee for Children of New York City published the findings of its study on the impact of prescription drugs on the family budget. The committee found that patients suffering from arthritis, rheumatism, cancer, heart disease, and tuberculosis often had "great difficulty meeting the cost of medicines essential to survival or alleviation of pain." Citing the cost of brand-name drugs as presenting a "real hardship" to low- and middle-income families, the committee claimed that patients would be able to buy drugs "at a fraction of the cost" if physicians used a drug's generic name when writing a prescription. Later that year, the consumer magazine *Consumer Reports* published two articles on the high cost of prescription drugs, in which it was similarly critical of brand-name manufacturers and called for the federal government to launch a comprehensive investigation of the industry.[18]

In December 1959, in response to these concerns, Senator Estes Kefauver (D-TN), chair of the Senate Subcommittee on Antitrust and Monopoly, launched a congressional investigation into the business and pricing practices of American pharmaceutical companies. At the center of Kefauver's concern was the purportedly high price of prescription drugs and what appeared to him to be an absence of price competition in the industry.[19]

As part of the investigative hearings, Kefauver's subcommittee heard testimony from consumer and patient groups condemning the prices patients—particularly elderly patients—were forced to pay for vital prescription drugs. Ethel Andrus, president of the American Association of Retired Persons (AARP) and the National Retired Teachers Association (NRTA), testified that numerous members of the AARP and the NRTA were struggling with prescription drug prices: "Their monthly costs for drugs prescribed for treatment left them little to live on. They estimated their drug costs somewhether [*sic*] between $15 and $30 a month." Reiterating the Citizens' Committee for Children's call the previous year, Andrus urged Kefauver to institute legislation that would require physicians to use the generic name instead of the brand name when prescribing drugs. As Andrus regarded it, this might be "the only way in which we can meet the problem of big pharmaceuticals."[20]

Over the next two years, Andrus's call for prescription reform was echoed by representatives from state welfare agencies, consumer groups, hospital pharmacists, and a handful of academic physicians who regarded the prescription pad as an appropriate site for legislative intervention. By the end of the investigative phase of his subcommittee's hearings, Kefauver agreed, having concluded that brand-name manufacturers held unfair monopolies in the pharmaceutical marketplace in large part because they were able—through their marketing activities—to persuade physicians to prescribe using brand names rather than generic drugs. In an effort to reduce physicians' dependence on the brand name and increase competition in the pharmaceutical marketplace, Kefauver introduced a bill that, among other provisions, mandated that physicians prescribe by generic name rather than by brand name. In this way, Kefauver's bill aimed squarely at reforming physicians' prescription practices and their autonomy over the prescription.

Kefauver's attempt to introduce mandatory generic prescribing faced staunch opposition from brand-name manufacturers, organized medicine, and a cadre of academic physicians who, while supporting the *concept* of generic prescribing, challenged the scientific and regulatory rationale on which it was premised. Their collective opposition to Kefauver's prescription reform rested on three key grounds.

First, brand-name manufacturers argued, the brand name was associated with the quality of the drug and the reputation of the manufacturer producing that drug. As Theodore Klumpp, president of Winthrop Laboratories, explained, when the physician prescribed by a brand name, he could be sure he was choosing a quality product. "If his only way of identifying a drug is by generic name, rather than by brand-name," as Kefauver would have it, Klumpp continued, "he will have no way of knowing the true quality of the drug which he prescribed." In other words, when physicians prescribed Schering's brand of prednisone, Meticorten, they were not just writing an order for a specific active ingredient, they were writing an order for a complete pharmaceutical product manufactured to the highest quality control standards by a reputable and trustworthy firm. In contrast, brand-name manufacturers contended, because generic drugs were not as rigorously evaluated as brand-name drugs and because they were usually produced by small drug firms whose reputation could not be guaranteed, physicians could be certain of neither their quality (and therefore safety) nor their efficacy. Moreover, Klumpp argued, conforming with the principles of the free market, the use of brand names ensured that only high-quality prescription drugs would succeed in the marketplace: "Brand-names enable the consumer to reward the product which is proved good . . . through repurchases of the product. If a product proves unsatisfactory, the consumer has the means of punishing it—by refusing to buy it again." Because "brand-names facilitate[d] reward or punishment," they were "a prime factor in stimulating reputable manufacturers to produce the best product."[21]

Second, brand-name manufacturers and leading academic physicians argued that generic drugs were not necessarily therapeutically equivalent to their brand-name counterparts. As Austin Smith, former editor of the *Journal of the American Medical Association* and current Pharmaceutical Manufacturers Association (PMA) president testified, the suggestion that "generic drugs could be readily substituted for brand-name drugs at huge savings to the drug buying public . . . with no danger to the patient and no lessening in therapeutic results . . . is a potentially dangerous action." Smith and his allies argued that it was inaccurate and extremely dangerous to assume, as the reformers did, that generic drugs had the same therapeutic effect as their brand-name equivalents. Rather, as PMA chairman Eugene Beesley testified, advocates of generic prescribing "overlook the fact that there can be important variations between drugs with the same generic therapeutic agent. They can have different inactive ingredients and special excipients, vehicles, and bases." Because physicians could "identify these important variations by the manufacturer's brand-name," Beesley argued that it was "most desirable that the physician know exactly what his patient has taken so that he can assess its value and chart his future treatment of the patient."[22]

Finally, brand manufacturers and the AMA argued that mandatory generic prescribing threatened the sanctity of physicians' prescribing autonomy. The NPC's Newall Stewart testified, "It is in the public interest to require that the prescribed drug must be dispensed . . . because the prescribing prerogative of the physician should remain inviolate."[23] Any effort by the government to intervene in physicians' prescription practices would undermine that prerogative. On this point in particular, the industry found much support in the AMA. In June 1961, the AMA's House of Delegates unanimously approved a resolution that put the AMA "in opposition to legislative and administrative mandates which would compel physicians to prescribe drugs, or require pharmaceuticals to be sold, by generic name only."[24] The AMA maintained, however, that this did not mean the AMA opposed the use of generic names by physicians. Rather, the AMA opposed "the compulsion inherent in" proposals for mandatory generic prescribing and preferred "that physicians should have the privilege of prescribing drugs by either generic or brand-name."[25]

At the same time, however, that the AMA professed staunch opposition to Kefauver's attempts to mandate generic prescribing, it encouraged physicians to prescribe only generic drugs to welfare patients. Since the late 1950s, several states had worked with the AMA to encourage physicians to write generic rather than brand-name prescriptions for welfare patients. In 1960 the AMA's House of Delegates endorsed the recommendations of a report by its Council on Medical Service in collaboration with the American Public Welfare Association titled "Guides for Drug Expenditures for Welfare Recipients." The report urged physicians to use "counterpart [read: generic] drugs of equal therapeutic effectiveness when available, when the quality of the product is assured, and when a price differential exists." In doing so, the AMA made clear that the physician's right to prescribe as he saw fit would be preserved: "Whatever methods are used to reduce expenditures, leeway must be left for exceptions, when medically justifiable, and for appeal of agency decisions by the attending physician." In this way, the AMA placed the onus on the physician to consider economic factors when deciding which drug to prescribe, and *not* on the state to restrict what the physician was permitted to prescribe.[26]

Although at first pass the AMA's support for welfare prescription reform seems to contradict its opposition to Kefauver's prescription legislation, this support should instead be read as fitting within the association's broader mandate of stymieing federal health care reform. By working with state welfare agencies, the AMA was able to craft policy solutions that preserved physicians' professional and economic interests and staved off proscriptive federal interventions that threatened to curtail both. Indeed, by providing support to state welfare agencies, the AMA, in essence, took the wind out of Kefauver's proposal for mandatory generic prescribing. After all, if, through local state-based negotiations, welfare agencies and physicians could agree to work together to control prescription drug costs, much of the economic incentive for Kefauver's proposal would be lost.

Although Kefauver's efforts to institute mandatory generic prescribing failed, the Kefauver hearings did culminate in passage of legislation that reformed—to some degree—the practices of the drug industry. Yet the reforms embedded in the 1962 Drug Amendments emphasized improving the safety and efficacy of drugs while leaving the economic and professional underpinnings of the pharmaceutical enterprise unchanged. In the end, both the "brand-name" drug and physicians' prescribing autonomy had been successfully defended. At the same time, the social networks that characterized the pharmaceutical and medical enterprise had taken on a distinct political quality. Certainly the drug industry had drawn upon the AMA's support to oppose the efforts by the Food and Drug Administration (FDA) to secure greater authority over prescription drug labeling in the 1940s, but during the Kefauver hearings, the industry and the medical profession developed a full-blown political alliance against any attempt by the federal government to intervene in prescription practice.[27]

The Prescription as a Site of Legislative and Regulatory Contest

As health care reformers battled, through the mid-1960s, to secure passage of Medicare and Medicaid, congressional and public debate about the cost of prescription drugs continued to mount. In 1964 Senator Philip A. Hart (D-MI) chaired a senate subcommittee hearing into charges that U.S. drug firms were fixing the price of the antibiotic tetracycline. These hearings resulted in the filing of antitrust charges against Pfizer, Bristol, and American

Cyanamid and reignited concerns among patient advocates that brand-name manufacturers were out to exploit the American patient's wallet.[28]

Although Kefauver had failed to secure passage of legislation requiring that physicians prescribe by generic name only, Senator Gaylord Nelson (D-WI) continued the push for prescription drug reform. When Nelson launched his investigation into "competitive problems in the pharmaceutical industry" in 1967, he placed the issue of generic drug prescription at its center.[29] Like Kefauver, Nelson regarded mandatory generic prescribing as the solution to rising prescription and health care costs. Nelson joined with others in Congress and in state legislatures to push for the use of generic drugs for all welfare patients and to secure prescription drug benefits for seniors under Medicare.[30] As during the Kefauver hearings, brand-name manufacturers and their physician allies vigorously opposed passage of generic prescription legislation.[31] They did so on two familiar grounds: first, that generic drugs were not necessarily therapeutically equivalent to their brand-name counterparts, and second, that generic prescription legislation would abolish the physician's autonomy.

Brand-name manufacturers had initially raised concerns about the therapeutic equivalence of generic drugs during the antisubstitution campaigns of the 1950s. But in the years since, pharmacists and pharmacologists working in government and academic laboratories had become engaged in the equivalence question and had found evidence that chemical equivalence (the current official criteria of a generic drug) did not always produce therapeutic effects equivalent to those of their brand-name counterparts. During the 1950s and early 1960s, for example, researchers at the Food and Drug Directorate of Canada's Department of National Health and Welfare had found that for several drug products, in vitro dissolution and disintegration tests—the existing pharmacopoeial standards of *chemical* equivalence—were an inadequate measure of biological availability and thus did not guarantee the therapeutic equivalence of the drug products.[32] These and similar studies had led officials at the United States Pharmacopeia (USP)—the organization responsible for setting drug standards in the United States—to worry that there was a need for "a new kind of official standard."[33]

Thus, as Senator Nelson pushed forward his proposal for generic prescribing, brand-name manufacturers were able to draw upon the mounting evidence and growing awareness among academic pharmacists, pharmacologists, and officials from the FDA and the USP that not all versions of a pharmaceutical agent produced the same therapeutic effects. For example, the PMA's president, C. Joseph Stetler, during his testimony before Nelson's subcommittee, cited a recent Parke, Davis study that had compared the equivalence of its own brand of chloramphenicol (Chloromycetin) against several generic versions of the drug and found that some of the generic drugs were absorbed into the blood stream of human volunteers at rates remarkably different from the Parke, Davis brand. Subsequently, the FDA had withdrawn nine of the generic chloramphenicol drugs from the market. Here, Stetler argued, was clear evidence that "although two products may contain, or are supposed to contain, the same amount of the same active ingredient, this provides no assurance that both products will produce the same clinical effect in any particular patient."[34]

In the absence of comparative data, brand-name manufacturers held up the regulatory process as a guarantee of brand-name superiority. Under the 1962 Drug Amendments, all new chemical entities (and therefore brand-name drugs) had to undergo rigorous clinical evaluation before the FDA could grant marketing approval. Generic drugs, however, named for the very fact that they were *not* new chemical entities, fell out of the purview of any such requirement and were granted approval merely upon demonstration of their chemical equivalence to their brand-name counterpart. Academic physicians and industry representatives alike argued before Nelson's subcommittee that this double standard in the regulatory process made hazardous any attempt to institute mandatory generic prescribing. The preeminent pharmacologist Alfred Gilman warned, "Without clinical data" on generic drugs "we have no idea of the efficacy in man of many of the generic drugs that we have declared to be effective."[35] Gilman contended, emphasizing it as an imperative: "All producers, and certainly the generic houses, should be required to submit proof of the performance of their drugs in human patients before they are permitted to market them. Once that is required, and this double standard is eliminated, I believe many of the problems facing us will be reduced."[36]

For many academic physicians, then, the proposed legislation reflected a fundamental lack of understanding among policymakers and the public concerning the intricacies of pharmaceutical practice. For Boston physician Dale Friend, the "widespread belief" among "a great many . . . union leaders, many of the laity, [and] a good many . . . congressional leaders . . . that a solution to at least some part of the high cost of drugs to patients could be brought about by the simple expedient of using generic name drugs" was a cause of serious concern, not least because he considered most of it to be "based on a failure to understand the technicalities of drug therapy." The consequent move by congressional reformers "to legislate the use of generic drugs" would, he expected, "rather seriously interfere with the physician in his practice of Medicine."[37]

For the majority of nonacademic physicians, whose exposure to the scientific and regulatory debates over therapeutic equivalence was limited to the pages of the major medical publications, the key issue raised by generic prescription legislation was—as Friend implied—the threat it posed to their autonomy. For example, one Memphis physician asserted to readers of the *Journal of the American Medical Association* in 1966, "The physician should be allowed to prescribe drugs from a pharmaceutical house that he knows is ethical and holds to high sanitary standards, even though it might cost his patient a little more money for the prescription at his pharmacy."[38]

In the wake of the passage of Medicaid and Medicare, the AMA leadership was particularly attuned to any legislative effort that sought, in their eyes, to grant the federal government any further control over medical practice.[39] Mandatory generic prescribing looked to do just that. In an editorial in the *AMA News* in 1967, the physician group argued that physicians prescribed drugs by brand name so that they could be confident in the drugs' quality. "Legislation that would nullify this knowledge [about quality] by removing the decision-making power from" physicians, the AMA asserted, "clearly is not in the public interest."[40]

Just a year earlier, the AMA had gone on the offensive to protect the physician's prescribing autonomy from what the group perceived to be an increasingly interventionist FDA.[41] Following reports of severe side effects among patients taking the long-acting sulfa drug Madricin, newly appointed FDA commissioner James Goddard had pulled the drug from the market and ordered the relabeling of other long-acting sulfa drugs. More alarming for the AMA, Goddard had also insisted that physicians prescribe a short-acting sulfonamide before considering the use of a long-acting sulfonamide.[42] As John Adriani of the AMA's Council on Drugs wrote to Jean Weston, director of the AMA Department of Drugs, Goddard's action was "anything but a small matter." He continued, "It involves a very far reaching, fundamental principle. I feel sure that Dr. Goddard did not intend to usurp the prerogative of the physician's right to choose the drug he thinks best for his patient, but his edict certainly does legislate by a regulation and not by a Congressional act the manner in which we practice medicine."[43]

Later that year, the AMA saw the "fundamental principle" of physician autonomy under further attack when Senators Russell Long (D-LA) and Joseph Montoya (D-NM) introduced amendments to the Social Security Act calling for the use of generic prescriptions for all patients receiving Medicaid benefits and adding a (generic) prescription drug benefit to Medicare.[44] Adriani, a Louisianan, wrote to Senator Long on several occasions offering his suggestions on the drug bill. Adriani was sympathetic to Long's goal of making prescription drugs more affordable and thus accessible to all those who needed them; "The idea of requiring drugs to be prescribed by generic instead of brand-names," he wrote, "has considerable merit." Yet Adriani thought Long's proposal unworkable "for two main reasons." He explained, "The first of these reasons is that doctors resent being told how to practice medicine, not only by laymen but also by other physicians. I know this from experience as one of the associate administrators at the hospital. I also know it from my own personal reactions. I want freedom of choice in the treatment and management of a patient who has entrusted his life and welfare to me." The second reason was that many physicians simply did not know the generic names of drugs. Rather than attempting to "tell a physician how he should practice medicine" by mandating the use of generic prescriptions, Adriani urged instead educating physicians on the generic names of drugs and requiring manufacturers "to dispense the drug by its generic name." In other words, Adriani opposed legislative attempts to restrict the physician's right to *prescribe* whichever drug he wanted but supported changing the way manufacturers *labeled* their drugs.[45]

Brand-name manufacturers later became vehement in their opposition to what they perceived as Adriani's attempt to abolish the brand name. And subsequently, as Adriani became increasingly vocal in his efforts to promote generic labeling, the AMA leadership distanced itself from Adriani's position. Yet Adriani's assertion that physicians would, on principle, oppose any attempt to encroach upon their prescribing autonomy reflected the overwhelming sentiment of the medical profession. For example, Adriani recounted the story of Louisiana's former commissioner of welfare, who had, several years earlier, attempted "to force" generic prescriptions on physicians in Louisiana but "was taken care of via the route of medical politics." That commissioner was "now practicing law somewhere in West Louisiana."[46]

In spite of his pronouncements to Senator Long, in personal correspondence with his medical and industry colleagues, Adriani continued to express his misgivings about generic prescription legislation. Indeed, in May 1967, he wrote to Jean Weston, who had since left the AMA to serve as executive director of the NPC, that although many physicians were "sympathetic to the idea of generic prescribing," it was not "practical under all circumstances." He summarized, "We can't make a size 8 shoe fit everybody." Two years later he restated his concern, distinguishing the pressing need "for a clarification of the matter on nomenclature—one name for one drug," while maintaining that physicians should still be able to prescribe a brand-name drug if they chose: "The integrity of the brand means much to me."[47]

The FDA and the Department of Health, Education, and Welfare (HEW) were equally unwilling to endorse Senators Long and Montoya's proposals because of the uncertainty over the therapeutic equivalence of generic drugs. At a hearing before the House Interstate Committee in March 1967, Representative Archie Nelsen (R-MN) asked Commissioner Goddard "whether there may be a variation as to the effectiveness of a drug of a similar generic name, ignoring the trade or brand-name?" Goddard responded, "This is unfortunately true. I say unfortunately, because it means we are not performing our functions as well as we have to. We [the FDA] view our goal as being one where the physician can prescribe any drug that is in the marketplace on any basis he wishes in terms of whether he uses brand-names or generic names, and be assured that those drugs are all effective and they are all safe."[48]

Despite the repeated efforts of Long, Montoya, and Nelson through the late 1960s, neither Long's nor Montoya's bill passed (the Long and Montoya amendments related specifically to Nelson's own prescription reform agenda and as such received considerable attention during the hearings of Nelson's subcommittee). Nelson's generic-prescription legislation suffered the same fate. In each case, the prescription-reform legislation rested on two problematic assumptions: that generic drugs were therapeutically equivalent (and, indeed, cheaper) than their brand-name counterparts and that the prescription was an appropriate site for legislative intervention. Given their shared foundation, it is no surprise that all failed to secure congressional passage. In the end, at least three factors account for their legislative failure.

First, concerns about the equivalence of generic drugs raised profound challenges to the drug regulatory system, forcing the FDA to reconsider the standards by which generic drugs were granted market approval. Until the FDA was able to create new regulatory standards to measure biological equivalence (which it did in the 1970s), policymakers seeking to introduce legislation mandating generic prescribing faced a significant administrative hurdle. Second, the economic ground on which Long, Montoya, and Nelson predicated their calls for generic prescribing was shaky. Certainly, in the late 1960s health care costs were rising and presented a serious financial burden to American senior citizens, the indigent, and working poor. Yet generic-prescription advocates had little political rope with which to work: prescription drug costs represented a relatively small percentage of total health care costs (around 10%), and prescription drug prices had actually declined slightly (0.1%) between 1965 and 1970, while total health care expenditures had increased 6.1 percent during the same time period.[49]

Finally, the lack of political traction accorded prescription drug legislation was all the more significant as policymakers confronted the AMA's and the drug industry's powerful and persuasive campaigns against generic legislation. The PMA, for example, circulated among physicians' offices, pharmacies, and congressional members and staffers two pamphlets that warned of the dangers of making generic prescribing mandatory: *Compulsory Generic Prescribing—A Peril to the Health Care System* and *Drugs Anonymous?* The latter warned that such action "could and probably would bring about deterioration in the quality of medical care—through the wide sale of substandard products—by discouraging the struggle for excellence which has marked the astounding progress in the pharmaceutical field—and by impeding drug research on which future progress depends."[50] The industry also invested significantly in lobbying and forging legislative alliances with congressional members. In 1968, for example, the PMA was accused of "pour[ing] money into Wisconsin in an attempt to defeat Senator Gaylord Nelson" in his reelection bid that fall.[51] And a year later it was accused of pressuring the White House to withdraw its appointment of John Adriani to the position of director of the FDA's Bureau of Medicine (because of Adriani's support for generic drug labeling).[52] So, too, the AMA had for decades proclaimed the sanctity of physicians' professional autonomy; indeed, this very argument had helped the AMA secure from health care reformers the promise that under Medicare patients would retain the "freedom of choice" to select their physicians.[53]

By the 1970s, however, the physician's autonomy and the hierarchical system of health care on which it was predicated were under threat. The thalidomide crisis of the early 1960s, the growing incidence of adverse drug reactions, and reports of physicians, clinical researchers, and pharmaceutical companies conducting unethical experiments on human subjects led many Americans to question whether physicians alone should make health care decisions for patients. The women's health movement, for example, held up the failure of the medical profession, the FDA, and pharmaceutical firms to adequately protect women from the dangers of taking oral contraceptives, diethylstilbestrol, and Depo-Provera, as evidence of the oppressive and tragic consequences of medical paternalism.[54] The mounting distrust of physicians was part of a much broader critique of the role of government, institutions, and the professions in American society that grew out of the civil rights, student, and countercultural movements of the 1960s and continued through the rights-based movements of the 1970s. These events led to a profound weakening of American physicians' cultural authority and called into question the medical profession's decades-long claims to professional autonomy.[55]

Changes made by the FDA in the scientific and regulatory foundations of the drug approval process in the early 1970s further compounded physicians' weakening claims to autonomy. As more and more studies documented the inadequacy and clinical consequences of current generic drug standards (based as they were on chemical equivalence), the FDA began requiring generic drug manufacturers to submit evidence of the biological equivalence of their generic drug to its brand-name counterpart before granting marketing approval for the generic.[56] As the FDA did so, bioequivalence data assumed new significance in physicians' therapeutic decision-making, which raised concerns among medical leaders, government officials, and academic pharmacists about the abilities of physicians to interpret such data. Out of this uncertainty, pharmacists moved to claim a new professional role for themselves. Arguing that physicians lacked sufficient pharmacological training with which to evaluate bioequivalence data, pharmacists sought to identify themselves instead as the pharmaceutical experts best able to evaluate equivalence claims and hence best qualified to determine the most appropriate version of a prescription drug to be dispensed to a patient. To assert this expertise, pharmacists called, once more, for the authority to substitute another drug product for the one prescribed.

Recrafting Professional Expertise, Recasting Autonomy

In April 1970, the largest association of American pharmacists, the American Pharmaceutical Association (APhA), passed a resolution calling for repeal of the state antisubstitution laws. The APhA's about-face on the issue of substitution reflected the tensions and separations that had grown between organized pharmacy, organized medicine, and the PMA during the 1960s. The APhA had become increasingly frustrated by the tendency of physicians and industry representatives to suggest—in their congressional testimony and public speeches—that prescription drug costs were high in part because they incorporated the professional fees and markups applied by retail pharmacists. Because pharmacists' fees and pricing practices had been excluded from Kefauver's (and later Nelson's) investigation of the pharmaceutical industry, industry officials and physicians had been able to shift some of the blame for prescription drug prices onto pharmacists without congressional recourse.[57]

To pharmacy leaders, this blame-shifting strategy reflected a lack of respect among physicians for the pharmacists' professional abilities. William Apple, president of the APhA, accused the AMA of "relegat[ing the pharmacist] to the role of a merchant." He explained, "I don't think the AMA fully appreciates the extent to which physicians are relying on pharmacists to help patients purchase their prescriptions more economically. I don't think AMA is aware of the informal understandings that many physicians have with pharmacists regarding brand interchange." According to Apple, the pharmacy profession was "fighting for liberation" from medicine, the path to which depended on the profession's "eliminat[ing] the old concept that the pharmacist is merely the handmaiden of the physician." It would also be necessary to "wipe out the last vestiges of the view that the pharmacist is the final link—the end-of-the-line—in what is essentially a marketing system for prescription drug manufacturers."[58]

To be sure, pharmacists had been struggling since the 1950s to craft a professional identity for themselves. The compounding pharmacist of the first half of the twentieth century had been superseded, in the 1950s, by the vertically integrated pharmaceutical firms that compounded and packaged their own drug products. In the same decade, regional and national supermarket chains had begun offering pharmacy services to their customers, supplanting the unique services once provided by the community pharmacist.[59] Through the 1960s, pharmacists—and their professional organizations—were relatively content to accept their subordinate position within the health care team. But as Elizabeth Siegel Watkins describes in chapter 4, during this decade a clinical pharmacy movement arose within the profession, seeking to create a new clinical role for pharmacists and establish them as critical members of the health care team. In the early 1970s, recognition of the clinical importance of bioequivalence offered pharmacists a new foundation upon which to craft this new *clinical* professional identity. Indeed, as practicing physicians struggled to make sense of the ever-growing number of new drugs on the market, pharmacists argued that they—not physicians—possessed the appropriate pharmaceutical knowledge and expertise to make sense of the plethora of me-too drugs and identify the most appropriate version of a prescription drug for each patient. The APhA hoped that through repeal of the antisubstitution laws and the assertion of pharmacists' authority in drug product selection, pharmacists would finally be regarded "as bona fide members of the health care team," with equal professional standing to physicians.[60]

Pharmacists' claims for greater professional authority were also couched in economic terms. In the first half of the 1970s, health care costs—especially under Medicare and Medicaid—were expanding at unanticipated rates. In 1973, in an effort to rein in the federal government's prescription drug expenditures, the secretary of the Department of HEW proposed limiting reimbursement for drug purchases under federal programs to "the lowest price at which the drug is generally available, unless there is demonstrated difference in the therapeutic effect."[61] Secretary Caspar Weinberger's Maximum Allowable Cost program placed the issues of therapeutic equivalence and physicians' autonomy over the prescription in the legislative spotlight and highlighted the economic rationale behind generic substitution: the expectation that generic-name drugs were cheaper than their brand-name counterparts. Doing so revealed another aspect of prescription practice in which the pharmacist could claim superior knowledge. While physicians were largely ignorant of prescription drug prices (legislative attempts to include drug prices on all advertising and in all drug compendia had fallen flat in part because of opposition by pharmacists), retail pharmacists were fully knowledgeable about prescription drug pricing. In making the case for a new professional role, pharmacists argued that they possessed both the technical and the economic expertise with which to incorporate any legislative changes into prescription practice.

The APhA's resolution to repeal the antisubstitution laws provoked a stern response from the AMA. In an editorial in the *Journal of the American Medical Association*, the AMA expressed "alarm" at "such ill-considered action by a segment of the pharmacy profession." Said the writer, "[It] denigrates the profession itself and indicates a disrespect for the patients the profession serves." As far as the AMA was concerned, the antisubstitution laws were working perfectly because they left "completely to the physician's judgment the final decision about the drug product his patient" would receive. Repeal of the antisubstitution laws, by contrast, "would permit a pharmacist . . . to dispense substitutes in defiance of the physician's best therapeutic judgment," so that the physician would have "no control over the drug product to be dispensed."[62]

The AMA's Council on Drugs was equally disturbed by the APhA's actions. The NPC's Jean Weston wrote to John Adriani, chair of the Council on Drugs, expressing concern that "pharmacists, by self-confession, have suddenly become therapeutic experts capable of taking over the physician's responsibilities in that area once he has made a diagnosis." Weston urged Adriani and his colleagues on the council to make "some pretty vigorous comment" opposing the pharmacists' actions. Adriani was all too happy to oblige: "I agree that a statement saying *why* is important because many people do not understand what a 'Pandora's box' this [repeal of the antisubstitution laws] would open and that the public's best interest would not be served." The following May, the Council on Drugs passed two resolutions requesting the AMA "to resist repeal or modification of existing anti-substitution laws and vigorously to support the prerogatives of the physician to designate the drug of choice for his patient."[63]

While the APhA argued that the pharmacist was best qualified to evaluate the bioequivalence of prescription drug products, physicians countered that the uncertainty over equivalence made it all the more important that the physician's judgment serve as the basis for sound therapeutic decision making. D. N. Goldstein, editor of the *Wisconsin Medical Journal,* reflected some of the medical profession's concern that the pharmacist's ability to substitute a physician's prescription would lead to poor clinical outcomes. Goldstein argued that the physician "must be certain that what he prescribes is dispensed." "It is conceivable," he continued, "that when a patient is stabilized on a particular product, [the physician] is assured that the same degree of absorption, blood level and duration of action will continue throughout the course of treatment, whereas legalized brand-switching without the doctor's knowledge . . . could be hazardous to [the patient's] health." For Goldstein, this was "not a matter of questioning the druggists [*sic*] knowledge or his judgment"; it was "simply a matter of sound medical practice." To be sure, a series of studies published in Canada, the United States, and Britain between 1971 and 1973 documented the inequivalence—and resulting clinical effects—of different commercial preparations of digoxin, tetracycline, ampicillin, and warfarin.[64] These studies made it clear that for drugs with a relatively narrow range between their effective concentration and their toxic level, differences in bioequivalence could have profound clinical effects.[65]

Philadelphia physician Milton M. Perloff expressed similar reservations about giving pharmacists responsibility for making product selections. While acknowledging that physicians were not always able to adequately evaluate equivalence claims, he felt sure that "so far," no one was "better qualified than the physician to make this determination." In case the division of labor and the relative expertise of physicians and pharmacists were unclear, Perloff delineated the basis for the physician's autonomy over the prescription: "The pharmacist who fills the prescription is not trained for this clinical evaluation, and he is not exposed to the clinical atmosphere that exists between the patient and physician (where the actual evaluation of what the drug does for the patient is made)." In other words, clinical considerations—the appropriate domain of the physician—were paramount in all acts leading to the dispensing of the physician's prescription. No matter the pharmacist's knowledge about pharmacological effects and interactions, only the physician had all the necessary information available upon which to make therapeutic decisions. The AMA reaffirmed this position in a *Journal of the American Medical Association* editorial in August 1971, arguing that without such clinical knowledge, it was "difficult to believe that any amount of technical knowledge about drugs per se would place the pharmacist in a better position than the physician to determine which drug product would be best."[66]

In the end, the AMA and brand manufacturers lost the battle over substitution, and pharmacists succeeded in securing a new professional role in prescription practice. By mid-decade, driven by rising medical costs, state legislatures around the country were acting upon the APhA's recommendation to repeal the antisubstitution laws. In February 1975, for example, New Jersey state senator Frank J. Dodd and Assembly member Martin A. Herman each introduced a bill that would authorize pharmacists to substitute a generic drug for a specific brand-name drug prescribed by a physician. Signaling the state's escalating Medicaid costs, Senator Dodd argued, "We could save as little as $3 million or as much as $5.7 million" by authorizing generic substitution, a measure that "would stabilize the ever-increasing drug costs to Medicaid in the future."[67] The following month, California assembly member Barry Keene introduced a similar measure authorizing substitution amid claims that it "could save Californians $40 million in drug costs the first year and more in future years." The California legislature passed the generic-substitution bill later that year.[68] In New Jersey in February 1977, after two years of legislative apathy on the matter (not surprising, since New Jersey is home to several of the world's largest brand-name manufacturers), the legislature finally approved a bill authorizing generic substitution.[69] And by 1979, forty states and the District of Columbia had modified their antisubstitution laws so as to permit pharmacists to substitute generically equivalent drugs for the brand-names prescribed.[70]

Conclusion

What explains the changed regulatory context and altered professional boundaries of prescription practice between the early 1950s and the late 1970s? How did physicians—previously autonomous in their control over prescription practice—come to cede some of that control to pharmacists? Part of the explanation certainly lies in the declining social and cultural authority of physicians brought about by the activism of health feminists and the revelations about the Public Health Service's Tuskegee syphilis study. But to account specifically for the loss of physician control over the prescription, we must look also to the changes in the regulatory and scientific context of prescription practice.

Since the 1950s, brand-name manufacturers had predicated their opposition to prescription reform—be it the state-mandated use of generic drugs or pharmacy-controlled generic substitution—on the grounds that the FDA could not adequately regulate generic drugs and thus could provide no guarantee that a generic drug was therapeutically equivalent to its brand-name counterpart. Therefore, any legislation that usurped the physician's autonomy over the prescription by mandating the physician's use of generic drugs or by allowing the pharmacist to practice substitution put the patient at risk of receiving inferior—and worse, unsafe—prescription drugs. For brand-name manufacturers and their physician allies, this meant that "until all similar drug preparations can be equated meaningfully in terms of their bioavailability to permit the interchange of different forms of a drug on a rational basis, legalistic maneuvering designed to weaken or revoke drug anti-substitution laws should be opposed vigorously."[71]

The flip side of this defense, though, was to invite the possibility that once the "equivalence problem" was resolved and the FDA was able to guarantee the therapeutic equivalence of generic drugs, the rationale for opposing substitution and generic prescribing would be lost. And this is exactly what happened. Once the FDA began approving generic drugs on the basis that their manufacturers had demonstrated the biological equivalence of the generic to a brand-name counterpart, the PMA and the AMA lost a major leg on which to stand in their opposition to substitution.

Although the AMA bemoaned the loss of physician autonomy over the prescription, in the end, states crafted the repeal or modification of their antisubstitution laws in ways that preserved much of the physician's control. Certainly, the new substitution laws permitted pharmacists to substitute a different drug for the one prescribed by the physician, but they did so only if the physician specifically *authorized* the pharmacist to make such a substitution. Under Missouri's new substitution law, for example, a prescription form was valid only if it contained "two signature lines at opposite ends at the bottom of the form, with 'dispense as written' under one, and 'substitution permitted' under the other." When writing a prescription, the physician was then required to sign on one line or the other according to her preference. In Michigan, the pharmacist was prohibited from substituting unless the patient specifically requested a generic drug or unless the physician actively authorized the substitution. And in California, as in the majority of other states, physicians were required to check a "no substitution" box on the prescription form if they did not want the pharmacist to substitute.[72]

Rather than undermining the physician's prescription autonomy, the new substitution laws added a check on it. Physicians were still autonomous in determining the type of prescription drug their patients needed, and if necessary, the physician could still specify the brand of that prescription drug to be dispensed. What had changed was the degree to which physicians exercised unchecked authority over the therapeutic decision-making process. In the new regulatory context in which bioequivalence standards had become a core aspect of prescription practice, the new substitution laws acknowledged that physicians were not always in the best position to identify which version of a specific drug best met the pharmaceutical *and* economic needs of their patients. In this regard, the new substitution laws created a new clinical space in which pharmacists—as the health care professionals best equipped to evaluate the technical and economic aspects of drug selection—could stake their claims to expertise and authority over prescription practice.

CHAPTER 4

Deciphering the Prescription

Pharmacists and the Patient Package Insert

Elizabeth Siegel Watkins

In September 1980, after five years of spirited debate, the U.S. Food and Drug Administration (FDA) announced new regulations requiring patient package inserts (PPIs) to be included with prescription drugs. Three months later, *American Druggist* published an editorial in which the journal's editor, Dan Kushner, bemoaned the mandatory PPI. He conceded that consumers had "every right to information about every drug they are instructed to take," but he objected to placing the burden on pharmacists to distribute this information. His concern that this new role would jeopardize the pharmacist's relationships with both doctors and patients was made clear in the editorial's title: "Are You Set to Become a Punching Bag between the Physician and the Patient?"[1]

The first patient package insert was mandated in 1970 for oral contraceptives. This little leaflet was to be included with each filled prescription to warn patients of the increased risk of blood clots associated with taking the pill. The PPI raised important questions about the information coded within the prescription. Up to this point, the physician had controlled how much and what kind of information to reveal to patients about the prescription drugs they were directed to take. As Dominique A. Tobbell shows in chapter 3, physicians' authority over whether to prescribe brand-name or generic drugs had been challenged in the 1950s and again in the 1970s, but the substitution controversy pitted pharmacists against physicians; it did not include patients directly. In the 1970s, the women's health movement and the consumers' movement joined forces to reject the privileging of information and to demand that *patients* be given access to all available information so that they could make their own informed decisions about whether to take prescribed drugs. Whereas the doctor had traditionally communicated orally with his patients, feminist and consumer activists advocated for patients to receive written information about their prescription medications. The ensuing deliberations revealed widely discrepant opinions about the content and dissemination of drug information to patients, as well as the wisdom of sharing this information at all. Should the PPI be identical to the information provided to physicians by drug manufacturers, or should it provide an abridged version? Who should be responsible for drafting its language? Who should distribute the information to patients? Where and when should the patient receive the leaflet: at the doctor's office or at the pharmacy? Before or after she paid for the prescription? How would patients use this new knowledge? What would be the effect on so-called compliance with prescribed drug regimens? It was clear that the prescription stood as a cipher for a vast amount of expert knowledge about a drug product; what was unclear was how or how much, if at all, to decipher this information for the layperson.

The debates over the PPI in the 1970s also exploded the notion that medical communication took place along a single axis between doctor and patient. Some groups contributed to the flow of information about drugs and therapies not through face-to-face contact with prescribers and consumers but via indirect routes. Pharmaceutical manufacturers influenced knowledge and shaped opinions about drugs through advertising to the medical profession and public relations campaigns that targeted everyone else.[2] Journalists wrote magazine and newspaper articles about drug therapies, and broadcast news regularly featured coverage of medical topics. The actors who communicated most directly with both doctors and patients were pharmacists, who played an essential role in the interpretation, transmission, and translation of prescription information. Interactions occurred along the three sides of a communication triangle, with pharmacists, physicians, and patients each occupying one of the vertices. The PPI generated fresh controversy over the appropriate scope and content of information about prescriptions that pharmacists should convey to their customers. Also, as indicated by the editorial mentioned above, it highlighted anxieties about the role of the pharmacist as intermediary between doctor and patient.

The proposal for the PPI came at a time when the pharmacy profession was undergoing a sort of identity crisis. In the 1960s, a movement arose to expand the role of the pharmacist from drug dispenser to drug consultant. Clinical pharmacy, as the movement was called, sought to increase contact between pharmacists and patients and to include pharmacists on health care teams with physicians and nurses. The advocates of clinical pharmacy claimed that pharmacists should be responsible for providing information about the safe use of drugs to both the public and other health care professionals. They called for a revised curriculum in pharmacy schools, one that would focus on patient care through drug therapy. For these modernizers, clinical pharmacy would integrate pharmacy more fully into the health care system and, in so doing, advance the professionalization of pharmacy. Many retail pharmacists, however, viewed these proposed changes with skepticism. They were in the business of pharmacy, where success was measured by sales of drugs and other merchandise. It was true that pharmacists no longer spent their time compounding drugs; the medicines they sold were manufactured and, in many cases, prepackaged, by the pharmaceutical companies. But as doctors prescribed more and more drugs, pharmacists had more and more prescriptions to fill, and many pharmacists understood their place to be behind the counter dispensing drugs, not on the pharmacy floor counseling patients. In the 1970s, the PPI served as a lightning rod for these tensions within the profession of pharmacy.

In spite of pharmacists' key role in the circulation of prescriptions and pharmaceuticals in the late twentieth century, historians of American medicine have not paid them much attention. Although historians of pharmacy have traced the evolution of the profession from the apothecary to the modern drugstore, pharmacists have not yet been integrated as participants into narratives and analyses of the delivery of modern health care or the structure of the American consumer culture.[3] One of the goals of this chapter is to move pharmacists from the margins to the center of our historical understanding of how an economy of information developed alongside an economy of prescription drugs. By expanding this historiography, we gain a fuller understanding of how and by whom the information encoded in the prescription was transmitted and translated.

The controversy over the mandatory distribution of PPIs with prescription drugs illuminates just one of the challenges to therapeutic authority in the late 1970s. In this case, it was not the power of physicians to write prescriptions that came into question, but rather their authority to withhold information about those prescriptions. When consumer advocates and government regulators sought to claim for patients the right to information about prescriptions, pharmacists found themselves caught in the middle of the contest. The PPI compelled pharmacists to reconsider their own professional identities as purveyors of medicines, providers of drug information, and intermediaries in the flow of prescriptions between physicians and patients.

Efforts to Expand Access to Drug Information with Patient Package Inserts

Initially, any information about drug products was addressed to doctors. The 1938 Federal Food, Drug, and Cosmetic Act required manufacturers to provide prescribing information on the package labels of drugs, including indications, effects, dosages, methods of administration, hazards, side effects, contraindications, and precautions. A manufacturer could get around the requirement of printing all of this information if the package label stated that "literature would be sent on request." In 1961 the FDA closed this loophole with a new regulation that mandated "full disclosure" on or inside prescription drug packages, which led to the package insert. Manufacturers were responsible for writing these leaflets, based on the data submitted to the FDA as part of new drug applications, and distributing them with the drug products.[4] Although the inserts were intended for physicians, in practice they were most often received (if not necessarily read) by pharmacists, the ones who handled the drugs. Pharmacists did not include these leaflets when repackaging prescriptions for patients. It was entirely up to the physician to decide what, if any, information to share.

In 1968 the FDA quietly mandated the first patient package labeling requirement. The asthma inhaler isoproterenol had the paradoxical reaction of constricting, instead of expanding, the bronchial airways if overused. Since the drug was self-administered, the FDA decided that patients should be informed directly about this hazardous effect, and a two-sentence warning was printed right on the container. However, relatively few people were affected by this mandate, so it garnered little attention.

As I have discussed at length elsewhere, much more attention was focused on the first PPIs, the information leaflets ordered in 1970 for oral contraceptives, which were used by some 9 million women in the United States and another 10 million worldwide.[5] Concerns about the safety of birth control pills had escalated in the late 1960s, when several studies demonstrated a higher risk of abnormal blood clotting in women who took the pill. Other controversial studies suggested a relationship between the pill and various types of cancer, and dozens of metabolic reactions to the pill had been documented, but the implications of these effects remained unclear. After a highly publicized set of senate hearings on the safety of the pill, conducted as part of Gaylord Nelson's wider investigation into the pharmaceutical industry, the commissioner of the FDA, Dr. Charles Edwards, announced his agency's intention to require the distribution of a PPI with every prescription for oral contraceptives. With this proposed solution, Edwards hoped to accomplish two goals: first, to improve the status and power of the FDA as a regulatory agency, and second, to respond to the demand from legislators and the public to do something about the perceived problem of oral contraceptives.[6]

The first draft of the proposed insert described in layman's terms the health risks, side effects, and contraindications of oral contraceptives. Opposition to this version from the medical profession, the pharmaceutical industry, and the Department of Health, Education, and Welfare forced the FDA to rewrite the insert. The revised text was a scant one hundred words, one-sixth as long as the original, and it mentioned only five possible symptoms of side effects, whereas the earlier draft had listed more than twenty-five. When the abridged version was published in the *Federal Register* with an invitation for public comments, more than eight hundred individuals and groups wrote letters to the FDA in response.[7]

About half of the letters were copies of form letters distributed by women's and consumer's groups protesting the abridged text on the grounds that it did not provide full disclosure on all possible adverse health effects of the pill. Individual men and women also wrote letters in their own words objecting to the insert's reduction in length and scope and calling for greater access to information about medical information. Patients' access to medical information was exactly what most physicians opposed. Doctors' organizations (such as the American Medical Association and the American College of Obstetricians and Gynecologists) and individual physicians wrote to the FDA to protest the very idea of the PPI. They expressed outrage at the government's efforts to regulate their profession and argued that the mandated insert would interfere in the relationship between doctor and patient.[8]

Pharmacists did not weigh in on this particular debate, perhaps because the oral contraceptive PPI did not seem to impact their day-to-day operations. In the first five years after the FDA announced its final ruling in June 1970, the inserts were mailed by manufacturers to physicians to hand out with each birth control prescription. After 1975 the PPIs were included inside individual packages of oral contraceptives. In both cases, pharmacists did not have to do any additional work to provide patients with the mandated inserts.[9]

The PPI that was finally approved by the FDA was even weaker than the draft published in the *Federal Register.* It described no symptoms but merely told readers, "Notify your doctor if you notice any discomfort." Although it did mention that abnormal blood clotting was the most serious known side effect, it gave no indication of the vast amount of data that had been collected on the pill and its effects on health. The brief one-hundred-word insert referred to an information booklet that a patient could request from her physician. This eight-hundred-word booklet, written by the American Medical Association in consultation with the FDA and the American College of Obstetricians and Gynecologists, provided more information in greater detail. However, the onus was on the patient to ask the doctor, not the pharmacist, for the booklet, and in practice, very few women ever received it.[10]

The first PPI did little to reveal the wealth of information hidden behind the prescription. Women knew that the pills would prevent pregnancy, but they were not told how the pills worked, which symptoms they might experience, or what the relative risks of using oral contraceptives were as compared to pregnancy or other birth control methods. The disparity in knowledge between patients and physicians remained great, and in spite of their dire predictions about the disruptive potential of the PPI, physicians maintained their authority in matters medical and pharmaceutical.

Nevertheless, the oral contraceptive PPI whetted the appetites of health feminists and consumer activists for patients to receive increased access to information about their prescription medications. By the early 1970s, the consumer movement was in full swing and had achieved several legislative successes, most notably the establishment of new regulatory agencies (the Occupational Safety and Health Administration, the Environmental Protection

Agency, and the Consumer Product Safety Commission) and an expansion of the Freedom of Information Act. In the realm of health care, the American Hospital Association had been persuaded to adopt a patient's bill of rights, which declared, among other provisions, "The patient has the right to obtain from his physician complete current information concerning his diagnosis, treatment, and prognosis in terms the patient can be reasonably expected to understand."[11]

Consumer advocates wanted to extend the right of patients to receive full disclosure about medical care beyond the hospital walls. They found a sympathetic ear in the new commissioner of the FDA, Dr. Alexander M. Schmidt. When Schmidt took office in the summer of 1973, he affirmed his support for patients' right to know about the drugs they took and the PPI as the vehicle to provide that information. In 1974 he formed the Patient Prescription Drug Labeling Project within the agency to investigate the possibility of a wide mandate for PPIs. The head of the project, Dr. Vincent J. Gagliardi, was charged with convening meetings with interested parties, including representatives from the medical profession, the pharmaceutical industry, consumer groups, and pharmacists' associations.[12]

Although meetings were held in the fall of 1974 with the Pharmaceutical Manufacturers Association, the National Association of Retail Druggists, and the American Medical Association, consumer advocates grew frustrated with the glacial pace of the FDA's proceedings. On March 31, 1975, the Center for Law and Social Policy filed a formal petition with the FDA calling for patient labeling of prescription drugs, on behalf of the Consumers Union, the Consumer Action for Improved Food and Drugs, the National Organization for Women, the Women's Equity Action League, and the Women's Legal Defense Fund. The FDA took this petition seriously, publishing it in the *Federal Register* and inviting comments from interested parties.[13] In an effort to encourage the general public to weigh in, the FDA issued a press release that was picked up by the Associated Press. The publicity campaign worked; more than one thousand groups and individuals wrote letters expressing opinions on patient labeling for prescription medications.[14]

Perhaps predictably, the AMA opposed the petition for mandatory written patient package inserts, arguing, "The physician is the one who is able to ascertain the patient's physical and emotional characteristics and whether a particular patient has need of such information."[15] The Pharmaceutical Manufacturers Association (PMA) and the American Pharmaceutical Association

(APhA), the largest professional organization of pharmacists in the United States, raised questions about the feasibility and advisability of requiring PPIs for all drugs. The PMA worried about manufacturers' liability; the APhA worried about distribution of the inserts. Both groups suggested a pilot study of just a few classes of drugs, to evaluate the effectiveness of the PPI, instead of wholesale adoption of a program covering all prescription drugs.[16] But these skeptics and opponents were in the minority. The vast majority of the comments—93 percent—backed the petition. The FDA marshaled these letters as evidence of popular support for patient information, but it hesitated to authorize the requirement, opting instead to hold more meetings and gather more data and opinions from all interested parties.

To this end, the agency organized a joint symposium titled "Drug Information for Patients: The Patient Package Insert," cosponsored by the AMA, the PMA, and the Drug Information Association, a nonprofit organization of professionals involved in pharmaceutical research, regulation, and marketing that published the peer-reviewed *Drug Information Journal*. The conference, held in Washington, DC, in November 1976, attracted almost seven hundred participants representing medicine, pharmacy, nursing, hospital administration, the pharmaceutical industry, consumer groups, and the media.

Pharmacists occupied a prominent position on the program. Eight of the twenty-two presentations, more than one-third, were given by pharmacists or faculty from schools of pharmacy. Their participation in this conference signaled the recognition by both organized pharmacy and academic pharmacy that pharmacists must be engaged in the politics of decision making about drugs and drug information. Unwilling to be the silent partners of physicians, pharmacists had to ensure that their voices would be heard in matters directly relevant to the practice of their profession. The symposium began with a series of "viewpoints," from the perspectives of the FDA, the physician, the drug manufacturer, the consumer, the malpractice lawyer, the hospital nurse, the hospital pharmacist, and the community pharmacist. Pharmacists also contributed their opinions on information to be included in the PPIs, potential distribution problems, potential effects on the patient, and potential effects on the health care system.

It is worth pausing here to note the distance between hospital pharmacists and community pharmacists, two groups who represented very different poles of the professional spectrum. Hospital pharmacists worked in hospitals and similar clinical settings (e.g., nursing homes); community pharmacists owned or were employed in freestanding pharmacies and drugstores. The hospital pharmacy provided medications to inpatients, whereas community pharmacies catered to walk-in customers and their representatives. Community pharmacists, unlike their counterparts in the hospitals, had to be concerned about customer relations and other issues related to running a retail business.

Pierre S. Del Prato gave the community pharmacist's viewpoint.[17] He observed that advocates of the PPI program promoted several incommensurate goals that could not be attained with a single one-size-fits-all patient package insert. If the objective was to provide patients with full disclosure, then the PPI should include information about side effects, potential complications, and alternative therapies. The physician should have the patient read the PPI before the prescription was written, so that the patient could participate in the decision to use the drug. To illustrate, Del Prato drew an analogy with surgery: "Using the commonly accepted practice of requiring informed consent prior to surgery as an example, we might well ask, do patients generally give their informed consent for surgery just before the anesthetic is administered? The answer, of course, is no. Generally, patients are informed of the potential consequences of their surgery at the time that the decision is being made whether or not to undergo a given surgical procedure and that usually takes place in the physician's office."[18] On the other hand, he noted, if the aim was to improve compliance with drug therapy, then the PPI should give explicit instructions about how to take the drug and an explanation of why the drug is necessary. This sort of PPI would be best distributed by the pharmacist at the point of sale, so the patient would have written information to take home with the filled prescription. In this scenario, pharmacists would be faced with additional burdens of space and time: space to store the leaflets and time to affix them to each prescription and to respond to patients' questions and concerns. Pharmacists would have to be compensated for these added duties, Del Prato warned, and consumers would bear the brunt by paying higher prices for their medications.

The speaker representing hospital pharmacists had a somewhat different perspective, due to the location and nature of their practice.[19] Mary Jo Reilly, executive director of the American Society of Hospital Pharmacists, identified hospital pharmacists as professional members of the health care team responsible for patient care; she and her colleagues did not share the commercial concerns of their counterparts in community pharmacy. Reilly aligned pharmacists with physicians as arbiters of how much information to divulge to patients and in what form. "The pharmacist as well as the physician," she insisted, "must be given the flexibility of *not* dispensing to the patient the preprinted, standardized insert."[20] She agreed with Del Prato (as did most of the other presenters at the symposium) that patient education was essential to ensure the safe use of medications. But she made a distinction between hospital inpatients and all other recipients of prescription drugs, and she argued that the former should be exempt from any PPI mandate, because inpatients were under constant medical surveillance and because the sheer volume of prescriptions dispensed daily in a hospital setting precluded the practicality of also dispensing PPIs. The exceptional circumstance of the hospital allowed Reilly to pay lip service in support of patient package inserts, while at the same time excusing hospital pharmacists from having to deal with them.

Nobody at the symposium was openly opposed to better educating patients about prescription medications, but opinions on the mechanics of providing such information were widely divergent. Lacking consensus on how best to implement a large-scale system of patient package inserts and in the absence of evidence that such a system would actually be beneficial to patients, the FDA opted to continue studying the issue instead of taking immediate action.

While the Patient Prescription Drug Labeling Project inched forward (or stalled, depending on one's perspective), the FDA made a decisive move on one particular class of drugs, the estrogens prescribed for menopausal and postmenopausal women, mandating a PPI in September 1976.[21] Like the decision made in 1970 about oral contraceptives, this one followed the publication of epidemiological studies and a senate hearing. By 1975 the Premarin brand of estrogen had become the second-most-frequently-prescribed drug in America; 28 million estrogen prescriptions were dispensed that year. But the studies showed that estrogen was associated with a higher risk of endometrial cancer, and the senate hearing revealed that estrogen was being marketed and prescribed for a wide range of vaguely defined symptoms beyond the hot flashes for which it was indicated. Indeed, while many women took estrogen to relieve the hot flashes brought on by menopause, many others took the hormone because they believed it made them look and feel younger and more feminine. Well into its second year of deliberations on the idea of a broader mandate for patient package inserts, the FDA decided in the case of estrogens that patients needed full disclosure on these drugs to be able to make informed decisions about whether or not to take them.[22]

Once again, the FDA commissioner announced the agency's intention in the *Federal Register* and invited the public to comment on the proposed PPI. Almost four hundred individuals and organizations sent letters. About half of the correspondents supported the FDA's proposal, and many of them wrote in the language of informed consent and the patient's right to full disclosure that had developed in the six years since the oral-contraceptive PPI mandate. Physicians contested the patient package insert for the same reasons they gave in 1970: they objected to government regulation of medical authority and intrusion into the doctor-patient relationship. This time, pharmacists joined the debate. Those who wrote letters to the FDA clearly sided with the physicians.

These pharmacists expressed objections to the PPI that fell into two categories, one ideological and the other distinctly more practical. Some were opposed on the same grounds as their medical brethren: they did not want the government to meddle in the doctor-patient relationship. One correspondent wrote: "I have been a licensed pharmacist for nearly half a century and your recent regulation concerning the distribution of literature to patients with their prescriptions on estrogen fills me with horror . . . This interference in the physician-patient relationship is very dangerous and intolerable."[23] This pharmacist upheld the belief that decisions to divulge information about prescriptions should remain in the hands of the prescribers. The more practical objections were based on the everyday matters of time and money. The National Wholesale Druggists' Association wrote to the FDA on behalf of its members to oppose the PPI because of the "inflammatory effects of these proposals on drug prices and the serious distribution problems that may be created."[24] These concerns reflected the commercial aspects of pharmacy; many pharmacists were businessmen who had to pay close attention to the bottom line.

The absence of letters from pharmacists in support of the estrogen patient package insert does not imply that the profession was united in opposition. On the contrary, there were pharmacists who applauded an increase in patient information. These differences in opinion reflected a much larger division among pharmacists as the profession grappled with existential questions of what pharmacy was at present and what it should become in the future.

Pharmacy's Identity Crisis in the 1970s

In 1976, the year when the FDA announced the proposed estrogen patient package insert, the much-anticipated *Pharmacists for the Future: The Report of the Study Commission on Pharmacy* was being read and discussed by pharmacists around the country. The Millis Commission, named for its chair, John S. Millis, was formed in the fall of 1972 by the American Association of Colleges of Pharmacy, and the final report was published at the end of 1975. It was the twentieth in a string of reports, surveys, and conferences on pharmacy since the end of World War II.[25]

Each of these studies addressed some aspect of the concern expressed by pharmacy leaders about the character and status of their profession. The first major study undertaken in the immediate postwar years was focused "by pharmacy, on pharmacy and for pharmacy."[26] It examined the practice of pharmacy and the education of pharmacists and made recommendations as to how the latter could best meet the needs of the former. In the next few decades, pharmacy underwent numerous changes, resulting from forces both internal and external to the profession, that led by the 1960s to more outward-directed inquiries into the role and relevance of the pharmacist in society. Those changes included a shift in the number of years of study (from four to five or six) and the recommended degree for pharmacists (BS versus PharmD), the development and adoption of a national licensure examination, the passage of state and federal legislation regulating aspects of pharmacy practice, the rise of third-party prescription payment plans, and growing alarm in American society about the misuse and abuse of all kinds of drugs, licit and illicit, prescription and nonprescription.[27]

Unsure of their role in the health care system, pharmacists felt themselves to be "over educated and under utilized."[28] Almost all of the postwar studies pointed out that pharmacists were anomalous as health care professionals in two key ways. First, their compensation depended on the sale of products, not the provision of services. The goods sold by pharmacists were not limited to drugs but included a vast array of sundry items, which detracted from the image of a pharmacist as a health care professional. Second, they were isolated from the rest of the health care system, both in their education and in their practice. Not all pharmacy schools were affiliated with medical schools and medical centers; students who attended freestanding schools lacked exposure to patients and other health care providers. Most pharmacists worked in community pharmacies, not hospitals; their minimal contact with doctors and nurses contributed to the sense of geographical, intellectual, and professional isolation.[29]

Pharmacy leaders also worried about their profession's public image. To evaluate popular perceptions of pharmacy and pharmacists, the American Pharmaceutical Association commissioned a study by the Dichter Institute for Motivational Research, a market research company that applied Freudian concepts to help companies influence consumers to purchase more goods, from Jell-O to Barbie dolls to flights on American Airlines. The report, published in 1973 and based on a survey of more than five hundred consumers in twenty-three states, found that a lack of communication between pharmacists and patients left patients feeling alienated.[30] The pharmacy counter was more than just a physical barrier; it served as a symbolic obstacle to the establishment of a professional relationship.

Better communication between pharmacists and patients was one of the key tenets of the new clinical pharmacy movement, which endeavored to shift the focus from the drug product to the drug patient. Historical accounts of the precise origins of clinical pharmacy differ, but sources generally agree that the movement really took off in 1966 with the establishment of the 9th Floor Pharmaceutical Service Project at Moffitt Hospital at the University of California, San Francisco (UCSF).[31] This project aimed to expand the role of pharmacists by charging them with, among other duties, interviewing admitted patients and taking their "drug histories," instructing discharged patients on medications to be taken at home, and providing in-service education to doctors and nurses. Within a few years, a room in the campus library had been dedicated as the Drug Information and Analysis Center, where pharmacists developed detailed clinical opinions in response to specific questions about drugs and patients, as well as critical evaluations of new drugs being considered for the hospital formulary. At the same time, the UCSF School of Pharmacy began to experiment with its curriculum to incorporate clinical training for its students, launching a new era in patient-oriented pharmacy education.[32]

The founders of the clinical pharmacy movement were frustrated by the disconnect between their training and the realities of pharmacy practice. Most curricula still spent a significant amount of time on the how-tos of drug dosage preparation, when in fact this function had been taken over by the pharmaceutical industry long before. As early as 1946, 75 percent of the prescriptions dispensed were formulations prepared by manufacturers, to which the pharmacist had only to attach a label.[33] By 1970 just 1 percent of prescriptions required compounding by the pharmacist.[34] Clinical pharmacy advocates wanted to expand the role of the pharmacist from merely counting out pills and handing filled prescriptions to customers; they wanted pharmacists to become more involved in the counseling of both patients and physicians on the nuances of drug therapy. Clinical pharmacy was called "a new area of truly professional service to prescription patrons." Pharmacists who accepted this responsibility would find "their image as sellers of goods . . . altered significantly."[35]

Part of the movement for change was to integrate pharmacy more fully into the health care delivery system. Amid concerns about rises in the incidence of drug-therapy-related complications (estimated to be about 10%) and the number of drug reactions that required hospitalization (accounting for 5% of all hospital admissions, or 1.5 million per year), the pharmacist rose to the fore as the health care professional whose training could be best suited to handle this crisis. "We have a major problem of drug misuse in the country," announced Henry Simmons, director of the Bureau of Drugs at the FDA, speaking about licit prescription medications at a conference on pharmacy manpower at UCSF in 1970. "The pharmacy profession, if properly utilized, can be a significant factor in its solution . . . The medical profession will have to accept this new expanded role on the part of this fellow group of health professionals."[36] Simmons's remarks implicitly referred to medicine's perception of itself at the top of the health care hierarchy, with pharmacists, nurses, and others in subordinate positions below, and acknowledged that such an attitude could confound efforts to establish the practice of teamwork in hospitals and clinics. However, pharmacists, not physicians, were the subject of this symposium, and what became clear was that pharmacy education would have to change if the pharmacist was to be qualified to assume these new roles of counseling patients and collaborating with doctors and nurses.

It was in this context of uncertainty about the status, role, and education of pharmacists and unease about the use, abuse, and misuse of drugs in American society that the Millis Commission began its work. Over the course of two years, the commission met eleven times, conferred with eighty-one consultants, and generated more than six thousand transcribed pages of discussion. Mindful of the new discourse stimulated by the clinical pharmacy movement, the group began its report with a consideration of the question "What is pharmacy?" Eschewing the dictionary definition, "the art and science of compounding and dispensing drugs or medicines," as too narrow, the commission proclaimed, "Pharmacy is a health service." It went on to describe all health services—pharmacy, medicine, nursing, dentistry—as "knowledge systems." In the case of pharmacy, the knowledge centered on the interactions between humans and drugs.[37]

The commission believed that dissemination of knowledge was what pharmacists ought to be doing. It made no mention of the role of the written prescription in encoding and transmitting this knowledge, focusing instead on oral communication between the health care provider and the consumer. The commission found fault with the existing system of information distribution and called for the pharmacists to be information providers as well as service providers. "This will require more effective communication between professionals and patients," the report affirmed, as well as "new interprofessional cooperation and collaboration." The notion of service provision alone was radical to the old guard who saw themselves as product providers. The commission recognized this tension: "Not only is there no economic incentive to communicate with the patient[;] there is the reverse incentive *not* to communicate," in that advice given to patients (about, e.g., drug interactions) could result in the loss of a potential sale.[38] However, it made no suggestions about how to rectify the economics of the practice of pharmacy; it restricted its recommendations to the education of new pharmacists. The final report concluded that the old ways of training pharmacists in chemical knowledge and mechanical skills must yield to a new model of education that would incorporate knowledge about people and their behavior as well as about drugs and their actions. People skills would be essential to the pharmacist of the future, who would be a communicator and an educator as well as a dispenser of drugs and drug information.[39]

The commission acknowledged that a key force in effecting change in all aspects of the health care delivery system was what it called "the right to know." Brought about by consumer activists, patient advocates, and ordinary citizens, this movement arose "from a desire on the part of the public for more information."[40] The commission acknowledged further that the health professions had been more than remiss in imparting their knowledge; indeed, they had actively protected it as privileged information. The members of this study commission predicted the advent of a sea change in the relationships between consumers and providers of health care, based on new demands for complete information and full disclosure, and they urged pharmacists of the mid-1970s to transform themselves and their profession to meet these demands.

Pharmacists and the Patient Package Insert

With its emphasis on knowledge, information, and communication, *Pharmacists for the Future: The Report of the Study Commission on Pharmacy* spoke, albeit indirectly, to the crux of the patient package insert debate. How could patients gain access to information to help them understand and make decisions about prescription medications? Historically, the two groups who held that knowledge, physicians and pharmacists, had been loath to divulge it. Physicians thought patients were unable to comprehend expert knowledge; possessing privileged knowledge also helped doctors to maintain a privileged position in their relationships with patients. Pharmacists' reasons for nondisclosure stemmed from traditional restraint, as spelled out in the American Pharmaceutical Association's Code of Ethics of 1952: "The pharmacist does not discuss the therapeutic effect or composition of a prescription with the patient. When such questions are asked he suggests that the qualified practitioner is the proper person with whom such matters should be discussed." However, as we have seen, the pioneers of the clinical pharmacy movement bridled under such restriction and brought about a revision of the APhA Code of Ethics in 1969. The new version read, "A pharmacist should . . . render to each patient the full measure of his ability as an essential health practitioner."[41]

Given the profession's proclaimed shift from reticence to involvement, it might be expected that pharmacists would support the PPI as a tool to facilitate greater communication with patients. While some did adopt this position, others resisted the PPI, and a wide spectrum of opinion was expressed in the pages of pharmacy journals from the mid-1970s into the early 1980s. The published literature—editorials, articles, and letters to editors—presents evidence of the array of opinions among American pharmacists.

In an early editorial on the subject of PPIs, William A. Zellmer wrote in the June 1976 issue of the *American Journal of Hospital Pharmacy*, "In view of the limitations of our current method of physician-pharmacist-patient communications, the FDA is correct in pursuing a requirement for more patient package inserts." An article in *Drug Topics* was more circumspect. Titled "Patient Package Inserts: Will They Replace the Pharmacist's Advice?" it laid out the potential pros and cons—from pharmacists' perspectives—of written information about prescription drugs. On the pro side, it cited the opinion of one pharmacist that "they [PPIs] could push pharmacists into the very role they want to assume: that of a drug coordinator."[42] On the con side was presented this unsettling speculation: "Your role as a professional counselor on drugs could be usurped by a sheet of paper."[43] A survey of pharmacists' attitudes toward PPIs, presented at the annual meeting of the Drug Information Association in 1976 and published in that organization's journal the following year, yielded mixed results. While a majority of the relatively small sample (63 of 95) said they favored the development of PPIs, they also enumerated several possible disadvantages to patient labeling.[44]

At this point, PPIs were largely hypothetical. Since they were ordered only for asthma inhalers, which were not widely prescribed, and oral contraceptives, which had the PPI packaged by the manufacturer within each unit of sale, pharmacists did not have much actual experience with the provision of written information to their customers. This situation changed in July 1977, when the FDA issued its final ruling on PPIs for products containing estrogen. Once the order took effect, pharmacists were required to distribute a lengthy leaflet with each prescription for estrogen. The following year, the FDA added a mandate for PPIs to be distributed with progestins (synthetic forms of progesterone) as well. In 1977 the number of estrogen prescriptions dispensed at retail pharmacies in the United States was 22 million, and the number of noncontraceptive progestin prescriptions was 2.5 million; clearly, these rulings would have an effect on the nation's eighty thousand community pharmacists and fifteen thousand hospital pharmacists.[45]

Legislators also called for more information for drug consumers. The proposed Drug Regulation Reform Act of 1978, sponsored in the Senate by Ted Kennedy and in the House by Paul G. Rogers, included a provision to require "patient information labeling in layperson's language of the risks, benefits, side effects and so forth of any drug entity or product." Although the bill failed to pass in 1978 and again in 1979, it stimulated a great deal of discussion among pharmacists about government regulation of pharmaceuticals and medicine. Then, on July 6, 1979, the FDA finally published its formal proposal to implement PPIs for all prescription drugs.[46] As usual, the notice in the *Federal Register* invited public comment; this time, some fifteen hundred letters poured into the FDA headquarters in Rockville, Maryland.

According to historian Barbara Troetel, who studied these responses, organized opposition to this PPI proposal was stronger and better coordinated than it had been three years earlier, when the FDA published the Center for Law and Social Policy's petition, "because they [professional and industry groups] now faced an actual threat to their functioning and autonomy, rather than a theoretical one." Organized medicine maintained its hard-and-fast stance against the PPI, and organized pharmacy joined in the opposition. The members of the state associations of pharmacists in Nebraska, North Carolina, and Indiana, for example, sent letters protesting the FDA mandate because it would escalate the cost of prescription drugs and because it threatened the conviction that the doctor (and the pharmacist) knew what was best for the patient.[47] As both Congress and the FDA appeared to be accelerating toward a mandatory PPI regulation, pharmacists around the country turned to the pages of their professional journals to share their concerns, express their opinions, and speculate about their future.

American Pharmacy devoted much of its November 1979 issue to the topic of PPIs. The cover featured a photograph of a bewildered pharmacist being showered from above with information leaflets (fig. 4.1). The lead editorial depicted pharmacy at a crossroads: "The current consumer clamor for drug information and the proposed FDA rule come at a time for pharmacy which may be either propitious or awkward . . . A patient labeling program mandated by FDA will probably accomplish little, but the threat of such a regulation can be the stimulus to move the profession out of its tradition-bound inactivity." Companion articles echoed the irritation with government intervention. "The thought of mandatory patient labeling," one began, "touches a raw nerve in pharmacists who—justifiably—consider their profession already overregulated." A report on existing commercial information services noted glumly, "These highly successful operations may be doomed if FDA's proposed patient labeling regulations are implemented." And an article called "A Hard Look at FDA's Patient Labeling Project" argued that the FDA's cost estimate of $270 million over five years was much too low; APhA calculated a price tag closer to $1.8 billion.[48]

Editors of the nation's leading pharmacy journals made clear their dissatisfaction with the proposed PPI mandate. George F. Archambault, writing to an audience of hospital pharmacists in *Hospital Formulary,* titled his editorial "I'm Worried—Aren't You?" Although he professed his endorsement of the principle of PPIs, he limited it to outpatients and railed against the inclusion of hospital inpatients. In another issue of the same journal, the editor-in-chief took for his subject the FDA's proposal in the *Federal Register;* he excerpted quotes from the announcement and added his comments, which included "humbug! . . . meddlesome mischief! . . . Are you crazy, or am I?" Irving Rubin,

Figure 4.1. The cover of the November 1979 issue of *American Pharmacy*. Copyright American Pharmacists Association (APhA). Reprinted by permission of APhA.

the editor of *Pharmacy Times,* wrote a series of adamant polemics, in which he criticized the mandatory PPI as "overkill," "not in the patient's best interest," and "*not* the answer."[49] These commentators sided with physicians in defense of preserving the health professional's prerogative to decide how much information to share with the patient. They buttressed their position with arguments about the inflated cost, in terms of both time and money, of putting a comprehensive PPI program in place and speculation about the potential legal liability of pharmacists in malpractice suits.[50]

Individual pharmacists who took the opposite stance were not shy about voicing their opinions. The author of a letter to the editor of the *Journal of the American Pharmaceutical Association* asserted, "As a community pharmacist I hope to see PPIs become a reality." Another wrote to *American Pharmacy,* "The time is out for being an overeducated drug clerk/drug overseer. The pharmacist should be a utilized professional who is knowledgeable and willing to share that knowledge with his or her patients." One of the strongest supporters of the PPI was perhaps the nation's most prominent pharmacist, Jere Goyan. Plucked from his position as dean of the School of Pharmacy at the University of California, San Francisco, where he was an active proponent of the clinical pharmacy movement, Goyan became commissioner of the FDA in October 1979. He was the only pharmacist ever to serve as the head of the nation's food and drug regulatory agency (his predecessors tended to be either chemists or physicians; his successors were mostly physicians). In a 1982 speech, Goyan ruminated in retrospect that PPIs "seemed to be a logical extrapolation of the pharmacist's role as drug advisor, only a *slight* expansion of something the professional claimed to be doing already." He continued, "I anticipated some opposition from organized medicine but only a little from pharmacy. Was I wrong! . . . Almost every pharmacy association opposed PPIs."[51] Goyan and other PPI advocates sympathized with patients' desires for supplementary sources of information to help them make informed decisions as consumers; moreover, they saw the PPI as a "ticket for pharmacists to come out from behind the prescription counter that separate[d] them from the people they serve[d]."[52]

Although opponents claimed that the FDA had not collected adequate evidence to support the hypothesis that patients would benefit from written information about prescription drugs, an influential study proved otherwise. In 1979–80, the Rand Corporation conducted a prospective study of prototype PPIs for three kinds of drugs. Its survey of 1,821 prescription drug customers at sixty-nine pharmacies in Los Angeles County found that patients did indeed find the written information to be helpful. Of those consumers, 70 percent read the leaflet supplied to them, 45–56 percent kept it for future reference, and 22–32 percent reported reading it more than once. Contrary to the fears of pharmacists and physicians, the PPI did not increase the proportion of returned prescriptions, nor did it encourage patients to report more side effects.[53]

The Pharmaceutical Manufacturers Association tried a different tactic in its battle against the PPI. In 1977, when the estrogen PPI was ordered, this association took the offensive by filing suit in federal court against the FDA, arguing, first, that the FDA did not have the authority to mandate drug labeling and, second, that the requirement to dispense PPIs along with prescription drugs interfered with physicians' constitutional right to practice medicine as they saw fit. In February 1980, the U.S. District Court in Delaware ruled against the pharmaceutical manufacturers. The judge recognized the authority of the FDA and declared, "There is simply no constitutional basis for recognition of a right on the part of physicians to control patient access to information concerning the possible side effects of prescription drugs."[54] The judicial system had given its blessing to the PPI.

On September 12, 1980, six years after Commissioner Schmidt had formed the Patient Prescription Drug Labeling Project, the FDA published its final rule on PPI requirements for prescription drug products. Considerably less comprehensive than the original July 1979 proposal, which would have required PPIs initially for fifty to seventy-five drug classes and then for all prescription drugs, the mandate had shrunk to a three-year pilot program for only ten classes of drugs.[55] After the three years, the FDA would decide whether to expand it to include more drugs, based on data collected during the trial period. Recall that a pilot program of reduced scope was exactly what the American Pharmaceutical Association had proposed in 1976 in response to the *Federal Register* notice of the Center for Law and Social Policy's petition.

In general, the responses of many pharmacists to the final rule can be characterized as reluctant acceptance. Of course, there were those who still clung to their disapproval, such as the alarmist who asked, "Are You Set to Become a Punching Bag between the Physician and the Patient?" But the mainstream reaction was "It could have been worse." An article on the regulations quoted the executive vice president of the National Association of Retail Druggists, who commented, "10 is better than 375," referring to the earlier proposal to cover all prescription drugs. "The chain drug field," the article reported, "probably offers the best example of ambivalent reaction to FDA's final regulation—ideological opposition to a mandatory PPI program coupled with pragmatic preparation for compliance." An editorial in the *American Journal of Hospital Pharmacy* was more optimistic. Titled "Thinking Positive about PPIs," it concluded, "[The] patient package insert regulation balances fairly the interests of consumers and the concerns of health professionals." Given the program's limited scope, the author predicted "minimal burdens on pharmacists." As one of the potential outcomes of the PPI program, he said, "more pharmacists might gain confidence, and reap satisfaction, in discussing medications with patients, and, as a result, allocate greater time to counseling."[56] These more optimistic writers imagined that the patient package insert might, after all, help to bring pharmacists off the sidelines of the health care delivery system.

Denouement: The Revocation of the Patient Package Insert Mandate

While pharmacists, physicians, and pharmaceutical manufacturers readied themselves for the start of the patient package insert program, scheduled to go into effect on April 12, 1981, newly elected President Ronald Reagan—acting on his commitment to deregulation, privatization, and smaller government—issued an executive order for all federal agencies to review the necessity and cost-effectiveness of existing and proposed regulations. In rapid response, Arthur H. Hayes Jr., Jere Goyan's replacement as FDA commissioner in the new administration, deferred the implementation of the PPI mandate. One year later, in February 1982, the FDA announced that it would revoke its earlier ruling on patient package inserts, based on the rationale that a voluntary program using drug information and educational materials developed by the private sector would cover more products, reach more consumers, and be less costly than the mandatory federal pilot program.[57] The Reagan Revolution overturned a decade's worth of work toward government support of the patient's "right to know," leaving decisions about demystifying the prescription to market forces.

After the revocation, written information about prescription drugs was produced by a patchwork of organizations, including the American Medical Association, the United States Pharmacopeia, and the American Association of Retired Persons. A commercial enterprise, the PHARMEX division of the

Automatic Business Products Company, produced and sold to pharmacists what it called Patient Advisory Leaflets for frequently prescribed drugs. It also published a compendium called *Family Guide to Prescription Drugs.*[58] While written information was available for those who actively sought to learn more, the provision of drug information to consumers was not an incorporated part of the prescription process.

Some pharmaceutical companies began to develop their own patient package inserts, or patient information sheets, to comply with the FDA's drug advertising requirements, so they could promote their products directly to consumers. Starting in 1985, FDA allowed drug companies to advertise in mass market magazines and newspapers so long as they included a "fair balance of information," as was required for ads in medical journals. Most companies fulfilled this obligation by simply reprinting the text of the FDA-sanctioned physician package insert. This professional labeling served as the basis for developing PPIs, to be distributed along with the medication. The medical terminology, dense verbiage, and tiny fonts of these inserts made them inscrutable to the average consumer, thus rendering them virtually useless as informational sources.[59] By the mid-1990s, written information had been voluntarily developed for some thirty to forty medications,[60] a small fraction of the armamentarium of prescription drugs sold in pharmacies.

Pharmacists' acceptance of the wide distribution of PPIs became a moot issue after the revocation. However, they still had to contend with PPIs for birth control pills, estrogens, and progestins. As mentioned earlier, the oral contraceptive PPI did not present much of a burden to the pharmacist, because it was packaged by the manufacturer with each unit of sale. Two studies of the estrogen PPI found, however, that a majority of pharmacists did not comply with the regulation to distribute the information leaflet along with the filled prescription. In the larger of the two studies, FDA researchers sent undercover agents, armed with prescriptions for estrogen, into chain drugstores and independent pharmacies in twenty cities across the country; their results showed that in only 39 percent of interactions did the pharmacist automatically hand out the patient package information.[61] This finding was corroborated by a smaller local study of a single midwestern city, in which only 42 percent of customers received the PPI. The authors of the latter study hesitated to explain pharmacists' behavior, calling instead for "additional study . . . to ascertain the basis for pharmacist compliance or non-compliance with the PPI regulations."[62] The FDA researchers were less circumspect: "It is . . . possible that pharmacists' attitudes contribute toward the failure to deliver the PPI . . . Some pharmacists may feel that the PPI is an unwarranted government intrusion into their profession."[63]

The FDA study also reported troubling findings about pharmacists' verbal communication with patients. In almost three out of every four cases (72%), the customer received no counseling from the pharmacist. What had happened to the goals of the clinical pharmacy movement? Apparently, there was still resistance among the rank and file to expending time and effort on counseling. While pharmacy leaders debated the merits of mandating patient counseling and proposals for its compensation, many community pharmacists circumscribed their responsibility to the dispensing of drugs only. Over the course of the 1980s, the dispensing of information along with the dispensing of medicines did not—as was so heavily feared by pharmacy groups—become a required or an automatic aspect of the prescription process. Instead, the consumer bore the onus of requesting information. The profession of pharmacy remained, as one observer put it, "in transition."[64]

Conclusion

The saga of the patient package insert reveals how access to prescription information shaped the professional identities of pharmacists in the 1970s. The profession was split over whether or not to share with patients that privileged information—knowledge about the indications and contraindications for, side effects of, and alternatives to prescription medications. Some of the profession's leadership, those who worked as faculty in schools of pharmacy and participated in the clinical pharmacy movement, called for pharmacists to provide the professional service of drug consultation. Other leaders, such as the editors of the profession's journals, spoke for those pharmacists who juggled professional and commercial roles. While they paid lip service to the notion of patient counseling as trained drug experts, they opposed the distribution of written patient information on economic grounds as small businessmen. Willing neither to ignore the fiscal realities of pharmacy operation to focus exclusively on the provision of health care service nor to forgo their education as licensed specialists to act solely as merchants, pharmacists remained in professional limbo.

Pharmacists' objections to the PPI upheld the function of the prescription as a boundary marker between experts and laypeople. Recall that pharmacists contested the PPI not only because of the added burdens of space, time, and expense, but also because they opposed the intrusion of the government into the practice of both pharmacy and medicine. Pharmacists supported physicians' prescriptive authority (even as they contested the prescription of brand-name over generic drugs), and they believed they shared with physicians the authority to selectively mete out information about the medications prescribed. The prescription—and the coded knowledge it contained—symbolized the professional status of pharmacists. As evidenced by the stance they took in the PPI debate, pharmacists had a vested interest—consistent with that of physicians—in retaining control over the information within the prescription, as a way to set themselves apart from their customers and patients.

Feminist health activists and consumer advocates sought to reduce what they regarded as paternalism in the practice of health care in the United States. For them, the prescription represented an undesirable gulf between the providers and the consumers of health care. They resented the privileging of prescription information and sought to break down traditional boundaries between the haves and have-nots, in terms of access to that information, and thus they seized on a federal mandate for written patient information about drugs as a way for patients to become more actively knowledgeable about and involved in their medical care. These health care reformers wanted to remake the prescription as a starting point for open discussion between health care providers and consumers about the pros and cons of a therapeutic regimen. Pharmacists and physicians, in contrast, preferred to maintain the prescription's coded language and limited access.

Through the 1970s, the FDA appeared to side with consumers in efforts to decode the prescription through the mechanism of the PPI. As the FDA tried to increase its regulatory power and exercise greater control over the pharmaceutical industry, it also sought to decipher the cryptic prescription. The proposed PPI acted as a locus of government intervention in what had been until then the relatively autonomous practices of medicine and pharmacy. However, the FDA is an agency of the executive branch of government, and as such, its goals and actions are determined by the administration in office. Under presidents Ford and Carter, the FDA was authorized to pursue efforts to ameliorate the flow of information about prescription drugs, but once President Reagan entered office, that sanction was quickly revoked.

The FDA (before 1981) and consumer advocates were at odds with physicians and pharmacists about how to improve patient safety with regard to prescription medications. For the FDA, the PPI would reduce the misuse of drugs by providing consumers with better information about how, if at all, to take the drugs prescribed for them. For physicians and pharmacists, the prescription itself served as a safety valve, by restricting knowledge of potentially harmful information to those equipped to handle it, namely, themselves. This episode in the history of the prescription shows how access to prescription information was interpreted by the opposing sides as either beneficial or detrimental to consumer safety.

By revoking the ruling on PPIs in 1982, the FDA preserved the veil of secrecy that enshrouded the prescription. Consumer information did not become an obligatory part of the action of writing and filling prescriptions. To be sure, some physicians and pharmacists took it upon themselves to provide information to their patients and customers, but this remained a wholly voluntary aspect of the prescription process. Thus, the prescription continued to function as a gradient of knowledge and information, with patients at the low end of the slope and pharmacists and physicians arrayed higher above. Those with the greatest access to the information contained within the prescription—the physicians—maintained the highest rank of power, with pharmacists in a somewhat subordinate position but still superior to the public, whose access to drug information was severely limited. The PPI had threatened to upend this order by forcing pharmacists and physicians to divulge the well-guarded information, but the curtailment of the PPI program, before it even began, preserved the privileged position of physicians and postponed the need for pharmacists to change their ways.

CHAPTER 5

The Right to Write

Prescription and Nurse Practitioners

Julie A. Fairman

In 2006 the state of Georgia became the last state to allow nurse practitioners to prescribe, slightly more than twenty-five years after North Carolina became the first state to do so. After a sixteen-year battle between the Medical Association of Georgia and various nursing groups and their allies, nurse practitioners gained the "right to write."[1] Nurses' and physicians' groups framed the new regulations in different ways, according to their own political standpoint. Nursing groups and their physician allies in medical schools and private practices applauded the newly gained rights but lamented the restrictive physician-supervision clause included in the regulations. This restriction, they believed, decreased patient access to care and placed an undue burden on already overloaded practitioners, many of whom worked solo in poor rural areas where physicians were scarce or absent.

In contrast, the medical association noted it was "pleased with the outcome for physicians and patients." It stated, "While there were numerous attempts to broadly expand the scope of practice of mid-level practitioners, the physician's association was successful in halting those efforts."[2] Its political message implied that nurse-practitioner prescribing without significant physician oversight would endanger patients and that the restrictions were appropriate within a rational system that offered differentiation between clinical training models as the proxy for safe patient care. According to the Georgia physicians, the limits on nurse-practitioner prescribing were a rational response to the variance in length and content of nurses' and physicians' training programs.

The process of achieving nurse practitioners' legal right to prescribe and the response from nurses, physicians, and others to this process were, however, anything but rational. Instead, the battles over the right of nurse practitioners (nurses with advanced knowledge and training) to prescribe illustrate the arbitrary and contested boundaries of clinical practice set by both states and professional organizations, as well as the constructed nature of clinical practice that shaped modern health care in the last decades of the twentieth century. These battles entailed challenges to entrenched medical-practice boundaries that were set in physicians' hearts and minds long before the questions of who could and should have prescription authority arose.

Nurses and Prescribing

To understand nurse-practitioner prescribing of medications and treatments, the practice must be acknowledged as part of the continuity of nurses' patient care decision making. Historically, before the nurse-practitioner role emerged in the mid-1960s, prescribing by nurses was not unusual, if traditional remedies are included and the analytical lens is focused closely on what was actually happening in clinical practice. But it is also important to understand that a group of nurses who gained additional knowledge and skills that were traditionally part of physicians' exclusive territory differentiated and formalized nurse prescribing. The differentiation did not mean the end of prescribing by the generalist nurse, who continued to prescribe as patient need dictated, informally, without the regulatory support and protection eventually gained by nurse practitioners.

Nurses had a long, perhaps subversive, history of circumventing the usual prescribing channels, both with physicians and independent of them. Prescribing activities were sustained because patient circumstances demanded them and because nurses prescribed effectively. Historian Arlene Keeling describes the work of the Henry Street District Nurses in early-twentieth-century New York City; they carried various types of solutions and ointments, in addition to narcotics, caustics, and mustard, in their black bags. Most of these items were part of the medical armamentarium, similar to medications found in middle-class homes, and often the subjects of advertisements in popular magazines of the time. During their patient visits, the nurses frequently used a process of "de facto diagnosing" to make decisions about treatments, including dispensing and prescribing the appropriate therapy.[3] Perhaps after the visit they might have informed a physician about their actions or referred the patient for further treatment. Nevertheless, nurses made decisions and acted to prescribe treatment even when they had no formal authority to do so.

In general, what nurses could and did do in terms of prescribing differed somewhat from what they were legally allowed to do. Although much of what nurses prescribed in these circumstances was informally tendered (in the absence of explicit authority), nurses at the turn of the twentieth century certainly knew how and what to prescribe for a growing set of patient complaints. Most nurse training schools at the time included courses on materia medica that were similar to medical school courses. Public health nurses, in one of the most elite nursing specialties, acquired additional pharmaceutical knowledge during their postgraduate training, and their expertise in diagnosing and prescribing approximated that of physicians.[4] Patients also expected nurses to prescribe. Physicians sometimes made mistakes, were not always available, and were not always interested in indigent cases. Nonetheless, as dictated by the realities of their political position in the medical hierarchy, nurses would not typically draw upon the language of diagnosis or prescription, as these powerful words were already fully part of the medical lexicon and used in newspapers, medical journals, and texts of the time in reference to physicians.

Lillian Wald, the founder of the Henry Street Settlement on New York's Lower East Side, described how one visiting nurse chose language to relay her diagnosis and prescription suggestions to the physician who had "left a prescription" but had not returned to treat a febrile woman. "With fine diplomacy," Wald noted, "an excuse was made to call upon the doctor . . . My colleague [the nurse] presented her credentials and offered to accompany him to the case immediately, as she was sure conditions must have changed since his last visit or he doubtless would have ordered so and so, suggesting the treatment the distinguished specialists were then using."[5] In this case, the nurse contested the prescription left by the physician, hinting that it was not appropriate for the patient's current status and that she knew from her assessment and diagnosis of the patient what was needed in this case. She saw to it that the patient did not fill the physician's prescription, and she had probably already dispensed the treatment she believed the patient needed. But the limits of her ability were clear, as Wald's narrative suggested, as was the political field she had to navigate to get the appropriate treatment for the patient.

There are many more examples of prescribing by nurses, including the nurses of the Frontier Nursing Service and the Indian Health Service, who carried syringes, various drugs, suture kits, and dressings in their saddlebags and nursing kits and employed them to treat patients when needed.[6] In poor rural areas, in cities with large poor immigrant populations, and in areas where physicians were scarce or unavailable, nurses diagnosed and prescribed drugs and treatments for patients as an essential part of their practices. Many physicians, in fact, agreed that nurses should prescribe, realizing they themselves could never provide care to certain populations in certain areas, nor did they want to do so. In busy hospitals in the mid-twentieth century, as Leonard Stein explained in his classic description of the doctor-nurse game, physicians relied upon nurses to diagnose and prescribe. A nurse might call a physician and describe a patient's condition and then suggest the treatment ("Doctor, the patient is short of breath, do you want me to give him 40 mg of Lasix [a diuretic]? Yes, nurse, please give him 40 mg of Lasix"). Or, to allow a physician to get some rest at night, a nurse might treat the patient and tell the physician later. The common thread in these examples is that the nurse decided when (and how) to treat and when to call for help.[7]

Experienced nurses knew which physicians would expect them to diagnose and prescribe or would support their doing so and which ones might report them to their supervisors for practicing medicine and thus overstepping their boundaries. By the 1960s, nurses in hospitals and clinics informally prescribed any number of drugs to revive stopped hearts, relieve pain, or treat high blood pressure or blood glucose levels. For example, nurses in the newly organized intensive care units of the 1950s and 1960s typically started resuscitation procedures and determined which drugs and how much to administer before a physician arrived. There was no one else who could start the process to save the patient.[8] In these situations nurses relied on "situational credentialing," literally having the temporary authority to prescribe at any place and at any time until a physician could be consulted to approve the treatment decisions made.[9] "Whatever I did, I let the doctor know," one nurse reported, "but I was there and he wasn't and sometimes decisions have to be made right then. If you found the physicians, it might be too late . . . I didn't want to lose the patient."[10] Sometimes nurses took (and still take) these actions based on protocols, and sometimes they acted independently. But even when using protocols (e.g., an insulin scale that adjusted dosage based on blood glucose level), nurses operating under their own licenses used clinical judgment to determine when and how to use the protocol parameters and then what amount of medication to prescribe. Situated credentialing occurred in rural and urban hospitals and public and private institutions and depended upon the relationships negotiated between nurses and physicians. Its application was also influenced by time, place, and personal relationships, appearing and disappearing as the clinical scenario changed and illustrating the contingent nature of prescriptive rights.[11]

Nurse Practitioners and Prescription

Differentiation of the generalist nurse, who prescribed in particular situations, from the more highly educated nurse, who formally (or at first informally) prescribed, occurred fairly quickly within the first few years after the emergence of the nurse practitioner. In 1965 Colorado nurse Loretta Ford and pediatrician Henry Silver came together to address the health care problems of rural residents. Trained as a public health nurse, Ford was already expanding her practice boundaries when she first met Silver, but she knew she needed additional skills to provide even more of the care her patients needed. Ford and Silver developed the nurse-practitioner certificate program at the University of Colorado. At the same time, nurse Barbara Resnick and physician Charles Lewis were developing new models of care in the outpatient clinics at the University of Kansas in Kansas City. Plagued by poor patient follow-up and outcomes, Resnick and Lewis gave nurses additional training in physical assessment and clinical thinking and integrated them into the clinic structures to provide more continuous and sustainable patient care.[12]

As reports of these experiments were published in professional journals, the number of training programs for nurse practitioners grew rapidly, supported in part by federal funding through the Nurse Training acts in the next two decades and in part through the largesse of private foundations such as the Robert Wood Johnson Foundation, which supported new types of service models and training programs through demonstration projects. Nursing education responded both to the opportunity for funding new programs and to the increased number of nurses demanding access to the skills and knowledge necessary to expand their practices. As these programs grew, so did the boundaries of nursing practice as both nurses and physicians experimented with the limits of collaborative care. This flexibility was particularly evident in clinics and offices outside of hospitals, where institutional oversight was scarce or nonexistent. Nurse practitioners inserted intrauterine devices (IUDs), read x-rays, sutured wounds, ordered and administered immunizations, and prescribed birth control pills, to name just a few of the ways practice boundaries expanded. These were skills that nurses had already unofficially claimed, but now, in an officially expanded capacity, their work became more visible to other practitioners and was seen as a challenge by some of them.

Perhaps the most controversial public expansion of nursing practice was into the realm of prescribing medication and treatments. Besides surgery, this was the last piece of medical terrain to which other allied professions laid claim. To be sure, what physicians traditionally assumed as their own did not always remain within the medical domain.[13] But history and disciplinary custom suggest that the crossing of tasks between professional boundaries moved mainly in one direction, from doctors to nurses, in part because nursing knowledge and skills had lower status than those of medicine.[14] That is, physicians passed on tasks to nurses (such as taking temperatures, measuring blood pressures, and drawing blood) that no longer interested them because the skill had lost its complexity or "newness." But we can also find precedents for exceptions in cases where nurses and physicians worked closely together, usually in areas where physicians' jurisdictional claim was weaker or the clinical path uncharted.[15] For example, sharing of knowledge (e.g., technical, scientific, and communicative skills) between nurses and physicians occurred in the critical care units of the 1950s, in the dialysis units of the early 1960s, in the AIDS units of the 1980s, and during disasters and wars.[16] The nurse-practitioner movement followed this pattern of exceptions. The medical tasks (e.g., history taking, diagnostic reasoning, and prescription) and related knowledge (clinical thinking) assumed by nurse practitioners were distinctive because they were not discarded by physicians; physicians continued to perform them. Moreover, physicians took on the better communication skills they learned from nurses, in a reversal of the usual directional flow of tasks. In many instances, physicians themselves taught nurses how to expand their repertoire of skills to gain relief from busy or boring practices or the ability to cheaply expand their private practices.[17] In turn, nurses chose to learn the skills and use the knowledge, in the process gaining the status and power associated with them.[18]

Nurse practitioners understood that many patients did not need complicated treatments or drugs for most of the problems they brought to the clinics and offices and that prescribing the right medication, while a powerful symbol of medical authority and power, was not very complicated. In addition, this group of nurses was more vocal and independent, self-selecting a more autonomous role or looking to leave traditional nursing, which they considered boring and overly dependent.[19] Their broader knowledge of physical examination, diagnosis, and prescribing treatment, as well as their closer work with physicians in clinics or offices, also set them apart from generalist nurses, particularly those who worked in institutions like hospitals where most of nurses' work was rule-bound and beholden to the nursing, hospital, and medical administrations.

Nurse practitioners and their allied physicians were indeed colleagues, but to other physicians, nurse practitioners became competitors for patients and resources. Outside of the individual nurse-physician partnership, the nurse-practitioner profession faced many challenges to its boundary expansion. Physicians who were not familiar with what nurse practitioners could do, or who saw the expanded practice boundaries of nurse practitioners as encroachment on their professional prerogatives and their relationships with patients, created confrontational environments, which included legal challenges to state boards to investigate expanded practice boundaries.

For example, in the 1970s in Pennsylvania, a state with large swaths of poor rural areas and multiple counties that lacked physicians or had only a few, nurse practitioners were frequently reported to the State Board of Medicine for practicing medicine. In the state's northwest and north central counties, medical practices sometimes covered multiple clinic sites over large geographic areas, and physicians traveled between them by helicopter.[20] During their absence, physicians needed qualified partners to help expand their practices and provide continuous patient care. As many physicians learned, new medical partners were expensive and hard to find for rural communities, and so they began to work with nurse practitioners. As part of their practice, nurse practitioners prescribed medications, using protocols and prescription pads presigned by physicians. These were not new or extraordinary actions but were quite consistent with practices that occurred informally around the country.[21] But the nurse practitioners were accused of practicing medicine by the State Board of Medicine, perhaps in response to reports from local pharmacists who themselves wanted to prescribe, or at the least be able to substitute generics, or other physicians in the area who sought to disrupt the expansion of clinic services.

Physicians must have felt besieged as nurse practitioners, pharmacists, and physician assistants in the state (and nationally) all tried to gain prescription privileges at the same time during the 1970s and into the 1980s. But they may also have gained some respite when these groups, all less powerful than physicians and with fewer resources, temporarily shifted their focus from physician contestations to trying to prevent each other from gaining prescription privileges. Pharmacists in Pennsylvania, as in other states at the time, claimed greater authority than nurse practitioners or physician assistants because of their detailed pharmaceutical knowledge and immediate contact with patients after dispensing a prescription.[22] Physician assistants in Pennsylvania claimed that they were more qualified to prescribe because they practiced under the authority of physicians' licenses (in contrast to nurses, who worked under their own license) during the 1970s. Indeed, they were the first allied health profession to gain prescribing rights in the state, but not without a struggle. In particular, the State Board of Nursing campaigned against physician assistants with the rhetoric (but not the facts) that the state nurse practice act did not allow generalist nurses to follow written orders signed by physician assistants. The Board of Pharmacy also challenged the physician assistant prescription privilege. It gained regulatory oversight over their prescribing practices, by winning a seat on the Board of Medicine and permission to participate in generating the language for the statute. Pharmacists were less successful in their attempts to influence nurse-practitioner legislation, despite mounting a fierce lobbying campaign both to participate in nurse-practitioner regulation and to gain their own prescription rights. Pharmacists never gained this privilege, although they successfully challenged physicians in other arenas, such as the right to substitute generic medications. Physician assistants and nurse practitioners succeeded in part because they were known clinical partners, were tied professionally to physicians for patient access, and had gained the support of powerful legislators.[23]

Pharmacists, physicians, and nurse practitioners also battled each other in New Hampshire, as nurse historian Deborah Sampson found. There, the ability of nurse practitioners to legally prescribe rested on the convenience of the public and the intersecting interests of other professional groups, such as pharmacists. Nurses in New Hampshire prescribed fairly independently until about 1973, when their prescribing became the focus of medical board scrutiny and challenges from the Board of Pharmacy, which saw nurses' prescribing as hindering pharmacists' attempts to gain their own prescription rights. The opposition to nurse prescribing came not from erroneous prescriptions or evidence of patient harm but from actions meant to limit nurses' broader practice claims. After twelve years of constant lobbying and political maneuvering, New Hampshire nurse practitioners gained their full formal prescription rights by persevering through changes in the leadership of the medical and nursing boards (who brought varying political allegiances and preferences into the mix), by holding off the pharmacists, and by learning how to gain legislative support from sympathetic state representatives whose friends were patients of the nurse practitioners.[24]

Physician and pharmacist opposition to nurse prescribing involved numerous complexities, including economic opportunism and control over practice prerogatives that trumped any evidence of nurses' ability to safely prescribe. This was perhaps most evident in court challenges filed in response to nurse prescribing. Although the early nurse-practitioner programs were not officially structured to support nurse prescribing, by the mid-1970s most programs included courses in basic sciences, pathophysiology, and pharmacology that provided a solid knowledge base. One of the first direct court challenges came almost fifteen years after the nurse-practitioner role emerged, indicating perhaps the difficulties state medical organizations had in bringing such cases to court, the presence of an opponent who gained political allies, or the particular local politics of the state.[25] Whatever the case, in 1980 the Missouri Board of the Health Arts (which regulated nursing and the practice of medicine) voted to recommend criminal prosecution for nurse practitioners working in a rural family planning clinic who were prescribing birth control pills.

In the case *Sermchief v. Gonzales* (1983), the board contended that nurse practitioners were practicing medicine without a license when they inserted IUDs and prescribed birth control pills (all according to collaborative protocols) and argued that patients were receiving inferior care, despite the lack of evidence that patients had been harmed or were unsatisfied. The Missouri nurse practice act, which had been revised in 1975 to remove the need for direct physician supervision of nurse practitioners, broadly allowed nurse practitioners to practice in a way that was commensurate with their education and training. At the University of Missouri, which had the only formal nurse-practitioner training program in the state, pharmacology and training in IUD insertion were part of the course work. Both medical and nursing students at the University of Missouri also learned how to perform pelvic examinations, conduct physical assessments, and prescribe birth control pills from nurse-practitioner instructors.[26]

The nurse practitioners, as in many of the cases brought against them, received broad support from nursing groups and academic institutions, as well as from individual physicians and the dean of the state medical school. Thus the challenge to nurse prescribing was not fully supported by all physicians in the state, but perhaps just by those who remained rooted in traditional political hierarchies.[27] After an initial trial court defeat for the nurse practitioners, the Missouri State Supreme Court reversed the judgment, noting that the Missouri nurse practice act was similar to those found in more than forty states and that nurse-practitioner practice, including prescription authority, went unchallenged in these states.

This case was important because it presented a critical illustration of how challenges to nurse prescribing influenced patient care. In Missouri at the time of the trial, nurse practitioners provided much of the care to the state's poorer rural citizens, since most of the private physicians refused Medicaid patients and more than forty-two counties were designated federal manpower shortage areas.[28] If the nurse practitioners had lost the case, most of the patients using the family planning clinics in the state would have lost their providers and access to the full scope of services available to other, more affluent patients.

The Politics of Regulation and Professions

As part of its regulatory function, each state decides who can prescribe medicines and medical devices within the health care realm, from over-the-counter drugs (e.g., Tylenol) to durable products (assist devices such as splints or canes) to narcotics. In the 1980s (as today), regulatory acts differed from state to state, thus making prescription an inconsistent boundary marker of professional practice. The powers of the state medical societies and their political connections and resources, as well as the strength and resources of the state nursing associations, were powerful influences on nurse-practitioner prescribing regulation (as seen in the New Hampshire example), although by this time, all states gave nurse practitioners some prescription rights. The variability, unrelated as it was to quality and safety issues, raised questions concerning the interests of the state professional boards. Why, for example, could nurses prescribe independently in Alaska but not in Georgia? Why were nurse-practitioner prescribers considered safer in certain states than others at certain points in time? Did the boards exist to protect the public or the political interests of the professions they regulated? The answers reflected both the aim of public protection and the influence of politics across time and place. By the mid-1970s, some states recognized nurse practitioners' ability to safely prescribe controlled substances and other drugs by formulary (i.e., a list of medications nurse practitioners were or were not allowed to prescribe) while supervised by physicians. By 2011, eighteen states and the District of Columbia allowed nurse practitioners to independently prescribe all non–Schedule I drugs (every drug category except illegal drugs like heroin) without physician supervision.[29] By authorizing independent prescribing, particularly of the more complex and dangerous types of medications and treatments, regulators in these states acknowledged that nurse practitioners had the knowledge and skills to accurately and safely prescribe these medications as part of their clinical role. These more permissive states tended to have a rural character, a high proportion of poor or immigrant populations, places and populations with very low physician-to-population ratios, and low visibility of the state medical societies.

State regulators who did not grant broader prescribing privileges to nurse practitioners were convinced by the resources and rhetoric of other professional groups (or by the absence of stronger resources and arguments by nurse organizations) who portrayed nurse practitioners as unsafe unless they were supervised by practitioners who had more extensive education and who had decades ago established and normalized their role as exclusive prescribers. Absent from most of the discussions surrounding nurse-practitioner prescribing was any reflection by regulators on the changing parameters of how much and what types of knowledge were actually needed by health care professionals of all kinds to safely prescribe for most of the patients seeking care at the end of the twentieth century.

The cases noted in the previous section serve as late-twentieth-century examples of century-old processes heavily influenced by history, tradition, and politics as each state regulated its health care providers. By the late 1880s, most states had passed mandatory medical practice acts, with the medical profession designating the areas they believed were or could be part of medical practice. The medical acts were broadly written, reserving to physicians exclusive clinical practice rights, and they were notable for leaving little for others to claim that was of high status or for which fees could be collected.[30] Prescription was included as part of the terrain of medical practice, but not as a compulsory activity. Until 1914, with the passage of the Harrison Narcotic Act, which made prescription for certain narcotics mandatory, the prescription was more of a symbol than a required means for patients to obtain dispensed medications.[31]

Throughout the twentieth century, the American Medical Association actively worked to develop expansive medical practice definitions within state practice acts, ensuring a continued broad scope of practice for physicians. The foundational knowledge a physician needed to safely prescribe was assumed by the public and by state professional boards to be constantly updated and responsive to scientific discoveries and new therapies. This was a reasonable assumption, but it did not take into account the changes in practice that occurred as population demographics and patient needs changed in the last four decades of the century. Older adults and patients with chronic illnesses needing health promotion and prevention services made up a growing part of a provider's practice, and they required of the provider less specialized but broader, more comprehensive skills to safely prescribe, as well as the ability to manage all of the various components of care within the fragmented health care system. A common refrain from the 1970s and 1980s is still heard today: "If nurses want to practice medicine they should go to medical school."[32] This phrase carries the assumption of physicians' putative ownership of the ability to prescribe medications and treatments, and it captures the momentum medicine, as practiced by physicians, has had in the past century to consistently define its practice boundaries across states without constraint despite the changing practice arena.

In contrast, nurse practice acts were fairly broad regulations in the first three decades of the twentieth century, protecting the title "Registered Nurse" against challenges from untrained attendants rather than defining specific practices. But change came in 1938, when New York became the first state to establish mandatory licensure, marking the first time nursing practice was legally defined. This event was notable as the beginning of more codified restrictions on nursing practice. The New York regulations positioned nursing practice as complementary to physician practice, and they included a requirement for direct physician supervision of patient care functions, an atypical expectation up until that point.[33] Constructed by hospital administrators and physicians, with the consultation of nurse training school supervisors who believed the new regulations would keep untrained and poorly educated nurses out of the patient care market, the highly politicized regulations were out of date from their inception. Private duty nurses, for example, although receiving and taking orders from physicians, rarely worked under direct physician supervision and were known for their independence and sometimes blatant disregard for medical orders.[34]

Although they created baseline parameters for nursing practice, the New York regulations specified what nurses were not allowed to do (e.g., diagnose or prescribe) as opposed to what they were authorized to do in private duty, public health, and hospitals. To be sure, nursing education in the years before World War II was not designed to support nurse prescribing, nor was there great support for formal prescription privilege from the nursing leadership or nurses themselves, at least until they entered practice, where the reality of patient care overtook training socialization. The regulations also created hierarchies of trained and untrained nurses, putting into effect a strict and complicated structure for the nursing profession as it tried to improve its standards for entry into practice and to increase the legitimacy and social standing of professional nursing.

The main professional nursing organization, the American Nurses' Association (ANA), contributed very little to the discussion surrounding the definition of nursing practice and nurses' prescribing activities until 1955, when its House of Delegates approved a legal definition of nursing practice. Here, finally, should have been a national template for states to explicate the functions, standards, and qualifications of nursing practice.[35] The new definition, the ANA believed, supported its call for mandatory state licensure of nurses, and it did accomplish this aim, since most states had cooperated by the early 1960s.[36] But the new definition was itself problematic because it prohibited "acts of diagnosis and prescription of therapeutic or corrective measures."[37] These were areas into which nursing practice was already unofficially expanding in intensive care units, public health, and outpatient clinics.

The language used in the definition was based on recommendations by the ANA leadership, mostly hospital supervisors, and it may be that this group was unable to envision or acknowledge the changes occurring in practice both inside and outside hospitals. An admittedly admirable mission to bring about some sort of standardization for legislative and consumer protection purposes, the definition was obsolescent and restrictive from the start, yet many states adopted it over the next decade. Despite the inconsistencies between practice on the ground and the ANA definition, most state nurse practice acts included the prohibitions against diagnosing and prescribing well into the 1960s and 1970s, changing only when state medical societies challenged the broader reach of nurse practitioners and tried to further reduce their ability to practice, with varied success.

The Constructed Nature of Practice Boundaries and Prescription

That nurses prescribed is a historically situated fact, but states' authorization and acknowledgment of nurse practitioners' legal right to prescribe varied over time and place, depending as much upon the political, social, and economic milieu as on the acquisition of particular skills or knowledge or patient need.[38] The political and arbitrary nature of states' determination of prescription rights set up a challenge to the state boards themselves as the best place for these decisions to be made, highlighting the oppositional forces of state regulation, physician authority, and the growing voices of other providers and patients.

In fact, prescription may not form as clear a professional boundary marker as is typically claimed. Physicians built their proprietary prescription authority on their claims to exclusive relationships with patients, claims that were based in part on their knowledge of basic sciences and pharmacology but also on their exquisite understanding of the individual patient. Physicians' claims were strengthened by their efforts to control the conditions and possibility of other providers' involvement in the patient relationship, which they fortified by their strong influence with state regulators through class, gender, and professional alliances. But physicians' exclusive relationship with the patient was perhaps not as exceptional as many described it. There were others, including nurses, families, and later on, physician assistants and pharmacists, who also developed therapeutic relationships with patients, thus rendering suspect the idea, as Elizabeth Siegel Watkins notes in chapter 4, that communications with patients "took place along a single axis."

An important point to consider is that nurse practitioners did not threaten physicians' prescribing autonomy, but they did threaten physicians' monopoly and exclusivity, as well as the particular relationships physicians claimed with patients. Nor did nurse practitioners' efforts illustrate government interference in medical practice. Like the generic-drug designation sought by pharmacists (as discussed by Dominique A. Tobbell in chapter 3), nurse practitioners' prescribing represented a challenge to physicians' normative and seemingly timeless status as the only professional group who could safely and accurately prescribe, a status, as historian of medicine Harry Marks has shown, that was manufactured and substantiated only since the late 1930s.[39] Furthermore, these challenges came from providers who were educated for shorter periods of time and during a period of important changes in medicine's social and cultural authority.

As authors of other chapters note, physicians' proprietary claims to prescription authority were characterized by cold war politics of fortification and defense, but they were challenged by groups such as pharmacists and feminists and attempts by the federal government to put physician prescribing under surveillance (via the Food and Drug Administration). Organized medicine's inability to stop new government entitlement programs such as Medicare and Medicaid also resulted in fundamental changes in fee-setting and government oversight of practice. Physicians' claims to exclusive rights to prescribe were confronted during this time by growing public mistrust resulting from reports of mistaken prescribing and overprescribing of barbiturates, antianxiety drugs, and sleeping pills, as well as newly publicized accusations of physicians who did harm by using patients as research material without their consent. By the early 1970s, nurse practitioners' efforts to gain prescription rights compounded the onslaught of these challenges to medicine's authority, along with contemporary incursions into the medical marketplace by chiropractors and osteopaths. In this milieu, organized medicine and its individual members were perhaps unusually vigilant and sensitive during the 1970s and 1980s to any formal infringement on what they understood to be their historical rights, even as the reality of busy practices for individual physicians ensured and demanded their collaboration with other types of providers.

For nurse practitioners, authorization and acknowledgment of their right to prescribe presented a double-edged sword. On one hand, what nurse practitioners did routinely on an informal basis (e.g., individual and situational prescribing of medications and treatments without legal protection and codification found in state practice acts) with and without tacit physician approval was suddenly "dangerous" when they sought to codify and formalize the activities. Informal prescribing was in some ways quite liberating for nurse practitioners. Informality allowed them a modicum of freedom from oversight at the margins of clinical practice for patients who needed their attention. On the other hand, authorization and acknowledgment also meant that the states recognized new nursing roles that came with new proficiencies acquired by nurses, such as advanced knowledge of pharmacology and physical assessment skills. State recognition helped nurses to claim the rights and responsibilities of licensed practitioners, including reimbursement, and to fully extend their services to patients. Indeed, improving patients' access to services was one of the primary reasons for the existence of nurse practitioners.

The prescription privilege also exemplified how the nursing profession differentiated itself by education level and place of practice, essentially firming up the political hierarchy that had percolated within the profession for decades. The nurse-practitioner role was both similar to and different from traditional generalist nursing. It incorporated knowledge and skills traditionally in the physician's realm but applied them through the lens of nursing. For example, when a nurse practitioner discovered an infant's infected ear and treated it, he was also educated to ask how the family was coping with the new infant and whether resources were available to pay for the antibiotics. The subtlety involved in perception of the expanded role and the increased authority that came with it was confusing to a nursing profession that practiced primarily in hospitals, saw its work as highly distinctive from physicians' medical work, and could only recently boast a better-educated workforce. The nurse-practitioner role and the right to prescribe accomplished something nursing leaders had been unable to do in the 1950s and 1960s, that is, to differentiate the profession by levels of knowledge and, indirectly, by education levels and independence of practice. But these new roles and rights also created dissonance within the nursing ranks as many generalists saw nurse practitioners as elitists or as medical imposters. Many nurses refused to follow patient orders issued by nurse practitioners and went out of their way to subvert their practice.[40]

Physicians' authority to normalize their broad practice and prescription purview rests in traditional assumptions of their cultural authority to make expansive claims to these powers through the physician-patient relationship, as well as the public's complacency when it comes to challenging them. It is very hard to argue against physicians' power to prescribe, given their long cultural authority to so. It is easier to argue for others to also have that right. The most common reason that people have accessed the medical system in the late twentieth and early twenty-first centuries has been to get advice or treatment for the management of chronic illness, and the lay public has a great tendency to self-prescribe, blurring even further the boundaries of professional practice. This is particularly true for complementary and alternative therapies, over-the-counter remedies (especially those that were previously available only by prescription), and self-dosing of prescribed drugs.[41] If patients prescribe their own treatments, why is there such pushback from physicians over prescribing by other providers? Patients themselves became quite effective at managing chronic illnesses such as diabetes, using protocols to administer insulin and to determine during acute situations when provider consultation is needed. Sometimes they need to visit a nurse practitioner for insulin adjustment; if they are experiencing more complicated and acute problems, they might need to see a physician. To take this scenario even further, the patient might sometimes visit providers other than doctors or nurses, such as nutritionists or physical therapists, who themselves prescribe different treatments. This is a very different paradigm of care from the traditional fee-for-service physician-based system, and it stands to redefine the processes by which social networks and professional identities are formed.

As nurse practitioners become more commonplace and normative in clinical practice settings, they challenge at least a portion of physicians' hegemonic claims to prescription as a professional boundary marker. If nurses can prescribe as well as physicians for problems in particular populations, and they learn to do so in shorter periods of time, what do these findings mean for the relevancy of physicians' claims?[42] Perhaps it is the fear of irrelevancy (in addition to economic threats and decreasing political power) that guides physician opponents to challenge nurse-practitioner prescribing, knowing that in some fields, such as primary care (both general and pediatrics) and family practice, the difference between the types and quality of services physicians and nurse practitioners offer is narrowing and sometimes indistinguishable. Or it could be that physicians find it very hard to overcome medical school socialization that instills early on the notion that physicians must control everything surrounding their patients, especially prescription with its implicit and explicit power and meaning. Whatever the reason, the prescription and the "right to write" offer multiple standpoints for examining the relationships that shape and define clinical practice. Part metaphor, part icon, prescription as a proxy for political power, concrete knowledge claims, and cultural authority serves to illustrate the constructedness of clinical practice boundaries in modern health care.

CHAPTER 6

The Best Prescription for Women's Health

Feminist Approaches to Well-Woman Care

Judith A. Houck

On the evening of September 20, 1972, police officers and investigators from the California Department of Consumer Affairs, armed with a search warrant, entered the Los Angeles Feminist Women's Health Center. They were looking for any evidence of the illegal practice of medicine, including "prophylactics," speculums, patient charts, menses removal equipment, and "yogurt."[1] They were also looking for Colleen Wilson and Carol Downer, lay health workers at the center, who turned themselves in to the police the next day. After their arraignment, Wilson, charged with eleven counts of practicing medicine without a license, pled guilty to one count, agreed to a suspended sentence and two years probation, and paid a $250 fine. Downer, accused of diagnosing a yeast infection and treating it with yogurt, pled not guilty. On December 6, 1972, a four-woman, eight-man jury found Downer innocent. In the aftermath, Downer called the verdict "a victory for all women." She proclaimed, "Now we and all women are free to learn about our bodies, to learn good health care, and to use common sense in caring for ourselves."[2]

Downer's defense insisted that the language of "diagnosing and treating" in the statute was vague and overbroad and that, if enforced to the letter, it would prevent friends from passing Kleenex and mothers from giving children chicken soup for a cold. Further, the defense insisted that the use of yogurt was a home remedy and thus exempted from the definition of treating.[3] These arguments apparently resonated with the jury. The charges against Wilson were more serious. They included fitting a diaphragm, performing pelvic examinations, supplying birth control pills, removing an intrauterine device (IUD), and extracting menses—procedures that required more invasive bodily contact. Furthermore, since birth control pills were available by prescription only, Wilson had clearly crossed over into medical territory. Her guilty plea precluded judicial consideration of the permeability of the border between medical practice and self-help. At issue in the burgeoning feminist self-help enterprise was the question of professional authority, not between professional groups, but between health care professionals and laypersons without specialized training. In the negotiation of this issue, the writing of prescriptions took on material and symbolic importance in drawing the line between what lay providers could and could not do.

The "great yogurt conspiracy," as this event has been dubbed, established the positions over which feminist health activists, particularly those involved in providing health services, and the medical establishment negotiated for at least the next decade. On one hand, health feminists claimed for women their right to see, understand, and care for their own bodies, at least their own well bodies. On the other hand, physicians and their allies reserved for medical personnel the right to diagnose, treat, and prescribe. Feminist health centers became the site for many battles over these alleged rights. Feminist activists sought to expand the services provided by nonmedical personnel, and physicians, often supported by the state, sought to defend the "practice of medicine" from feminist agitators and dilettantes.

This chapter looks at how feminist clinics provided health care to their clients while remaining true to their politics and practicing within the law. In particular, it examines how feminist health activists coped with the legal restrictions defining who could prescribe and dispense medications and insert and fit medical devices. But as the great yogurt conspiracy illustrates, the state's suspicion of these clinics and physicians' hostility toward them did not focus solely on whether feminist health activists were illegally procuring prescription medications for their clients. Critics were more concerned with these clinics' larger prescription for women's health, which was based on the premise that the natural workings of women's bodies did not require medical inspection, oversight, or control. Women themselves, sometimes trained through state-sanctioned programs, sometimes merely guided by more experienced women, could identify and observe their anatomy, understand their menopausal experiences, and treat their menstrual cramps, all without the help of physicians. As the women staffing these clinics saw it, the best prescription for what ailed many women was not a pharmaceutical prescription at all but rather a supportive, feminist environment where women could share their questions, experiences, and discoveries while abandoning their shame, fears, and passivity. By accepting this broader feminist vision of health care, which relied less on professional expertise and innovative therapeutics and more on women's self-knowledge and sympathy, women would be better prepared to evaluate their health and well-being and better equipped to demand their liberation.

I focus on two feminist health clinics in largely rural northern California, both of which emerged from larger social, political, and economic developments that increased the access of Americans to health care during the economic prosperity of the postwar years. In 1962 President John F. Kennedy signed into law the Migrant Health Act, created to address the unmet health needs of agricultural workers. In 1965, as part of President Lyndon Johnson's Great Society, Congress funded Medicaid to improve health care for low-income people and Medicare to finance health care for the elderly and the disabled. Also in 1965, two doctors dedicated to securing social and medical justice for all founded a community health center in a Boston housing project. When Jack Geiger and Count Gibson opened the Columbia Point Community Health Center, it sparked a community-health-clinic and free-clinic movement that spread quickly across the nation.[4] As health care access and government health spending increased, medical professionals and scholars of health care resources began to worry that the demand for health care would outstrip the supply of professionals equipped to meet it. Predicting a health "manpower" deficit, policymakers and health care professionals considered the need for more "midlevel" health professionals, who could take on some of the "routine" health care tasks and serve as primary health care providers in some geographically isolated areas.[5] Although nurses had been delivering primary health care to particular patient groups long before the 1960s, advanced-practice nurses, physician assistants, and other medical auxiliaries increased in number and scope in response to the perceived shortage of physicians.[6] The

Figure 6.1. California, showing locations of Arcata and Chico.

two feminist health clinics described here reflected and expanded on these larger changes in health care.

The first is the Chico Feminist Women's Health Center (FWHC). In 1975 Chico was a town of eighteen thousand at the northern end of California's agriculturally rich central valley (see fig. 6.1). In general, this primarily rural area of the state had long leaned distinctly to the political right.[7] Nevertheless, Chico, influenced by its California State College campus, also had a history of progressive politics and radical protest. Not easily reduced to town-versus-gown conflicts, the city and its surrounding communities frequently became a "battleground" for far left and far right politics.[8] The Chico health center was part of a well-known "federation" of centers—somewhat controversial, even within feminist networks—associated with health activists Carol Downer and Lorraine Rothman.

The second is the North Country Clinic for Women and Children in Arcata. A small town of nine thousand people, Arcata was known in the 1970s for its hippies, its college, and its marijuana, hallmarks that occasionally caused friction with other communities in the rest of this rural, mostly working-class area in the far northwestern part of the state.[9] This clinic began as the Women's Health Collective of the Humboldt Open Door Clinic, but in the midst of this history, the collective broke off from Open Door to found its own, explicitly feminist, North Country Clinic.

The women involved in these two clinics believed that physicians should not have exclusive control over women's health care. They doubted that a medical or a nursing degree conferred special skills or knowledge that women could not acquire from study with each other, at least in the realm of the routine health and reproductive needs of well women. Both clinics' founders and staff drew boundaries between routine health care and the treatment of minor conditions (which could be provided by lay persons) and the care of acute or chronic conditions (which was best left to doctors). But beyond these shared bedrock commitments, their political differences led them to assign different roles to lay health workers and medical and nursing personnel. The organizers of the Chico FWHC reacted more strongly to what they perceived as the incursion of licensed professionals into well women's routine health care and insisted that these services be provided only by lay persons. However, they did rely on nurse practitioners and occasionally physicians to consult about conditions outside of "wellness" and, significantly, to provide authority to prescribe medications and devices such as oral contraceptives, antifungal medications, and IUDs. The feminist clinic in Arcata, by contrast, was founded in part by two licensed health care practitioners, a physician and a nurse practitioner. The Arcata clinic's personnel did not necessarily see organized medicine's role in women's lives as problematic, but—like their colleagues in Chico—they did not believe that physicians or nurses had any exclusive authority over women's bodies. As a result, they challenged the medical monopoly by employing a different tactic: the creation of new categories of legal midlevel health care practitioners. The histories of these two clinics suggest that a distinction might be drawn between a narrow sense of prescription (the right to prescribe drugs) and a broader interpretation that includes the ability to diagnose, evaluate, and make recommendations about the health of a woman's body. Feminists certainly mounted a challenge to physician control over the latter, in contesting the expert-lay boundaries set up by medical professionals. The "prescriptions" that well-woman clinics offered outside of the arena of prescription drugs highlight important ideological and practical differences between preventive health care and medical therapeutics.

Feminist Self-Help Care in Chico

Although the details of the great yogurt conspiracy may have been unique, the larger political claims it staked over the control of women's bodies were part of a much broader social trend. Even before women's liberation was an identifiable force, women across the country were fighting to repeal abortion laws and taking drastic action to assist women in terminating their pregnancies. For example, in the early 1960s, Patricia Maginnis and other California women campaigned for the repeal of abortion laws and simultaneously helped to secure safe abortions, often sending women to Mexico. By 1970, as women in Boston talked and wrote about their bodies, as women in Washington, DC, demanded safe methods of birth control, as women in Chicago operated an illegal abortion service, their various activities formed a notable movement. Although the movement was not unified by methods or goals, it nurtured and developed Maginnis's claim that "a woman's body is her own and she has a right to it."[10]

Many health feminists in the 1970s became health activists through their involvement with the cervical self-exam and the self-help gynecology clinic. The self-help clinic was developed in 1971 after Carol Downer discovered that with a mirror and a speculum, women could peer into their own bodies, a view heretofore off limits to them. The self-help clinic was not merely a demonstration that showed women how to find and admire their cervixes. It was a series of meetings of a small group of women who together explored their own and each other's bodies. Relying on their own experiences and inspired by their own curiosity, these women developed the skills and knowledge to trust their own instincts about and assessments of their bodies. Over time, advanced self-help clinics emerged that included the use of menstrual extraction and encouraged women to learn even more by studying their own bodies and the bodies of other willing participants.[11]

The self-help clinic explicitly challenged the medical monopoly over all things bodily. Downer, for example, decried the medical profession's efforts to gather and "hoard" information about women's bodies, repackage it, and then "dispense it like a product to those who are able to pay for it."[12] By creating and sharing a woman-centered base of knowledge about women's bodies, the self-help clinic wrested control from the medical profession.

The self-help-clinic movement spread quickly across the country and beyond. It also served as part of the inspiration for the development of women's health centers staffed largely or entirely by lay health workers. At women's health centers, "hippie free clinics," and community health centers, women who had been involved with self-help gynecology stepped forward to provide well-woman care. If they could recognize their own yeast infections, they reasoned, they could help other women to do so as well. If a doctor without a uterus could learn a bimanual pelvic exam, women who had experienced the cold speculum and the rough palpitations could learn it too (and, they believed, perform it better).

The Chico Feminist Women's Health Center was founded in 1975 as part of a loosely affiliated network of health centers (the Feminist Women's Health Centers) inspired and guided by Carol Downer and the self-help movement. The Chico center offered a variety of reproductive services, including pregnancy screening, birth control methods, and well-woman gynecological clinics. These services differed notably from those offered in traditional medical settings. Most significantly, all the well-woman and contraceptive services, at least initially, were performed not by nurses or physicians but by lay health workers who had received in-house training over the course of six months. They learned a variety of procedures (e.g., how to perform a bimanual pelvic exam, a breast exam, a Pap smear, and a sickle cell screening), and they researched medical and alternative views of women's bodies. The use of lay health workers challenged the medicalization of women's bodies, insisting that through practice, well women could learn to take care of themselves and each other.[13]

The women of the Chico clinic did not reject medicine altogether. They believed that "disease and pathology and prescribing prescription drugs" were rightly medical practices and that professional medicine had its place. Still, they insisted that women had the right to gaze at their own genitals and to touch and explore their own bodies. Further, they believed that menstruation, menstrual cramps, menopause, pregnancy, and the like were "life processes" rather than medical events. The clinic's mission was therefore not to replace medical care. Instead, the women of the clinic hoped to reclaim women's well bodies from medical purview and control. They also believed that by providing information about disease and treatment options, the clinic could help women drive their own medical care in the event of illness or pathology.[14] They did not want to replicate the medical model with lay health workers. They wanted to replace professional medicine for well women and to make women informed and empowered medical consumers in case of illness, effectively separating the spheres of preventive care and medical therapeutics.

Although seeing themselves primarily apart from medicine, the women of the center never worked entirely free of medical and state oversight. Indeed, to open their doors, they needed a state license, and a state license required a medical director. Because of the general hostility of the local medical community, their first medical director was also the physician they hired to perform their abortions, Carl Watson, who lived and worked in Oakland, more than 150 miles away. While Watson was willing to provide paper legitimacy, he was not interested in being involved in the day-to-day administration of the clinic. Where, then, would the Chico FWHC find someone who was authorized to prescribe the needed medications and devices while allowing the lay health workers to perform all the necessary examinations and to call all of the shots?[15]

At first, the Chico FWHC relied on Richard McDowell, a physician who was "very into para-medics and health workers." Perhaps his greatest value to the center was that he "didn't give a shit." "He wanted to sit there and be told what to do and he'd write out a prescription for what you told him." McDowell was perfectly willing to let the lay providers decide on the correct prescription and to lend his signature to seal the deal. His tenure at the center was brief, however, and terminated when he was eventually "run out of town" because he could not get staff privileges at the local hospital.[16] The Chico feminists next worked with a "young upstart from Sacramento, who wanted total involvement." He worked for them for about ten weeks, but in the end, "he decided he didn't want to work anymore." Tired of dealing with physicians who resisted working for the center on the center's terms (and having run out of candidates), the women decided to hire a nurse practitioner and push "the limits of the nurse-practitioner position to its absolute limits." In Brenda

Hanson, they found one who was willing to push the boundaries and test the law. "She feels like it's her right," they said.[17]

Hanson was hired by the Chico FWHC in 1975 to provide "routine gyn stuff," including performing pelvic exams, prescribing birth control pills, fitting diaphragms, and placing IUDs. In addition, she conducted preabortion exams on Fridays for the Saturday afternoon abortion clinics and provided aftercare for the abortion patients. Carl Watson, the Oakland-based medical director for the clinic, performed the actual abortions. Watson also agreed to serve as Hanson's "supervising physician," at least on paper, but he wanted no further responsibility for her or for the health care provided at the clinic.[18] He was willing, however, to presign a prescription pad. The center also relied on a local doctor who worked for the Butte County Health Department, Julian Lorenz. On Friday nights, the staff would take all of Hanson's charts to Lorenz's house for his review. (They delivered them to his house rather than his office to protect his professional life from the taint of association with the FWHC.)[19]

When Hanson started at the center, many nurse practitioners were prescribing drugs and inserting IUDs, but few had the explicit and undisputed right to do so (see chapter 5, by Julie A. Fairman, for the history of nurse practitioners' efforts to gain the right to prescribe). Hanson understood from the beginning of her tenure at the Chico FWHC the fuzziness of her scope of practice under the 1974 Nurse Practice Act.[20] She was also aware that she was technically required to work in collaboration with supervising physicians, but she exploited that ambiguity of the supervisory requirements. She does not believe she actually "collaborated" with either Watson or Lorenz. She did not even meet Lorenz until a decade or so after she left the center (at which time he complimented her on her "impeccable" charts), and although she thinks it's "logical" that she would have called him once or twice, she has no memory of doing so. Given this distance from medical oversight, along with the nature of the work she was doing at the clinic, Hanson understood that she was practicing at the limits of her license. She also understood that the FWHC was marginal to and often despised by the medical establishment. Nevertheless, she firmly believed that she was toeing the legal line. "I thought I was out on the cutting edge of nursing, [and] I knew that it would be a new frontier." Although she believed she was "working out here at the edge," she also believed absolutely that she was "not working over the edge." She insisted that she had been trained precisely for the job she was doing and that she remained on the right side of the law. But Hanson was not naive; she knew the precarious position of the center and her place within it. She understood that she was "working at the edge as a nurse practitioner in a clinic that was [itself] at the edge."[21] By using Watson's presigned prescription pads and by writing prescriptions that Lorenz later verified, Hanson had appropriated the traditional physician's prerogative in prescribing drugs for patients. For the feminists who ran the clinic, Hanson was, by virtue of her nursing license, enough of a "health care professional" to take on that role.

Although the Chico FWHC employed Hanson and other nurse practitioners, the directors of the center did not relinquish all or even most of the routine health care to these "professionals." Indeed, it is unclear how much of the routine gynecological care—the breast exams, the pelvic exams, the diaphragm fittings—the center allowed her to do. Even after hiring Hanson, the women of the center remained committed to living and modeling their belief that women as women, without any special license, "should have control" over their "health care and lives." As a result, they remained committed to providing routine gynecological well-woman care in a "participatory clinic," where groups of women and lay health care workers would examine and learn about their own and each other's bodies.[22]

Hanson was hired in part to provide prescriptions for oral contraceptives and to insert IUDs, but the women associated with the Feminist Women's Health Centers worried about the risks these methods of birth control posed to women's health, and they insisted that there were other reliable methods with fewer potential side effects. Like the feminist health activists described by Heather Munro Prescott in chapter 7, who resisted efforts to make the pill more easily accessible, the Chico staffers insisted that women needed unbiased information about birth control before they could choose the best contraceptive for their own situations. At the center, lay health workers and contraceptive seekers could swap personal experiences and informational resources to "make an educated choice." Women—health care workers and health care consumers—together created a space in which women could decide for themselves whether to use the pill; Hanson, at least in her role as nurse practitioner, merely supplied the prescription.[23]

The directors of the Chico FWHC acknowledged that their commitment to lay care in a participatory clinic sometimes challenged the sensibilities of the medical professionals they hired. With Hanson, they knew that their refusal to relinquish well-woman care left her with ambiguous responsibilities, which made her uneasy. "The only frustrating part for her is that we do much of it, that she doesn't get to do very much. And that is a conflict for her, because she wants to be doing it too."[24] Hanson clearly supported the center's politics, believing that women had the right to health care on their own terms and that women—not the medical establishment or anyone else—owned their own bodies. As a result, Hanson embraced her status as employee, despite her much greater health care training and expertise. She felt that the group welcomed her and valued her contributions. They would often ask her opinion on health care issues and sometimes asked her to help them learn more. In the end, she was grateful for the experience of working at the center. "They were so kind and generous to let me be a part of that clinic." Hanson may also have willingly taken a back seat because she believed that the participatory clinic worked. "I thought excellent health care was provided . . . because women were empowered to learn about themselves and whenever women are given knowledge about themselves they are empowered and they choose to learn more. So excellent health care was given."[25]

Although the center always tried to provide health care within legal constraints (necessary to remain open and funded), their commitment to lay health workers and their rejection of the medical model for healthy women continuously challenged the boundaries of legal health care provision. Sometimes the center survived by stealth. Julian Lorenz, who signed Hanson's charts, remembers that the system worked because they were quiet about their actions. "You know, we were just smart enough to fly under the radar . . . We just knew how to do it, to fly under the radar."[26] There is plenty of evidence, however, that their actions did not go unnoticed.

For reasons that are not entirely clear, in the fall of 1975, Hanson and Watson were investigated by the Board of Medical Examiners and the Board of Registered Nursing to determine whether Hanson was "practicing medicine without a license." It is likely that Hanson was caught in various contemporary struggles between the health center, the local medical community, and the state, mostly via the Office of Family Planning (OFP). In December 1975, for example, a representative of the OFP visited the Chico center and discovered many ways in which the clinic was operating outside the "standards of care." In her report, Mabel W. Daley, RN, was distressed to discover that no physicians attended the gynecological clinics and that the "job descriptions and the protocol" for the nurse practitioner were not adequate. (The 1974 Nurse Practice Act required that each clinic employing nurse practitioners establish their own policies and protocols.)[27] This report suggests that the center courted trouble both by operating independently of medical oversight and by letting a nurse practitioner practice without setting strict boundaries around the content of that practice.

But something else was at play in this episode. The center had faced opposition from the local medical community before it even opened its doors. For example, an influential physician allegedly blustered to the Butte-Glenn Medical Society that the center would open "over [his] dead body." The medical society responded, perhaps with more impact, by recommending that its members not refer patients to the clinic. In addition, none of the local physicians agreed to work at the center or to provide backup for the abortion clinic in cases of emergency.[28] These physicians were not protesting the employment of a nurse practitioner to prescribe drugs. Instead, they moved beyond this narrow interpretation to protest the clinic's broader prescription for well women: "Take control of your body. Take control of your life." In essence, the medical community opposed the clinic's prescription of a body-centered feminism. Hanson's actual practice of writing and supplying prescriptions merely provided an excuse to investigate the clinic and disrupt its operations.

If the Board of Medical Examiners and the Board of Registered Nursing had determined that Hanson had practiced medicine illegally, she would have lost her license. In the end, however, no charges were filed, perhaps because Watson insisted that Hanson always worked in consultation with him, a point Hanson conceded was "stretching the truth to the utmost." In the aftermath of the hearing, at least according to Hanson's memory, her practice changed only slightly (or not at all). Perhaps the women of the center tightened their charting procedures. Even though Hanson survived the ordeal generally unscathed professionally, the threat of losing her license for providing the health care for which she had been trained shaped her outlook and her politics forever afterward. She realized how people with power deployed it to quash the already marginalized. After being threatened by the state for doing her job, she came to understand "power and the power trajectory that was not only affecting, not only the feminist clinic and me as an individual, but women's health." According to Hanson, "these guys" came to the "little town of Chico" and tried to use their power to make the FWHC "step in line" and do what they wanted it to do.[29]

Although Hanson would not have called herself a feminist when she began at FWHC, her experience with the Board of Registered Nursing led her to embrace feminism explicitly. "It's kinda like at some point you need to join the club and you need to be a stakeholder . . . It's important to be able to say, 'I'm a feminist' . . . It's important to be able to say that, if indeed that's what it takes to continue to get quality healthcare for women . . . So I am a feminist, and there's a whole bunch of us out there."[30]

After Hanson left the Chico FWHC in the late summer of 1976, the center faced an ongoing problem of finding and keeping nurse practitioners (or anyone else willing to take on prescription authority) in their employ. For example, in 1977, the center hired a male nurse practitioner who was initially "very supportive" but who quit abruptly and with no explanation after a few months. The clinic's problems were probably compounded by its continued refusal to give medical professionals jurisdiction over physical exams.

In August 1979 the clinic was still scrambling to find someone who could bring medical legitimacy to their clinic. Apparently the current nurse practitioner had gone into labor prematurely and thus started her maternity leave sooner than planned. Steven Pulverman, a new doctor of osteopathy (DO, as opposed to MD) in town, was willing to work at the well-woman clinic once or twice because they "were in a bind and maybe . . . [the clinic] could return the favor some day."[31] He was allegedly not willing to do more because of pressure from local doctors. Although he was initially supportive of the clinic, after feeling out the situation he decided that "if he were to have a relationship with the Health Center, he would not be able to get anywhere in the medical community." His wife Katherine Ljunquist, a nurse practitioner, was also approached about working at the clinic. She was receptive at first, but when she learned how the clinic worked—that is, how lay health workers did many of the tasks usually reserved for licensed personnel—she backed away, claiming it didn't "feel good" to her. "Although she believed in education (sharing information) she was very uncomfortable with the fact that lay people did it here." The center was apparently desperate for her help, offering her money and telling her that they needed her "if only for a couple of hours" or else the clinic would be forced to shut down. She replied that she was "still unconvinced" and advised the center to "just give everyone condoms and foam" and reschedule them for a later time.[32]

Over time, the compromise that the Chico FWHC reached with the state—using nurse practitioners and physicians for consulting, referral, and prescribing, in exchange for "permission" to use lay health workers for well-woman care—fell apart. As state administrations changed and new bureaucrats from various state agencies discovered anew the practices at the Chico center, the pressure mounted to change the clinic's ways. Initially, the center appeared to win its battles. For example, when the Office of Family Planning in 1977 threatened to cut off its funding if the clinic didn't turn over the gynecological exam to a physician or a nurse practitioner, the center stood firm. "To have the Nurse Practitioner conduct the entire exam would promote professional separation between women and their health care. We will continue to use Nurse Practitioners in the capacity of a redource [*sic*] and back-up in a well-woman clinic."[33] The Chico FWHC then insisted that well women without state credentials could take care of themselves and each other, but they accepted nurse practitioners as medical resources and sources of legal legitimacy for prescriptions. As the directors of the FWHC put it, physicians and nurse practitioners could best serve as a "resource" through "standing orders, chart reviews, and individual consultation."[34] This admission suggests that the power to prescribe, even if it was exercised only derivatively through a physician's orders, was a powerful tool that the nurse practitioner provided to the clinic.

By 1979 complaints from physicians and one alleged complaint from a client put the Chico center under increased scrutiny.[35] Concern about the quality of care at Feminist Women's Health Centers in general led the Office of Preventative Services to perform site visits at all the FWHC clinics. The investigators, two women physicians, evaluated the centers positively, suggesting that all of the clinics, particularly Chico, provided first-rate care that required only minor tweaking to make it superior to that provided in most medical settings. Furthermore, the investigators defended the centers' nontraditional approach, arguing that "clients have the right to choose their care provider," even if that provider was a lay health worker. Nevertheless, their medical training seemed to get the best of them in the end. In their May 1979 report, the investigators suggested that if these clinics were to remain eligible for state funding, they must have a nurse practitioner or a physician present at, though not involved in, the clinics.[36] Although this ruling seemed punitive, given the conclusions about the quality of care, the Chico center reluctantly agreed. They decided that this requirement did not challenge the central structure or mission of the clinic.[37]

By 1981 this compromise was no longer enough for the state, reflecting perhaps the increased power of Mike Curb, the antiabortion Republican lieutenant governor.[38] Curb and his office had been "investigating" the Feminist Women's Health Centers since August 1979, hoping to find reasons to close them down or to withhold state funding, but only after 1980 did Curb have the power (in his temporary capacity as acting governor) to impose his will.[39] Under his watch, state agencies acted to close the FWHCs on several fronts. In March 1981, for example, representatives of the Medi-Cal Fraud Division made an unannounced visit to the Chico center, asking workers and clients about the health care and billing procedures of the clinic. Investigators focused particular attention on the Medi-Cal-funded clients. They showed up at clients' homes, asking them who performed their abortions and threatening them with court appearances to coerce answers. Although the fraud investigation was ultimately dropped, the Office of Family Planning notified the Chico center in April that its funding would be cut because its "methods," most notably its use of lay health workers, thwarted the Standards for Family Planning.[40]

This reversal of state policy caused a crisis among the directors at the Chico center, and they considered closing their doors rather than changing their policy. With much despair, however, the directors decided that providing urgently needed health services was, at least for the short-term, more important than standing by principle and risking closure. In July 1981 Dido Hasper wrote a carefully phrased letter to the head of the Office of Family Planning, explaining the center's new policy. "We have had many discussions about how . . . we can still maintain the integrity of the well-woman clinic concept and at the same time come into compliance with the Standards for Family Planning. We have implemented a new clinic format that utilizes mid-level practitioners and physicians for all examinations, diagnosis, and treatment while still using trained lay health workers in the health education aspects of the clinic." In an optimistic tone that covered the deep resentment and disappointment the women likely felt, Hasper concluded: "Now when a woman comes into the well woman clinic she has the advantage of both group participatory-information sharing and learning self examination, as well as an examination by a medical professional with a trained patient advocate to insure the woman receives the best possible care and gets all of her questions answered."[41] Although it required a few more bureaucratic changes, the Office of Family Planning eventually restored Chico's funding, but the role of lay health workers remained diminished.[42]

The Chico FWHC relied on nursing and medical professionals to keep at least part of its medical care within legal limits. As Hanson's experience shows, however, even the center's "legitimating" players pushed against (and maybe through) the boundaries meant to constrain their practice. Nevertheless, by playing by the rules, mostly, with regard to prescribing and dispensing drugs and devices, the Chico FWHC deflected at least some attention from some of its other practices. But these other practices—the clinic's larger prescriptive agenda to convince women that they had a right to expose, examine, and explore their own bodies and its insistence that women did not need state authority to share information about contraceptives or menstrual extraction—proved more threatening to the state. What the Chico FWHC ultimately prescribed for women could not be secured with a prescription pad. Instead, the women of the center wanted to give women the courage to claim what was rightly theirs, control over their own bodies. Perhaps this pill was just too hard to swallow.

Creating New Health Care Workers in Arcata

The Feminist Women's Health Centers were clearly radical in their reliance on lay health workers to provide the bulk of the health care services, with licensed personnel serving primarily as backup, both legal and medical. Other women's health clinics, however, were more willing to place licensed health care personnel at the center of their services. Their approaches should not be seen as less feminist, because they, too, strove to empower women.

In 1971 members of a "young idealistic counter-culture" decided to put into practice their belief that "all people had the right to healthcare, not just those who could afford it" and founded Humboldt Open Door Clinic in Arcata, California. Initially, this clinic served largely as a referral service "where people could get information about anything from pregnancy testing to food stamps."[43] By 1973 the clinic had recruited a few local physicians who volunteered to provide medical services two days a week. One of those early physicians was Gena Pennington, a newly graduated MD, who one day "wandered into" the clinic and was immediately identified as someone who could staff the state-sponsored family planning clinic. Pennington, who worked part-time at the Humboldt County Health department in its women's clinic, was game.[44]

Pennington and a few other local doctors at the time were not generally proprietary with their medical knowledge. For example, Suzanne Willow, who worked as the clinic manager, had learned some women's health basics while volunteering at the county health department. One of the doctors there had just said to her one day, "Well, why don't I teach you?" and he started showing her how to do various procedures. He taught her how to do breast exams, Pap smears, and speculum exams, but he drew the line at bimanual exams. "It's just too hard to teach you bi-manual exams," he said without offering further explanation. Willow and some other women at the clinic were eager to learn more, and Pennington didn't see why they shouldn't. She too had needed to be trained to perform bimanual exams and other procedures after she began seeing women patients. Consequently she started providing a "crash course in office gynecology," teaching the interested women how to perform pelvic and other examinations. "We had pelvic parties. We would have a potluck, and then someone would lay down on the kitchen table and everybody would do a pelvic on 'em—including me."[45] At some point, the women who had been trained at these pelvic parties began performing the exams in the clinic. No one appeared too worried about this at first, but eventually an administrator suggested that they protect themselves and the clinic by taking advantage of California's Experimental Manpower Act, AB 1503.

California's Experimental Manpower Act was the state's response to the alleged shortage of physicians, especially in rural areas and among poor communities. Too many patients were demanding too many procedures from too few doctors. In 1966 the National Advisory Commission on Health Manpower concluded that there was "a crisis in American health care."[46] To cope with the crisis, individual physicians and hospitals began seriously considering which health care procedures required a physician's expertise and which procedures required specialized training but not a medical degree. At the most informal level and generally without any legal authorization, physicians and hospitals began to assign nurses technical tasks that had previously been done only by physicians. More formally, physicians, hospitals, and medical schools established experimental training programs that shifted or expanded the roles of people with some health care experience.

Developed in 1972 and enacted in 1973, California's Experimental Health Manpower Act addressed the large number of midlevel health professionals who were routinely working beyond their scope of practice and confronted the shortage of medical professionals in the state. The act simultaneously approved the training and licensing of a variety of midlevel health workers and provided a "legal umbrella" over a range of practices currently in use that were otherwise technically illegal (or at least ambiguous) when performed by trainees in those programs or their graduates.[47] These programs created new midlevel positions and gave other, more established positions (e.g., nurses, nurse practitioners, physician's assistants, and dental professionals) expanded responsibilities and opportunities. Although the Experimental Health Manpower Act invited creative and innovative alternatives to medical practice as usual, its reach, while significant, was limited. Although individuals trained under the programs enjoyed relaxed rules and wider practice boundaries, at least during their training and apprentice years, most midlevel health practitioners remained constrained, at least legally, by state laws dictating practice.[48]

Some of these programs focused on training practitioners to meet the routine needs of well women. Drs. Donald Ostergard and Duane Townsend at Harbor General Hospital in Los Angeles, for example, founded a pathbreaking program for the training of Women's Health Care Specialists (WHCS's) in 1969, even before such a program was legally sanctioned by the state. The program taught women to be the "first-line provider of patient care to well women." The trainees, who included registered nurses, vocational nurses, and women with no health care education at all before the training, learned to conduct breast and pelvic exams, insert IUDs, fit diaphragms, and perform routine prenatal and other well-woman tasks. The designers of this program argued that WHCS's could provide "total or nearly total care for the well or worried well patients." Significantly, however, these women were "not independent practitioners." They functioned under the authority delegated to them by physicians.[49] Furthermore, architects of this program insisted that while Women's Health Care Specialists could discover "abnormalities," they were not authorized to diagnose or treat illness; their authority to prescribe medicines such as Monistat and family planning basics such as oral contraceptives remained ambiguous.[50]

The architects of these programs, most of whom were physicians, believed that they addressed several pressing social and medical problems, including meeting the health care needs of otherwise underserved women, containing the costs of health care, and getting out in front of health care reform in the guise of "designers and implementers" rather than suffering as "unhappy, passive recipients."[51] At least some of these founders were also influenced by the women's health movement. Richard Briggs, for example, founder of the Gynecorps project in Washington State, acknowledged that "much of the impetus in women's health care [had come from] the women's liberation movement." He agreed that women deserved to know more about their bodies and conceded that physicians had not generally been forthcoming with information. He also understood that women were often uncomfortable undergoing a pelvic exam or talking about sexual issues with male physicians. He posed the WCHS as a legitimate compromise, addressing feminist demands, patients' discomfort, and medicine's need to control health care.[52]

The prescribing authority of graduates of these programs was never absolutely clear. Some program administrators and trainees assumed that at least some of the health professionals trained under AB 1503 had acquired the authority to "dispense, administer, and prescribe drugs." An April 1976 opinion from lawyers at the Department of Health, however, suggested that while the programs gave trainees a waiver from state restrictions, they did not override federal drug statutes. Further, the opinion noted that while the programs might be interpreted to allow midlevels to "administer" and "dispense" "upon a valid prescription," they could not themselves "prescribe" the drugs.[53] A 1977 amendment to the original health manpower experiment program confronted this issue and "provided for experimental projects allowing registered nurses, physicians' assistants and pharmacists to prescribe, dispense, and administer drugs," but this provision met with stiff opposition from the California Medical Association, which had until then largely ignored the program.[54] In the end, some professional groups retained the right to prescribe certain routine drugs but not others, and lay providers never earned the privilege of writing prescriptions.

The women at Open Door Clinic, by then named the Women's Health Collective of Open Door Clinic, drew inspiration from this model. Rather than sending interested women to Harbor General for training, Pennington, Nancy Henshell (a nurse practitioner), and other clinic staff decided to create their own training program. This move was unprecedented. In 1974 there were only eleven "paramedical" training programs in women's health in the country, most of these accepted only trainees who were already nurses or nurse midwives, and virtually all were connected to universities or hospitals in major cities. By contrast, the Open Door program had no hospital or university support, and its only requirement for admission was the desire to become a women's health care specialist. "[Harbor General] had a bunch of people and lotsa teachers and lotsa money, and there was us. And we were in this decrepit bank in California." According to Willow, "everyone was a little shocked" that they would try to make it happen in a tiny community clinic in rural California. "How could this little thing" aim so high?[55] The initial class comprised women already involved in the clinic and recruits who were already engaged in health care in some capacity. Three of the initial cohort, for example, were lay midwives in the local community. Pennington provided most of the didactic teaching, but Henshell did the bulk of the organizational and logistic work (bringing in guest speakers, devising a curriculum, organizing the classes, and winning state approval).[56] The program ran through three cycles, graduating about seven or eight students each time. But what could they do?

The limits of practice for the Women's Health Care Specialists were always a little hazy, "a little grey." In general, they were able to do "whatever" their supervising physicians would approve and designate in written orders or protocols. They were "definitely" able to prescribe basic medications and devices including oral contraceptives, antifungals, antibiotics, IUDs, and diaphragms (or, more technically, to order prescriptions on the supervising physician's license), as long as a physician eventually signed off. Open Door and its offshoot the North Country Clinic for Women and Children also stocked commonly prescribed medications, and the women's health specialists were able to dispense them, according to protocols.[57] Physicians elsewhere were sometimes unwilling to relinquish much patient care, especially prescribing power, to Women's Health Care Specialists. Although some doctors willingly granted them a great deal of autonomy, others never came to trust them, "protocols or no." Susan Anderson, a WHCS graduate, recalled that one physician "required that he personally okay every prescription (even for Monistat) before writing a scrip." This physician's distrust "was a source of frustration" for Anderson.[58]

The women at the Open Door Clinic and the North Country Clinic embraced and employed legal—though clearly cutting-edge—efforts to gain legitimacy for their otherwise lay health care providers. Their willingness to seek state approval highlights their generally friendly relationship with the medical profession. Indeed, medical doctors supported the WHCS's in their attempt to gain more health care power and authority. A formal training program and the state approval it conferred allowed and encouraged these women to learn and to do even more.

By contrast, the directors of the Chico FWHC had considered and ultimately rejected seeking state approval for their own lay health workers through AB 1503, because it seemed contrary to their philosophy and their mission.[59] They insisted, "We have the right to take control of our own health care through self-help. Self-help is not an experimental project and we do not need or want to acquire a special state sanction such as AB 1503 in order to be able to offer it to women who use our facility." They further declared that "all women could evaluate their own health care needs, if given all the information."[60]

In spite of its successes, the Women's Health Care Specialist training program at Arcata was short-lived. The program ran one session out of the Open Door Clinic, and when some of the women at Open Door left to form the North Country Clinic, they took the training program with them and administered two more classes. When the program came to the end of its two-year term, none of the principals appeared to be interested in seeking renewal. As Suzanne Willow put it, "there wasn't enough energy to continue it." After awhile it became "not so fun." There may have also been ideological differences between members of the second and third classes and the founders and initial trainees of the program (many of whom were associated with Open Door or women's health more broadly). According to Willow, a few women in the subsequent classes subscribed to a notion of feminism that insisted that the instructors and the students in the classes should share power equally. And yet "they knew nothing about women's health care." Exhausting power struggles ensued.[61]

The Women's Health Care Specialist title and training program provided legal cover and professional legitimacy to a handful of women, some who were already providing health care and others who hoped to do so. But the WHCS remained a professionally precarious position: trained, skilled, and "certified" but without a regulating body or an advocacy group as their physician assistant or nursing counterparts had. WHCS's also may have found it hard to find employment that valued and used their skills. Indeed, too many trainees in the manpower program more broadly were "unable to secure employment utilizing their training," in part because of physician resistance to the erosion of medical prerogatives.[62] As a result, many of the women trained in the Arcata clinics believed they needed to gain further state protection or "fade into history."[63] Many of the women chose the former; the Women's Health Care Specialist Physician's Assistant program at Harbor General Hospital, connected with UCLA Medical School, allowed "many" of the women from Arcata to take its exams. After a week or so of written and clinical exams, these women then became licensed (and a bit later board certified) physician assistants.

One of the goals of the manpower projects was to increase occupational options for individuals from "socially or economically disadvantaged backgrounds" and to increase the number of people from underrepresented minorities in health care careers.[64] It is unclear whether any of the women trained as WHCS's in Arcata were members of underrepresented minority groups (although the administrator of the clinic was Native American). Still, some of those women did parlay their initial training into health care careers. As of 2010, at least five were still providing health care.[65]

Conclusion

The women in the Arcata clinic saw themselves as feminists, and they shared with the Chico FWHC a commitment to demystifying women's bodies. Because the Arcata clinic was founded by health care professionals, however, they did not share the Chico FWHC's goal of well-women's health care free from medical oversight. The women staffing the North Country Clinic did believe, however, that physicians and nurses should not have the exclusive right to provide basic health care to well women. When the state opened the door to a variety of new health care practitioners, these women embraced the idea. They did not relinquish their training to strangers' hands, though. Instead, they created their own training program to ensure that it remained feminist and met the needs of the women in the local community. Although they took different paths, reflecting different political relationships with professional medicine and the state, both clinics sought to protect women's rights to knowledge about their own bodies.

The histories of the North Country Clinic and the Chico Feminist Women's Health Center illustrate widespread health care trends in California and the United States in the 1970s, such as the rise of community clinics, the role of state and federal funding, and the proliferation of different levels of health care professionals. They also showcase the ongoing power struggles waged by health care stakeholders—medical societies, nursing boards, health departments, and others—over the defining parameters of "medicine," and they present two fine-grained examples of the challenges to medical authority that took place in the 1970s.

The prescription played a paradoxical role in these clinics' efforts to throw off the mantle of physician control of well-woman care. On one hand, the prescription limited the extent to which lay persons could provide health care, because only physicians (and, less certainly, nurse practitioners) had the power to prescribe pharmaceuticals, even those for minor conditions (such as the antifungal agent Monistat) or for the ongoing prevention of pregnancy (such as birth control pills). On the other hand, physicians sympathetic to the feminist cause could extend the power to prescribe by providing the clinics with presigned prescription pads or signing a week's worth of prescriptions at a time. These prescribing methods generally fell within the letter, if not the spirit, of the law, and they enabled feminist health workers to expand the scope of their practice. Thus, the prescription—as a written order for a pharmaceutical product—served as both barrier and bridge to feminist efforts to gain some measure of control over women's bodies.

However, in the feminist health clinics of the 1970s, the prescription also took on a much broader meaning. The clinic directors prescribed, for themselves and for other women, a new approach to health care, one that privileged the bodily experience of the woman herself and the value of shared knowledge. The feminist prescription carved out a separate space for preventive health services, in which women could take care of each other. They were willing to leave the treatment of acute and chronic illness to medical professionals, but they insisted on reclaiming routine health care on behalf of all women.

Many women sought out health care at feminist clinics for practical reasons. They hoped for an ointment to fade a rash or a device to prevent pregnancy. These clinics generally obliged with the requisite pharmaceutical prescription. But the women who staffed these clinics believed that health feminism offered women a much broader prescription than could be contained on the Rx legend. Consequently, feminist health activists sought to do more than relieve an itch; by prescribing bodily self-awareness, by providing health education, and by challenging medical control over women's lives, these women's clinics strove to secure both women's health and women's liberation.

CHAPTER 7

"Safer Than Aspirin"

The Campaign for Over-the-Counter Oral Contraceptives and Emergency Contraceptive Pills

Heather Munro Prescott

On January 21, 1993, the U.S. Food and Drug Administration (FDA) announced in the *Federal Register* that the agency's Fertility and Maternal Health Drugs Advisory Committee would hold an open public hearing to discuss issues related to providing oral contraceptives without prescription.[1] The meeting was to be chaired by Philip A. Corfman, director of the FDA Center for Drug Evaluation and Research, who had announced at a conference on birth control the previous year, "I think the pill is safer than aspirin and aspirin is available over the counter."[2]

One week after the posted *Federal Register* notice of the open hearing, FDA officials canceled the session in response to public criticism from feminist health activists and consumer protection groups. At the heart of their objections was a report in the *Wall Street Journal* that the FDA open hearing was backed by the R.W. Johnson Pharmaceutical Research Institute, the research division of Ortho Pharmaceutical Corporation, which manufactured the top-selling oral contraceptive brand, Ortho-Novum. Sidney Wolfe, director of the Public Citizen's Health Research Group, charged that the oral contraceptive pill was "too potent a drug for over-the-counter."[3] The National Women's

Health Network and the National Black Women's Health Project also argued, "At this time it is not [a] good idea to make oral contraceptives available without prescription."[4]

This chapter discusses two attempts to revise the prescription-only status of oral contraceptives and emergency contraceptive pills as case studies of the politics of the prescription in the late twentieth and early twenty-first centuries. In some ways, these stories are different from those of other products that have made the switch to over-the-counter (OTC) status over the past three decades. As Jeremy Greene observes in *Prescribing by Numbers,* the symptom has typically played a central role in public debates over which drugs are appropriate for OTC use. The FDA made this policy explicit in 1997 by stating that drugs considered for OTC status must be used for "self-recognizable conditions that are symptomatic, require treatment of a short duration, and can be treated without the oversight and intervention of a health-care practitioner." Greene uses the case of the cholesterol-lowering drug Mevacor (lovastatin) to demonstrate how the FDA applied these criteria in practice. Because high cholesterol did not present symptoms to relieve, the agency reasoned, and because high cholesterol was something that could be deduced only by a medical professional, Mevacor and other statins were not suitable for OTC use.[5]

Neither oral contraceptives nor emergency contraceptive pills fit conventional criteria of symptom relief, although proponents of making emergency contraception available without prescription argue that regardless of the lack of symptoms, a woman does not need a health professional to tell her when or why she needs to take this drug. Like other OTC drugs, emergency contraception regimens were meant for short-term use, but birth control pills were designed to be used continuously over long periods of time. For oral contraceptives, health care professionals were often at the forefront of efforts to remove the prescription requirement—in contrast to many other OTC switches. Experts who supported the OTC switch often used the same language as feminist health activists from the 1970s, arguing that women should be freed from the "paternalistic" doctor-patient relationship embodied in the prescription. In fact, some of these proponents were female health practitioners and male allies who were active in fighting for reproductive rights and professional opportunities for women in the 1970s and 1980s. This was not the first time that certain health care professionals had criticized the tyranny of the prescription: in the case of Mevacor, minority physician advocacy groups had criticized the "paternalistic and insulting assumption that only physicians" were able to make sense of high cholesterol numbers that many members of the public already understood through widespread public health campaigns.[6]

What is more surprising is that some feminist health activists were among the leading opponents of removing the prescription requirement for oral contraceptive pills. As Judith A. Houck demonstrates in chapter 6, it was feminist health centers that established women's right to understand and treat their bodies without interference from the medical profession. They, along with health care consumers who criticized the hegemony of the medical profession, helped contribute to the growing demand to switch drugs from prescription to OTC in the 1980s and 1990s. The approval of the OTC products Monistat 7 and Gyne Lotrimin for vaginal yeast infections in 1991 was in some ways the culmination of feminist efforts to empower female patients by giving them more autonomy in matters affecting their own health.

To explain the objections toward OTC oral contraceptives raised by some feminist health organizations, I contrast the feminist critique of the prescription as a paternalistic barrier to women's health with the feminist critique of the prescription as a necessary safeguard to women's health, placing these discussions within the longer history of feminist activism on behalf of consumer safety. These groups often sought to balance patient empowerment with adequate regulatory oversight aimed at protecting consumers from dangerous products.

A Question of Safety

Since the early twentieth century, grassroots activism by women's organizations such as the National Consumers League has played a critical role in the passage of laws protecting consumers from harmful products, including the first Food and Drug Act in 1906 and the Federal Food, Drug, and Cosmetic Act of 1938.[7] This vigilance by female consumers became especially strident in the 1960s and 1970s, when feminist health activists began raising concerns about products designed and marketed specifically to address women's health issues, most notably oral contraceptives and drugs for emergency contraception.

Concerns about the safety of oral contraceptives emerged soon after the approval of the first birth control pill, Enovid, in 1960. One article from *Time* published in 1961 declared, "All hormone like drugs are as powerful as TNT in the ways they affect much of the body's chemistry." Most magazine articles focused on minor side effects, such as headache, nausea, bloating, and weight gain, but some also reported on the rare cases of women who had suffered thrombophlebitis and thromboembolism while taking oral contraceptives. Between August and December of 1962, the number of reported cases of blood clots in Enovid users grew from 28 to 272.[8]

Reports about the possible dangers of oral contraceptives prompted the FDA to create, in 1965, its first permanent advisory committee, the Obstetrics and Gynecology Advisory Committee, which concluded that it could find no adequate scientific data to justify removing oral contraceptives from the market. Instead, the FDA, drawing on new regulatory powers granted by the 1962 Kefauver-Harris Drug Amendments, attempted to protect human subjects by requiring investigators to obtain written consent from patients enrolled in experimental drug studies and required manufacturers to list contraindications and adverse effects in all drug advertising and in a package insert for physicians to use when advising patients.[9]

These reforms were insufficient for many critics. In 1969 journalist Barbara Seaman published the book-length expose *The Doctors' Case against the Pill*, which roundly condemned scientists, physicians, and the FDA for foisting a dangerous product on unwitting female consumers.[10] *Washington Post* columnists Drew Parsons and Jack Anderson accused the FDA of suppressing data to protect pharmaceutical companies. In 1970 the Senate Subcommittee on Monopoly of the Select Committee on Small Business, led by Senator Gaylord Nelson (D-WI), investigated concerns that the medical profession and the pharmaceutical industry were withholding important information about oral contraceptives from women.[11]

Following the Nelson hearings, feminist health activists continued to pressure the FDA to protect women from further abuses by the medical profession. As the early critique of the birth control pill illustrated, the prescription as a consumer safeguard had failed to protect women from dangerous side effects of approved drug products. This critique was extended in investigations into the use of diethylstilbestrol (DES) as a postcoital contraceptive, what became known colloquially as the "morning-after pill." In April 1971 cancer researchers at Massachusetts General Hospital in Boston published studies that found a high incidence of a rare form of vaginal cancer in the daughters of women who had consumed DES while pregnant.[12] But in October of the same year, University of Michigan Health Service senior physician Lucile Kirtland Kuchera published in the *Journal of the American Medical Association* a study of one thousand patients given DES as a postcoital contraceptive at the University Health Service. Kuchera noted, "Formerly in this clinic we had only told the patient to wait for her next menses—not very consoling advice to a woman who does not want to be pregnant." She reported that the pill was 100 percent effective and that no serious side effects were experienced by any of the one thousand women who had received the treatment. Some patients reported nausea or headache, but 45 percent had no side reactions.[13] *New York Times* health columnist Jane Brody noted the irony that a national controversy over the cancer-causing properties of DES was brewing at the same time that doctors were proclaiming the effectiveness of the drug as an "after-the-fact contraceptive."[14]

Shortly after Kuchera's article appeared, Belita Cowan, a master's degree student in English at the University of Michigan, began an investigation into the use of DES as a morning-after pill at the student health service and the University Medical Center. Cowan worked part-time at the Medical Center, where physicians routinely gave the drug to female students and other women in the community who wanted to prevent unwanted pregnancies. The student health service advertised this form of contraception in the *Michigan Daily*, the University of Michigan student newspaper. Following a 1969 article in *Medical World News*, health service director Robert E. Anderson confirmed that the drug was being used on the Ann Arbor campus for contraceptive purposes, even though its legality was "open to question." He added that the morning-after pill was also used at the University Medical Center to treat rape victims after doctors there determined the drug was safe for this purpose.[15]

Cowan was horrified that doctors continued to administer DES as a morning-after pill even *after* the cancer reports began to appear in the medical literature and popular press. Cowan soon joined with women who had been patients at the University of Michigan Health Service to form Advocates for Medical Information (AMI) to protest what they considered to be reckless treatment of women students at the university.[16] In September 1972, AMI published an article entitled "Cancer and the Morning-After Pill (Will You Be Mourning-After)" in the feminist newspaper *her-self.* AMI cofounder Kay Weiss spoke with a participant in Kuchera's study and asked if she had been informed that DES was a powerful carcinogen. The woman replied that "she had been told in the context of a moral lecture that it was 'dangerous' and she would be taking it at her own risk, but not that it was cancer causing."

Another woman said she had been told the pills would make her "very sick" for a few days but nothing about the cancer risk of taking DES.[17]

A follow-up article in the January 1973 issue of *her-self* reported that "thousands, perhaps millions of women were administered DES in pregnancy" and that presumably large numbers of female university students had been exposed to DES in utero. Yet no one was telling these young women they might be at risk for developing vaginal cancer. Instead, many university health services around the country were giving women more DES in the form of the morning-after pill. Furthermore, even though federal law required that physicians must inform patients of the dangers of a particular drug, "coeds [had] been told DES [was] 'safe and harmless.'" The article claimed that women who received the morning-after pill were not given pregnancy tests, nor were they asked whether they had a family history of vaginal cancer. When women who had received the drug requested an iodine stain test for vaginal cancer, they were told they were "worrying too much or imagining things because of press reports." This "paternalism," the article argued, usually meant the doctor did not know how to perform the test or had never heard of it.[18]

On December 15, 1975, a group of women's health activists of the Washington, DC, area held a memorial service on the steps of the FDA building "to commemorate the thousands of women who died needlessly because of the pill, DES, and menopausal estrogen replacement therapy." The service was the public face of the group's participation in the various congressional hearings on DES and other estrogens. The day after the memorial service, Belita Cowan and several other members of this group testified at the House hearings on the use of DES as a "morning-after" pill. The public demonstration and the congressional testimony grew out of the efforts of Cowan and other feminist health activists—including Barbara Seaman, Phyllis Chesler, Mary Howell, and Alice Wolfson—to enhance the lobbying efforts of feminist health activists in the nation's capital. In May 1976 the group organized a national conference of women activists from around the country, who elected a twelve-member board of directors for the National Women's Health Network (NWHN), an organization that would "monitor Federal health agencies and ensure that the voice of a national women's health movement would be heard on Capitol Hill."[19]

The creation of the NWHN represented the growing professionalization of the women's health movement in the United States. Feminist health activists realized that to enhance their efficacy as advocates for women health issues, they needed to become professional and political "insiders": this included earning the professional credentials and learning the scientific knowledge necessary to "speak the language" of other experts who testified on the safety and efficacy of various contraceptive methods during the 1970s and 1980s.[20]

The ability to play on the same terms as the scientific establishment is exemplified by the testimony of NWHN board of directors member Judy Norsigian before the House Select Committee on Population in 1978. The hearing was organized to weigh evidence questioning the safety of various contraceptive methods that were being explored by federally funded population research, nonprofit organizations such as the Population Council and Planned Parenthood, and drug companies. Norsigian stated, "Those of us active in the women's health movement are concerned that present funding is too heavily weighted toward drug and device research . . . too often, such research has exposed human subjects, mostly women, to serious adverse consequences." As examples of the serious side effects some women had suffered from the pill and the Dalkon Shield intrauterine device, Norsigian referred to hundreds of letters telling "nightmare stories" that had been sent to her and the other editors of *Our Bodies, Ourselves.* Norsigian observed that most contraceptive researchers were male and hence had "little direct understanding of the practical impact of their research on women." Norsigian said that the male bias was evident in the policy recommendations on federal population research, which gave priority almost exclusively to drugs and devices meant for women. "We doubt if a committee composed primarily of women—consumers as well as researchers and government administrators—would have presented a similar list of recommendations."[21]

As an alternative to the morning-after pill and other potentially dangerous drugs and devices, Norsigian and other NWHN members endorsed the self-help model developed by feminist women's health centers in the 1960s and 1970s. Norsigian argued that the medical profession placed "too much emphasis" on "the presumed passivity of women and on the desirability of methods requiring little or no active participation." This attitude was especially apparent in the case of teenagers and young women. In the zeal to do anything to prevent the "epidemic" of unplanned teenage pregnancies, Norsigian said, population experts had too easily disregarded issues of safety. Norsigian argued that "questions of safety" were just as important when addressing the contraceptive needs of adolescent girls. In fact, because the long-term effects of oral contraceptives on female reproductive development were unknown, prescribing hormonal contraception to adolescents was especially unwise. Norsigian argued that, based on her work with inner-city youth in the Boston area, "the self-help model used in many women-run health centers [improved] use-effectiveness of barrier methods as well as the ovulation method," even among teenagers. Yet, these safer birth control methods did not receive "priority by those who control[led] the research dollars, while potentially dangerous methods [did] attract the majority of funds." The position of the NWHN was that "women should be creating policy on behalf of women" and "at the very least, that all users of contraceptives should have a significant voice in determining what kind of research is funded." In particular, there needed to be more research "conducted by community-based women's health centers which have worked directly with those who are intended to benefit from this research." To that end, the NWHN was conducting a survey of more than one hundred women's health centers and women's health education groups "to establish what women's health organizations see as their contraceptive research priorities."[22]

Despite this outcry from feminist health activists about the dangers of DES, demand for the morning-after pill continued. Women read about this pill in the popular press and saw reports about it on major network news broadcasts. While these articles and news stories emphasized the safety issues raised by the congressional hearings on DES, they also helped to raise awareness about this method of contraception.[23] Workers at rape crisis centers observed that while the use of DES was controversial, "some patients ask[ed] for it even after they [had] been warned of its possible consequences."[24] Physicians in emergency medicine at the University of New Mexico also found that "almost all victims taking DES voiced a strong desire to rid themselves of the physical and emotional vestiges of sexual assault."[25] College health professionals reported that they still wrote hundreds of prescriptions for postcoital pills containing DES per year. A survey of 42 health institutions (including the student health services of 14 large universities and the outpatient gynecology services of 14 major hospitals) conducted by the Cancer Control Bureau of the New York State Health Department indicated that 12 of the institutions prescribed DES to women directly and 11 others referred them to a physician who would prescribe it. The study estimated that 450 women were treated or referred in New York State alone during 1973, with the heaviest use at college and university health centers.[26] This ongoing demand from young women desperate to avoid unwanted pregnancies played a key role in efforts to find a new morning-after pill that might be available by prescription.

A Delicate Political Issue

One of the leaders in this search for a new postcoital contraceptive was A. Albert Yuzpe at the University of Western Ontario Student Health Services. Like their counterparts in the United States, Canadian physicians were increasingly reluctant to use DES as a morning-after pill, owing to concerns about the link between the drug and cancer. Yuzpe was aware that researchers in Hong Kong had used the progestin compound levonorgestrel as a postcoital contraceptive and found that women had fewer side effects such as nausea and vomiting than they did with estrogen compounds. In Canada, the only pharmaceutical product on the market that contained levonorgestrel was the combination oral contraceptive Ovral.[27] During the 1972–73 academic year, Yuzpe launched a pilot study of Ovral as a postcoital contraceptive, treating a total of 143 women aged 18 to 42 years.[28]

In the United States, Yuzpe's findings were duplicated by Dr. Lee H. Schilling, staff gynecologist at the California State University, Fresno, student health service. The study, conducted during the spring 1978 semester, enrolled 115 students, none of whom became pregnant. Schilling observed that despite the growing use of other estrogens for postcoital contraception, to many, DES and the morning-after pill were "synonymous." The link between DES and cancer had "created confusion" about the safety not only of DES but of postcoital contraception more generally. As a result of "perceived dangers" about the use of estrogens, postcoital contraception was "a poorly understood and vastly underutilized tool."[29]

In 1980 *Contraceptive Technology Update*, a newsletter for professionals in reproductive health care, published a report on Yuzpe's and Schilling's findings. The article described the ease of their method: clinicians could simply punch out four tablets from a regular package of the oral contraceptive and give them to the patient while she was in the office. Instructions for how to take Ovral—two tablets immediately, the other two twelve hours later—were easier to explain to the patient than the five-day regimen for DES and other estrogens. Patients were also less likely to experience nausea and vomiting while taking Ovral and were thus more likely to complete the course of treatment. Most importantly, women and practitioners who were reluctant to use DES might feel more comfortable using Ovral. Both Yuzpe and Schilling found that, measured against the risks of pregnancy and abortion, the risk of using Ovral as a morning-after pill was lower since it entailed fewer medical complications and avoided the financial, social, and psychological costs of abortion.[30]

A follow-up editorial by Robert A. Hatcher, professor of gynecology and obstetrics at the Emory University School of Medicine, who had adopted Yuzpe's method as a rape treatment protocol at Grady Memorial Hospital in Atlanta, declared that the act of prescribing a morning-after pill could "unlock the health care door" for women "who may not realize they need regular health care or family planning." These women included not only rape victims but teenagers, college students, and separated, divorced, or widowed women whose sexual activities were infrequent. The reproductive health professional could not only offer help in a crisis situation but also reintroduce women to the health care delivery system and thereby extend "a greater awareness of [their] own health."[31]

The Ovral regime, nicknamed the "Yuzpe method" after Yuzpe's pioneering work in this area, was approved by drug regulatory agencies in Great Britain and West Germany in the early 1980s. Yuzpe hoped that the same would happen in the United States and Canada.[32] Yet, although North American physicians were free to prescribe Ovral for this purpose, pharmaceutical companies were unwilling to apply to either the U.S. Food and Drug Administration or the Canadian Marketed Health Products Directorate to have oral contraceptives relabeled for use as morning-after pills. In an article in *Contraceptive Technology Update* entitled "Postcoital Contraception: A Delicate Political Issue," the Wyeth marketing representative for Canada, Gert Juegenkit, said that the company was less than enthusiastic about marketing such a method. For years, the company had been telling women not to take oral contraceptives if they were pregnant. The morning-after regimen entailed telling women to take a higher dose of Ovral if they suspected they were pregnant. "Explaining the difference could be a problem," he said.[33] Wyeth's manager of public relations in the United States, Audrey Ashby, stated that the company did not advise practitioners to use Ovral as a morning-after pill and would not support health providers who did so anyway and consequently encountered legal problems.[34]

The lack of FDA approval for the Yuzpe method made it difficult for health care providers to learn about this technology, because the information was not available on the package insert for oral contraceptive pills, nor was it readily available in medical or nursing textbooks. A survey conducted by *Contraceptive Technology Update* in the mid-1980s indicated that the majority of respondents did not prescribe the Yuzpe method because they were unaware of it. Three-fourths of all nurse practitioners and nurse midwives had never prescribed postcoital pills, and only slightly more than half of all physicians had done so. Furthermore, liability concerns made it impossible for some providers to administer the Yuzpe method to their patients. Federally funded family planning services could not prescribe or distribute drug regimens that did not have FDA approval.[35] Planned Parenthood's insurer would not cover the organization for the morning-after pill. Eve Paul, director of legal services for Planned Parenthood, instructed affiliates to refer patients seeking post-coital contraceptives to outside agencies.[36] Thus, the Yuzpe method remained unknown to many health care providers and most consumers.

Conversely, the few reproductive health advocates who did know about the Yuzpe method did not always access it through a directed prescription, instead practicing what anthropologist Lisa Wynn refers to as the "do-it-yourself (DIY)" approach to emergency contraception. During the 1980s and early 1990s, health educators in college health centers, independent abortion providers, and other reproductive health clinics cut up packets of Ovral and other brands of ethinyl estradiol–levonorgestrel oral contraceptive pills and gave these to women to prevent pregnancy after sex.[37] Women also learned about this "DIY" method through word of mouth or by reading about it in feminist magazines and on the Internet.[38] An advantage of the DIY approach was that health services and providers could remain under the radar of antiabortion groups. In addition, women who learned about the DIY method could assemble the regimen on their own using existing packages of oral contraceptives. Meanwhile, other consumers were benefitting from another form of "do-it-yourself" medicine in the form of new over-the-counter (OTC) drug products.

OTCness and Patient Power

The last two decades of the twentieth century saw a dramatic revolution in medical self-help, as hundreds of formerly prescription-only drugs were made available OTC. This rise of what pharmacologist R. William Soller called

"OTCness" was made possible by two developments.[39] First, major changes in drug evaluation policy were instituted by the FDA during the 1970s in response to pressure from consumer protection groups. Although the Kefauver-Harris Drug Amendments of 1962 mandated that the FDA evaluate the safety and efficacy of OTC drugs as well as prescription drugs, it was not until the 1970s that pressure from consumer groups for greater oversight over product safety compelled the agency to examine the more than three hundred thousand OTC drugs then in existence. Second, consumer demands for greater control over health care decisions challenged the professional dominance of the medical profession and with it one of the key justifications for the boundary between prescription and OTC drugs.

The 1951 Durham-Humphrey Amendment stipulated that drugs must be designated as prescription-only if they were "potentially harmful" and safe for use only under the supervision of a licensed medical professional. The rise of an egalitarian model of "patient power" in the 1970s questioned medical authority and led to demands that patients be treated as equal partners in the doctor-patient relationship. These claims to increased consumer privileges included the right to self-medicate. Some advocates of this revolution in self-medication argued that consumers should be not simply equals but "the primary practitioners in the new health care system." Both of these changes accelerated the movement to switch a variety of drugs from prescription to OTC status. During the 1980s, more than two hundred drugs were switched from prescription to OTC, and by the early 1990s more than four hundred OTC products were in use that had formerly been available only by prescription.[40]

Drug manufacturers and their allies used a grammar of consumer empowerment to promote the prescription-to-OTC switch. Soller, for example, argued that the growth of OTCness had made health care consumers more "self-reliant" and "responsible" by giving them new therapeutic choices to self-treat various medical conditions ranging from prevention of heartburn to restoration of lost hair.[41] Yet, there were economic factors at work as well. In 1984 the Hatch-Waxman Act gave manufacturers an additional three years of marketing exclusivity for formerly prescription-only products when they were made available over the counter. This act caused a dramatic surge in the availability of nonprescription drug products, as drug manufacturers recognized the economic benefits of the prescription-to-OTC switch. Many manufacturers hoped to extend the patent protection and benefit from brand-name recognition they had already established with a prescription-only product.[42]

Proponents of making oral contraceptives available without a prescription used similar notions of patient empowerment. In a 1993 article in the *American Journal of Public Health*, James Trussell, Felicia Stewart, Malcolm Potts, Felicia Guest, and Charlotte Ellertson argued that if giving women the right to choose when and whether they wished to become pregnant was a primary goal of reproductive rights advocates, then such advocates "should endorse women's full and direct access to contraception. Indeed, if this goal is central," they said, "then only compelling health concerns could justify restrictions such as a prescription requirement." In their opinion, the "ostensible benefits of protecting women from the harmful health effects of hormonal contraception fall short of the costs of medicalization." These costs included the time and money needed to visit a physician to obtain a prescription, the "financial and human costs of unintended pregnancies," and the administrative costs to the health care system caused by doctors' visits. They argued that package inserts and instructions written in easily understood lay language would enhance patient compliance and "enable women to judge for themselves whether oral contraceptives are medically contraindicated."[43]

Feminist health activists did not necessarily share this model of patient empowerment. "Don't make the pill easier to acquire," Judy Norsigian pleaded, arguing that screening for contraindications for pill use should be *more* strenuous, not less so. Data on the long-term effects of pill use were still inconclusive, she noted, especially for women who started taking the pill during adolescence. "Do we want to encourage greater pill use in that group before we have better data?" she asked.[44] A chief concern for Norsigian and other activists was that the major pharmaceutical company Ortho was a key player in the drive to make oral contraceptives available without a prescription. Feminist health activists and representatives from consumer protection organizations charged that Ortho had pressured the FDA to organize the meeting so hastily that opponents of nonprescription status for oral contraceptives had insufficient time to prepare their arguments.

From a social justice perspective, Cindy Pearson, program director for the National Women's Health Network, opposed moving oral contraceptives to OTC status because "[a] birth control prescription is the poor woman's ticket to health care." Women who visited a birth control clinic not only received a prescription for contraceptives; they were also screened for sexually transmitted diseases, high blood pressure, cancer, and other health problems. Furthermore, Medicaid would not cover nonprescription birth control pills, forcing women to pay twenty to twenty-five dollars out-of-pocket for contraceptives they had previously received at little or no cost. Physicians and other health providers would lose the opportunity to counsel patients about the proper use and side effects of oral contraceptives. Adolescents were especially likely to discontinue use if they encountered uncomfortable side effects. Health care professionals also were concerned that making the pill available over the counter would make it less likely that women would be screened for sexually transmitted infections, leading to further spread of these diseases.[45]

These criticisms persuaded the FDA Center for Drug Evaluation and Research to cancel the workshop scheduled for February 1993 and to conclude that a "broader range of groups both inside and outside of the government" needed to be consulted before the agency would explore the matter any further.[46] In June the Henry J. Kaiser Family Foundation hosted a two-day forum that brought together experts in family planning and reproductive health, pharmaceutical industry representatives, and consumer activists for "an unfettered discussion of the ramifications" of making oral contraceptive pills available without a prescription.[47] As in previous discussions, opponents raised concerns about pill safety and the possible increase in cost that came with the switch to OTC status. Francine Coeytaux from the Pacific Institute for Women's Health and Amy Allina from the Reproductive Health Technologies Project examined the international experience with nonprescription oral contraceptives. They found that this method of distribution did not pose a risk to women's health as long as there were mechanisms to ensure screening for contraindications. The developing countries in which this strategy was employed, however, had community-based distribution programs that put women in touch with trained health workers who were available for ongoing advice and support. More importantly, in all countries where these programs were in place, contraceptives were highly subsidized and available at little to no cost to clients. Unless such a program was instituted in the United States, the barriers to access for low-income women might actually increase, because they would have to purchase oral contraceptives at full retail price rather than having them covered by government-funded family planning clinics.[48]

Supporters of OTC status claimed that keeping oral contraceptives under physician control was paternalistic. Trussell, Stewart, Guest, Ellertson, and Potts, authors of the 1993 *American Journal of Public Health* article mentioned above, challenged the "common public perception" that women needed gatekeepers to help them to choose and teach them to use oral contraceptives.

"This perception might result from prescription status itself: The current paternalistic model of the physician-patient relationship encourages patients to be trusting and docile." These authors acknowledged that OTC status might lead to a decline in efficacy because women would not need to see a clinician and therefore would not receive counseling on how to use the product. Yet they also argued that the prescription created "significant obstacles to access," including costly initial visits and follow-ups to see a physician or nurse practitioner. In fact, prescription status itself discouraged use of oral contraceptives by implying that they were "unsafe." Untethering contraception from the prescription, they argued, would eliminate this obstacle by signaling the drugs were indeed safe. Furthermore, OTC status "would also stimulate direct advertising to women, creating another channel for consumers to learn about safe and effective" use of oral contraceptives.[49]

These public health proponents could not make oral contraceptives available OTC on their own; for that to happen, they had to persuade a drug company to file a supplemental New Drug Application (sNDA) with the FDA requesting that a specific product be switched from prescription to OTC. In August 1993 Schering-Plough Healthcare told the *New York Times* that it would apply to have its oral contraceptive sold without prescription.[50] Yet the company never followed through on this announcement. Proponents of OTC oral contraceptives were left to focus their efforts on raising awareness of the Yuzpe method and persuading the FDA to endorse it as an appropriate use of oral contraceptive pills.

Science and Politics at FDA

Advocates of the Yuzpe method pointed out that while the new drug RU-486 (mifepristone), typically used to induce abortion during the first trimester, could also be used as a postcoital contraceptive agent, regular oral contraceptive pills that were already approved in the United States were just as effective at preventing pregnancy within seventy-two hours of intercourse. The chief barrier to use of the Yuzpe method was that few women were aware of it because the FDA had not approved oral contraceptives for this purpose, and therefore this information was not included in the written indications or the patient package insert for oral contraceptives.[51] However, efforts to mainstream emergency contraception intensified in the midst of the uphill battle to persuade the FDA to approve the drug RU-486.

Not surprisingly, antiabortion groups linked and strenuously opposed these two efforts. The conservative daily newspaper the *Washington Times* quickly picked up on this association, reporting, "While pro-lifers battle to keep the French abortion pill, RU-486, out of this country, it has largely gone unnoticed that other morning-after medications are being sold here legally." Bob Marshall, a researcher for the American Life League, accused proponents of the Yuzpe method of "abusing their medical knowledge" by advocating the use of oral contraceptives for emergency contraception when the FDA had not approved the drugs for such use.[52]

Although the National Women's Health Network remained opposed to making oral contraceptives available OTC, they were convinced that enough scientific evidence about the safety and effectiveness of the Yuzpe method of postcoital contraception had accumulated that the Network was willing to "cautiously support its use." More importantly, increasing restrictions on abortion and access to federally funded birth control under presidents Ronald Reagan and George H. W. Bush convinced the Network that they needed to help ensure that women had access to this technology when other birth control methods failed.[53] As I argue elsewhere, this technology became a "bridge issue" between an older liberal feminist position that tended to support technological innovations such as hormonal contraception and a more radical feminist position that criticized the use of hormones but was otherwise in favor of reproductive rights. Their efforts culminated in the development of the dedicated emergency contraceptive products Preven and Plan B, which were approved by the FDA for prescription use in the late 1990s. Many advocates believed that requiring a prescription for these products hindered public awareness and timely access to this birth control method. Soon after the approval of the two products, the manufacturers began discussions with FDA about making them available OTC.[54]

These efforts and similar appeals from industry to further expand the number of OTC products prompted the FDA to consider developing more specific policies on which classes of drugs should be made available without a prescription. These included not only emergency contraceptive pills, but also cholesterol-lowering drugs, diuretics, antihypertensive agents, asthma treatments, and regular oral contraceptives.[55] At a hearing on June 28–29, 2000, the FDA heard testimony from twenty-five speakers representing the drug industry, scientists, health insurance providers, drug law experts, physicians groups, nonprofit organizations, and consumers. Many of those who spoke were in favor of making certain categories of drugs available over the counter because it would allow patients to get these drugs more easily.[56]

The issue of access was especially important to advocates of emergency contraception. Kirsten Moore, from the Reproductive Health Technologies Project, observed that women on a fixed or low income might not be able to afford an OTC product if their health insurance did not reimburse them as it did for prescription medications. "For this reason," Moore said, "we strongly encourage the FDA to consider maintaining dual status for ECPs," that is, to allow the same products to be available both OTC and by prescription.[57] Jack Stover, from the pharmaceutical company Gynetics, claimed that OTC status for Preven would not only enhance access but also increase awareness. Although the company had spent almost $15 million since the drug's launch in 1998, it had only been able to increase public awareness of emergency contraception to less than 10 percent of its target market. Gynetics had participated in independent studies and evaluations indicating that emergency contraception could be a $100 million a year product "with the proper level of support and advertising." Stover also observed that off-label use of regular oral contraceptives cut up and repackaged as emergency contraceptive kits with handwritten instructions continued even with two FDA-approved products available. "This surely must confuse the consumer even more and certainly acts as a disincentive to legitimate pharmaceutical companies" that spent millions of dollars to test and get approval for their products.[58]

Beyond access, other advocates argued that the prescription served no meaningful function beyond obstruction. Tara Shochet, from the American Society for Emergency Contraception, argued that there were no medical reasons to keep emergency contraceptive pills on prescription status. Women did not need help diagnosing the need for emergency contraception. "The doctor does not need to do a physical exam," she argued, "and as with aspirin or decongestant, there's little harm done if a woman takes the pills when she doesn't actually need them." Therefore, "requiring a prescription for using emergency contraception is as foolish as requiring prescriptions for using fire extinguishers."[59] Similarly, Beverly Winikoff, program director of reproductive health for the Population Council, argued that the FDA should allow women to obtain regular oral contraceptives over the counter, under certain circumstances. "At the very least," she argued, women who were regular pill users should be able "to reorder, replenish, and resupply themselves without needing additional medical prescriptions." Winikoff further attributed to the prescription status a paradoxical decrease in effectiveness: "With respect to oral contraceptives, the FDA must face an important reality: making these drugs more accessible more immediately also makes them more effective. American women deserve, indeed, require easy, over-the-counter access to these important adjuncts to health, self-care and peace of mind."[60]

Amy Allina, from the National Women's Health Network, disagreed with this position and explained why her organization supported making emergency contraception available without prescription but opposed doing the same for regular oral contraceptive pills. Like the FDA, the NWHN was "striving for consistency" in its position on which drugs should be available over the counter. The NWHN shared the FDA's concern that OTC products must have "a low incidence of adverse reactions" and "be intended for use in conditions that can readily be recognized by a consumer without the assistance of a clinician." Although the NWHN wanted "to see more OTC contraceptive options made available to women," it opposed doing this for oral contraceptive pills for continuing, regular birth control. The Network believed that prescription status for oral contraceptive pills was "necessary to maintain effective use of this method and to protect the health of women who [chose] to use it." Without the prescription requirement, there would be "no opportunity for a health care provider to screen out users who should not be taking oral contraceptives over the long-term." Furthermore, the opportunity for preventive health care and disease detection would be lost, a matter "of particular concern when it comes to women of color and low-income women who are already likely to have decreased access to such health services." If a third alternative between the current prescription status and OTC distribution were available in the United States, such as distribution by pharmacists from behind the counter, Allina said, "the Network would support distribution of oral contraceptives in that way."[61]

In addition, the Network believed it was important that consumers have access to "sufficient, accurate, and unbiased information" about a product to ensure that they could make an informed decision about whether or not to take a drug. The Network did not believe that a drug company would "make an unbiased judgment about what information should be conveyed to women or how best to convey it." Allina was especially concerned about the spread of direct-to-consumer advertisements and argued that such campaigns created by pharmaceutical companies were "not adequate information sources for consumers."[62]

The logistic barriers to obtaining emergency contraception within the seventy-two-hour time period convinced the NWHN that prescription distribution of this product was unnecessary to ensure safe and effective use. The Network supported OTC distribution of emergency contraceptive pills under the following conditions. There must be "appropriate label warnings to protect the health of women with contraindications to the use of emergency contraceptive pills." While there was no evidence of adverse reactions to the short dose of hormones contained in emergency contraceptive pills, the Network believed that women with a history of blood clots should avoid using this method and that the label of an OTC product had to contain "a clear and prominent warning explaining this precaution." The patient package insert should be in multiple languages and also employ techniques for women who could not read, such as pictorial representations. The package insert should warn women that the product did not offer protection against sexually transmitted infections. The Network also wanted to preserve affordable access to emergency contraception. As long as emergency contraception was a prescription product, it was covered by some insurance plans, but once it was made OTC, women would have to pay out of pocket. Sometimes the switch to OTC status had led to an increase in the price of a product. While it was impossible to predict what would happen with emergency contraception, the Network argued that to prevent barriers based on the cost of an OTC product, health care providers should continue to prescribe regular oral contraceptives for the Yuzpe method already supported by the FDA. Finally, the Network and other advocates were concerned that making emergency contraceptive pills available without a prescription might make it less likely that women using this method in cases of sexual assault would receive counseling and medical follow-up. Therefore, the Network supported "efforts to make counseling and medical services accessible to sexual assault survivors who choose to pursue them" and opposed "policies which require or otherwise pressure survivors to obtain such services against the woman's own desire or will."[63]

The purpose of the June 2000 hearing was solely to solicit public comments on how to address future switches from prescription to nonprescription status, so the FDA did not take action on any specific drug products. Nevertheless, supporters of OTC status for emergency contraception felt confident that the FDA was opening the door even wider for the spread of OTCness. They were also encouraged by the work of a group of organizations in Washington State that used legislation permitting collaborative agreements between pharmacists and physicians as the basis for a demonstration project on emergency contraception in the Puget Sound area. The collaborative agreement laws allowed pharmacists to prescribe and dispense certain drug therapies. The two-year pharmacy access demonstration project, conducted between 1998 and 2000, was funded by the David and Lucile Packard Foundation and managed by the Program for Appropriate Technology in Health, in partnership with the Washington State Board of Pharmacy, the Washington State Pharmacists Association, the University of Washington Department of Pharmacy, and Elgin/DDB, a public relations firm that had worked with the Reproductive Health Technologies Project to publicize the Yuzpe method of emergency contraception. The project conducted training sessions for pharmacists to educate them about all aspects of emergency contraception and to help them set up collaborations with health care providers in their area. Elgin/DDB created print and radio advertisements to inform women that pharmacists were prescribing emergency contraception and to promote the national telephone hotline and Web site for emergency contraception operated by the Reproductive Health Technologies Project.[64] This "quiet revolution" started by the Washington State demonstration project encouraged other states with pharmacist collaborative agreements to begin experimenting with pharmacy access programs for emergency contraception.[65] It also prompted manufacturers of emergency contraception to explore further nonprescription status for their products.

The Morning-After Pill Conspiracy

In April 2003 Women's Capital Corporation (WCC) submitted an sNDA, seeking to switch Plan B from prescription to OTC status. On November 25, 2003, the FDA published a notice in the *Federal Register* announcing that a hearing for the sNDA for Plan B would be held at a joint committee meeting of the Nonprescription Drugs Advisory Committee and the Advisory Committee for Reproductive Health Drugs on December 16, 2003.[66] At the end of the hearing, the joint advisory committee voted 23-4 in favor of switching Plan B to nonprescription status. Supporters of the sNDA were delighted but also surprised that they had prevailed, given the generally unfavorable environment for reproductive choice under the George W. Bush administration. James Trussell, a member of the panel who voted for the recommendation, told the *New York Times*, "It's hard to believe it actually happened." Kirsten

Moore, of the Reproductive Health Technologies Project, told the same reporter, "I guess I just didn't have a lot of faith that people would let the facts speak for themselves."[67]

This guarded optimism soon turned to disappointment: on May 6, 2004, Dr. Steven Galson, then acting director of the FDA's Center for Drug Evaluation and Research, rejected the recommendation of the joint advisory committee and issued a "non-approvable" letter to Barr Laboratories (which had purchased the patent rights to Plan B from WCC). Galson argued that Barr had "not provided adequate data to support a conclusion that Plan B can be used safely by young adolescent women for emergency contraception without the professional supervision of a practitioner licensed by law to administer the drug." Galson told Barr that it had to either submit data demonstrating that the product could be used safely by women under age sixteen without professional supervision by a licensed practitioner or provide an alternate proposal to allow for marketing of Plan B as a prescription-only product for women under sixteen and as an OTC product for women age sixteen and older.[68] Galson's suggestion was the first time FDA had advised that a drug be assigned to a prescription status based on age.[69]

Because Galson and the four joint advisory committee members who had voted against the OTC switch were political appointees of President George W. Bush, advocates of nonprescription status for ECP declared that these FDA officials were allowing the president's antichoice agenda to override their responsibility to be unbiased guardians of the public health. "Politics trumps science at the U.S. Food and Drug Administration," declared David Grimes in the journal *Obstetrics and Gynecology*, coining a catch phrase that was widely repeated in other media reports of the FDA decision. Barr submitted a revised application on July 22, 2004, but the FDA delayed ruling on it for more than two years.[70]

Meanwhile, a group of women who had testified in favor of the OTC application for Plan B decided to take more direct action. On January 21, 2005, the Center for Reproductive Rights (CRR) filed a lawsuit in the U.S. District Court for the Eastern District of New York on behalf of the Association of Reproductive Health Professionals, the National Latina Institute for Reproductive Health, and individuals from a grassroots advocacy group called the Morning-After Pill Conspiracy (MAPC).[71]

The lead plaintiff in the case, Annie Tummino, worked for Realbirth, a childbirth education and postpartum support center in New York City, and was chair of the Women's Liberation Birth Control Project. Tummino formed the MAPC in 2004 out of a coalition of feminist organizations dedicated to making emergency contraception available over the counter for all women regardless of age. Tummino said MAPC was inspired by the grassroots activism of the women's liberation movement of the 1960s and 1970s: "We speak out and engage in civil disobedience. Our goal is to send the message that women are the experts on our bodies and lives."[72] The group's name was "a tongue-in-cheek reference to the fact that under such restrictive conditions, many women obtained these pills from a friend, thus conspiring to break the law just to get a safe form of birth control." Because getting a prescription from a doctor and filling it within the time frame was "virtually impossible for most women," many sought out friends "who worked in health clinics to gain access and to stockpile packages for friends."[73]

In addition to filing the lawsuit, the MAPC used a variety of direct-action techniques to protest the stance of the FDA and the Bush administration on emergency contraception. They held consciousness-raising sessions and speak-outs in major cities, and they committed various acts of civil disobedience, including passing along emergency contraceptive kits to women without a prescription. Most emblematic of their ties to Second Wave feminist organizing were their actions at the March for Women's Lives in Washington, DC, on April 25, 2004. The group held a mini-rally where a dozen women "testified about rushing around trying to get the Morning-After Pill after a condom broke during sex, about the prohibitive costs associated with a doctor's visit, and about the tragicomic idea that anyone can get a doctor's appointment in 24 hours, especially starting on a Friday or Saturday night." In defiance of "unjust" prescription laws, the group flung boxes of Plan B into the crowd. They also invited spectators "to join them in signing the Morning-After Pill Conspiracy pledge to defy the prescription requirement (and break the law) by giving a friend the Morning-After Pill whenever she needs it." A group of physicians from the Association of Reproductive Health Professionals' Reproductive Health Access Project contributed to this display of feminist direct action by bringing their prescription pads and "furiously" writing prescriptions for emergency contraception for any women who wanted them. According to MAPC member Jenny Brown, these doctors "were illustrating a point which was repeated over and over in the FDA's advisory hearings—no physical evaluation or instruction from medical professionals is needed to safely and effectively use this medication."[74]

Members of MAPC declared that they "were proud to follow in the footsteps of feminists like Margaret Sanger, who passed out information on birth control when it was illegal to do so, and suffragists who were arrested for voting, to showcase how unjust the laws were." In addition, more than four thousand women from all fifty states signed MAPC's pledge, "Give a Friend the Morning-After Pill," further defying the prescription requirement. Like the feminist activists who protested against the abuse of women subjects during the 1970s, members held a sit-in at FDA headquarters in January 2005, where nine of their members were arrested for blocking access to the FDA, "just like they were blocking women's access to birth control."[75]

By filing the lawsuit, Tummino hoped to show that the FDA had not been following normal procedures and had held the sNDA for Plan B "to a much higher standard than any other [OTC] application."[76] Tummino and the other plaintiffs charged that the FDA's suggestion that Barr file a dual prescription-nonprescription application for Plan B violated their rights and the rights of women who needed Plan B to privacy and equal protection under the Fifth Amendment. Plaintiffs also charged the agency with "failure to follow medical science in their decisions on the MAP." The lawsuit sought the removal of the age restriction and behind-the-counter status for Plan B.[77] In a position paper, MAPC member Jenny Brown argued, "The age restriction has bad effects for all women, not just women under 16 . . . If we allow them to pit us against each other by age we will lose the chance to get what we really want, Morning After Pill over the counter for all women." The organization would not "accept the insult that young women are irresponsible when we try to obtain after-sex birth control. That is us taking responsibility."[78]

Feminists on Capitol Hill aided the efforts of the MAPC to force the FDA to cease stalling on the Plan B decision. After President Bush nominated Lester Crawford as commissioner of FDA, senators Patty Murray (D-WA) and Hillary Rodham Clinton (D-NY) announced in April 2005 that they would block a vote on Crawford's confirmation until the FDA ruled on Barr's sNDA. Senators Murray and Clinton released their hold on Crawford's nomination after Health and Human Services secretary Michael Leavitt promised that he would make sure that the FDA would rule on the Plan B application by September 1. After being confirmed as FDA commissioner on July 18, Crawford announced that rather than rendering a final decision on Barr's sNDA, the FDA would invite public comment on the "novel regulatory issues" that would be raised should the agency take the unprecedented step of approving dual packaging for the same drug. Susan Wood, director of the FDA Office of Women's Health, resigned in protest against Crawford's decision, stating that she could no longer work at an agency where "scientific evidence and clinical evidence, fully evaluated and recommended for approval for the professional staff here, has been overruled." Wood added that morale among FDA staff was low following Crawford's actions, because they worried that this would "severely damage the agency's credibility." Senators Murray and Clinton stated that Wood's resignation was another casualty of the Bush administration's "suppressing science" when it didn't "fit their political agenda."[79]

In the wake of public outcry following Wood's resignation, Commissioner Crawford also resigned abruptly for undisclosed reasons, and President Bush named Andrew von Eschenbach, a family friend and head of the National Cancer Institute, as acting commissioner. The image of the FDA was further tarnished by a report from the Government Accountability Office (GAO) issued in November 2005, which found that four aspects of the agency's review of the Plan B OTC-switch application were unusual. First, although the directors of the Center for Drug Evaluation and Research (CDER) supported the advisory committee's recommendation on the OTC switch, they were pressured by higher-level management to draft a nonapproval letter, which the directors refused to sign. Second, the GAO report found that this involvement by senior management was much greater than was usually the case for OTC applications. Third, FDA officials had conflicting accounts about when the decision not to approve Plan B was made. Finally, the decision of CDER acting director Galson was "novel and did not follow FDA's traditional practices," in that the agency had never before considered differences in cognitive development between adolescents and adults when making a ruling on an OTC switch. The GAO report also observed that the decision on Plan B was the only prescription-to-OTC decision made between 1994 and 2004 in which the CDER director overruled the recommendations of a joint advisory committee.[80]

The FDA continued to delay action on the Plan B OTC decision despite threats from Senators Clinton and Murray that they would stall von Eschenbach's confirmation as permanent commissioner of the FDA. In June 2006 the FDA formally denied the citizen petition, again citing lack of adequate data on safe use by adolescent girls. Finally, on August 23, 2006, the agency approved sale of Plan B without prescription, but only to those age eighteen and over. To ensure enforcement of the age restriction, the drug could be sold only in pharmacies or other facilities staffed by a health care professional and kept "behind the counter" so that it could be dispensed only to those who could show proof of age.

Senators Murray and Clinton dropped their hold on von Eschenbach's confirmation, but Tummino and her fellow plaintiffs did not drop their lawsuit against FDA. Following the agency's rejection of the citizen petition filed by the CRR, the lead attorney in the case, Simon Heller, declared: "The FDA's rejection of our citizen's petition in the midst of this lawsuit simply confirms what we have believed all along. The FDA, in the thrall of the Bush administration's anti-science agenda, has put aside its mission to promote public health in favor of depriving women of easier access to this important drug."[81] The CRR attorneys requested depositions from key FDA officials related to the case, including former commissioners Lester Crawford and Mark McClellan, Deputy Commissioner Janet Woodcock, former Office of Women's Health director Susan Wood, and CDER director Steven Galson. The judge in the case also allowed the plaintiffs to subpoena White House documents for the lawsuit, citing "strong showing of bad faith" as grounds for granting further discovery in the case. On March 23, 2009, the U.S. District Court for the Eastern District of New York ruled that the FDA "had put politics before women's health when it decided to limit over-the-counter access to the emergency contraceptive Plan B to women over 18." The court ordered the agency to reconsider its decision and to act within thirty days to extend nonprescription status to seventeen-year-olds. On April 22, 2009, the FDA announced it would follow the court's orders and "clear the way" for Barr to make its product available to seventeen-year-olds without a prescription.[82]

The Morning-After Pill Conspiracy was jubilant. Its Web site declared this decision to be "a huge victory for women's liberation." It also demonstrated the power of feminist organizing: "All along, feminists have accused the FDA of toeing the anti–birth control line of the Bush Administration in their decision making on the Morning-After Pill." By taking to the streets and the courts, the MAPC and other feminist health organizations had forced the FDA to grant OTC access to the morning-after pill for women over age sixteen.[83]

Conclusion

The two case studies presented here give two examples of how feminist grassroots activism helped redefine the meaning of the prescription in the early twenty-first century and how such activism became redefined in the process. The civil disobedience of the MAPC against the "injustice" of the prescription as a barrier for emergency contraception is one way in which feminist health activists continued to assert women's rights to control their own bodies in opposition to institutions of medical authority. The participation of health care professionals in this activism suggests that at least in the case of emergency contraception, the threats to professional authority posed by the erosion of the prescription are on balance outweighed by the public health goal of preventing unwanted pregnancy. Conversely, the dual critique of the prescription by feminist health activists—it was simultaneously attacked as a vestige of a paternalistic medical system that obstructed access to vital therapeutic objects in women's health and as an insufficient measure for safeguarding the bodies of vulnerable women—is itself emblematic of the complex relationship of industry, medicine, and feminist health activism in the early twenty-first century.

In this light, it is perhaps not surprising that the approval of nonprescription sales for emergency contraception has not entirely fulfilled the expectations of activists that all women could access this product without any interference from health professionals. Emergency contraceptives are not literally available over the counter but are kept behind the counter; customers must show identification proving they are seventeen or older before they can purchase them. This is not the only OTC drug product sold this way. Numerous states have recently passed laws or regulations requiring cold medicines containing the decongestant pseudoephedrine to be placed behind the pharmacy counter to prevent the bulk purchase of the drug by those who would use it to manufacture crystal methamphetamine. These developments have in effect created a third class of drugs, whose restricted market is no longer defined in terms of the prescription but by the nature of the "counter." It has also meant that pharmacists have replaced physicians and nurse practitioners as the "gatekeepers" for certain classes of drugs. Although this recent enhancement of the pharmacists' professional authority stands in contrast to the episodes of pharmaceutical deprofessionalization described by Dominique A. Tobbell and Elizabeth Siegel Watkins in chapters 3 and 4, not all pharmacists have welcomed this new role as guardian of the public health. Some pharmacists, in the process, have created new barriers to medication access by asserting that they have a right to conscientiously refuse to sell emergency contraception on religious grounds.[84]

Moving a drug out from behind the aegis of the prescription clearly does not make it immediately accessible to all who might benefit from it. Nonetheless, many of the individuals involved in achieving the OTC status for Plan B have been eager to extend this precedent to the sphere of regular oral contraceptives. In 2004 several of these parties formed the Oral Contraceptives Over-the-Counter Working Group "to explore the potential of over-the-counter access to oral contraceptives to reduce disparities in reproductive health care access and outcomes, and to increase opportunities for women to access a safe, effective method of contraception, free of unnecessary control, as part of a healthy sexual and reproductive life."[85] In November 2007 the FDA held a public meeting on behind-the-counter (BTC) availability of certain drugs, at which the NWHN's Amy Allina argued that the prescription was a significant barrier to access and compliance, since some women discontinued pill use because they could not schedule an appointment with a health care provider before the prescription ran out. If the creation of a BTC class of drugs was "well executed," she argued, then it could address some of these problems with access and compliance. However, she said, a "poorly-executed BTC system could make existing problems worse or even create new barriers for women" if health insurance did not cover BTC contraceptives.[86]

The health reform legislation passed in 2010 does not look promising in this regard: health insurance plans are not required to cover nonprescription drugs, and the new law prohibits the use of flexible spending accounts for OTC products. Thus, while the rise of OTCness has weakened the prescription as a boundary between patients and the health care professions, it has done nothing to address the economic inequalities in the United States that continue to pose an insurmountable barrier to those without the means to pay for OTC products.

CHAPTER 8

The Prescription as Stigma

Opioid Pain Relievers and the Long Walk to the Pharmacy Counter

Marcia L. Meldrum

Prescriptions for narcotics, specifically for the opioids, have long carried a social and cultural weight, as well as a legal meaning, not attached to other drugs. The relationship of both physician and patient to the opioid prescription evolved over the course of the twentieth century. But once physician registration and records of narcotic prescriptions were required in the United States with the passage of the Harrison Narcotic Act in 1914, these scrips made the behaviors of both doctor and patient visible and measurable and carried for both the risk of stigma and marginalization within society.

Although American physicians, legislators, regulators, and policymakers alike have consistently recognized the therapeutic value of opioids in pain relief, they have found it difficult to separate the analgesic properties of these substances from their addictive potential in determining the legitimacy of their use. Physician Raymond Houde remarked in 1978: "Ironically, the capacity to induce a sense of well-being, or euphoria, is a property of the narcotics which we have been making a great effort to eliminate from the so-called 'ideal analgesic' of our aspirations. This unfortunate paradox has been the result of an association of euphoria with drug abuse."[1] Opioids relieved suffering, offering comfort and pleasant sensations, but too much comfort and pleasure were suspect. Linked in the popular mind to irresponsibility, lack of control, unrestrained sexuality, disruptive behavior, and crime, the drugs became easily identifiable foci of "moral panics" (as discussed by Nicolas Rasmussen in chapter 1) whenever rapid social change generated public unease. The dual identity of these drugs distorts the lens through which both the opioid prescriber and the patient, the opioid user, are perceived by society, government, and even themselves.

Whereas a physician prescribing other potent pharmaceuticals can maintain a self-image as a rational scientist (as described by Jeremy A. Greene in chapter 10 and Scott H. Podolsky in chapter 2), offering the patient the benefits of his expertise, the opioid prescriber has found it hard to play that role. If he titrates his practice to the expressed needs of his suffering patients, his scrip volume—amounts, dosages, and duration—will be highly visible to drug regulators and enforcers and to other medical observers. Because an opioid overdose represents a lack of individual control, a physician's "overprescribing" suggests a lack of professional control. Either he is himself a drug dealer, turning his office into a "gray market" (as elaborated by David Herzberg in chapter 9), or he is the hopeless dupe of dealers posing as pain patients. In either case, he has lost the mantle of the rational scientist, of professional ethics and expertise; he is either a criminal or a fool.

The physician who intends to prescribe opioids and maintain her professional identity must distance herself from the drug and the patient. The prescription must become not a transaction or a gift, but a tool of discipline and control. The current pain management literature provides clear guidelines for this translation.[2] A physician must select patients carefully, have them sign written contracts, and monitor their behavior for signs of drug diversion, misuse, or abuse. The prescription—the amount, the dosage, and the duration—is translated into a tool to control and discipline the patient. While many therapeutic relationships carry an element of paternalism, the opioid prescriber is often advised to pursue a deliberate strategy that infantilizes or marginalizes the patient and denies his ability to act rationally. The patient who receives sufficient opioids to relieve the pain may accept this strategy; or he may bend or break the rules to get stronger analgesia or just to retain some control over the situation. In either case, he has lost legitimacy, in his own eyes and often in those of others, as a social actor, as an adult. The opioid user is either a child or a "problem patient."

These elisions of identity and reconfigurations of the prescription exist because a serious medical debate over the nature of addiction and the management of opioid use was short-circuited by the Harrison Narcotic Act. The prescription opioid was branded an illegitimate and risky substance, the use of which would be tolerated only under the aegis of a prescribing physician. The humble prescription thus became the mechanism whereby the social danger of the opioid would be defined and contained. The opioid prescription doesn't just define how much of the drug the pharmacist is to dispense; it makes an implicit statement as to who in this transaction has lost control, where the risk lies, and who must bear the stigma. After recounting the background, how the opioid prescription evolved as a stigmatizing practice since 1914, I present two case studies—one of a patient and the other of a physician—that describe in intimate detail the historical burden still shouldered by both subject and object of the narcotic scrip.

The Specter of Morphinism and the Harrison Narcotic Act

Morphine, the most powerful and promising of the opiate alkaloids, became the physician's prescription analgesic of choice, for the pain of wounds suffered in combat or the chronic agonies of middle-class neuralgia, after the introduction of the hypodermic syringe in the 1850s.[3] Many physicians praised the benefits of morphine prescriptions, writing confidently of the drug's "exaltation of our better mental qualities" and reassuringly that "of danger there is absolutely none." In the 1870s, however, some professional leaders began to express concern about the overuse of morphine and laid much of the blame at the doors of their colleagues. Virgil Eaton stated emphatically in 1888, "The doctors are to blame for so large a consumption of opium, and they are the men who need reforming."[4]

By the early twentieth century, American physicians had reached a professional consensus that morphine use was injurious, except for "special medical purposes." However, they still argued about whether its habitual use, iatrogenic or recreational, was a moral failing or a physical disease. If drug addiction was the mark of an inherently depraved character, doctors who unknowingly prescribed opioids to such "inebriates" were not at fault. But if addiction, or "morphinism," was a pathophysiological response to the drug, then the profession had a responsibility to monitor and treat the condition. An articulate advocate of the latter position, Ernest Bishop, wrote in 1920, "It is much wiser to supply to the addict who is not a public menace the drug of his addiction to the extent of his physical needs," maintaining his habit until the drug could be humanely withdrawn.[5] Even as Bishop presented his case, however, the profession's jurisdiction over morphinism as a disease was about to be sharply curtailed by the Harrison Act and the federal surveillance and enforcement authority it engendered.

A significant factor in the Harrison Act's smooth passage was that most American physicians by 1914 were more than willing to disassociate themselves from the casual prescription of morphine and other opioids. The profession hoped to exorcise the specter of morphinism and to file away the recurring stories of iatrogenic addiction and medical abuse. They mounted no strong protest against the act and its requirements for physician registration and reporting of all opioid sales and prescriptions.[6]

A medical minority intended to continue monitored prescription of opioids for the treatment of addiction under Harrison, but their efforts were forestalled. The Treasury Department took the position that maintenance prescriptions were not medical treatment; the drugs would either exacerbate the recipient's moral corruption or would be diverted to the illegal market. Treasury's Narcotics Division challenged, put under surveillance, and finally closed down maintenance clinics in at least twelve states between 1919 and 1923.[7] Although the Supreme Court's endorsement of these actions was not consistent,[8] few physicians contested them openly by offering or advocating maintenance prescriptions for addicts. Those who did, and their patients, could find themselves marginalized and stigmatized by the opioid prescription. Journalist Samuel Hopkins Adams noted sadly in 1924, "Under the unintelligible, inequitable, and constantly shifting 'interpretations' of the law, the treatment of narcotic addiction is rigidly if uncertainly limited, so that no intelligent and honest practitioner can feel himself safe from prosecution in treating one of these cases."[9]

Physicians disassociated themselves from the stigma if they subscribed to the psychoneurotic model of addiction formulated in the 1920s by public health psychiatrist Lawrence Kolb. Kolb's studies of 230 addicts led him to the conclusion that only 14 percent of cases could be attributed to iatrogenic addiction. Eighty-six percent were of the "psychopathic delinquent type"; these were persons who, whether they began taking opioids for medical reasons or recreationally, found that the effects relieved feelings of inadequacy and inferiority and produced sensations of excitement and power. Kolb later characterized these individuals as "thrill seekers."[10] They could not give up the drug, would inevitably devote all their energies to obtaining it, and would become part of a criminal subculture that spent "a great deal of their lives in prison." In Kolb's view, one welcomed by the Narcotics Division and perhaps by many physicians, most opioid users were the responsibility of the criminal justice system, not the medical profession.[11]

Prescriptions for Pain

The enforcement actions taken by the Narcotics Division and its successors created a regulatory climate around the opioid prescription from which most physicians preferred to distance themselves, and the Kolb model, although since challenged and substantially revised, created an enduring stigma of the "true addict" as someone less than normal, as a person who could not benefit from and hardly deserved medical treatment. If in theory many patients could take morphine or other opioids safely, physicians dreaded the possibility of giving a prescription to a "thrill seeker" who hid his vicious tendencies under an facade of normalcy. Opioid prescriptions for pain had a recognized place within medical practice from the 1930s through the 1970s (codified on Schedules II and III in the 1970 Controlled Substances Act), but textbooks recommended that morphine and other opioid prescriptions be restricted in practice to short-term use in severe trauma or postoperative recovery or to ease the pain of terminal illness. Patients who requested repeat prescriptions or higher dosages for acute pain set off alarm bells. In cases where the patient's illness was not fatal, or where the duration of life appeared to be more than a few weeks, physicians were "often loathe to give liberal amounts of narcotics" lest "the drug addiction itself . . . become a hideous spectacle," wrote cancer surgeon Warren Cole in a standard cancer textbook in 1956.[12] Once death was foreseen within a relatively short period, the patient was finally free of suspicions of psychopathology. "The medical and social problems of addiction are unimportant when the few remaining days of life need the blessed relief provided by opiates," wrote anesthesiologist M. J. Schiffrin and pharmacologist E. J. Gross in that same textbook. "In the patient with long term chronic pain . . . every effort should be made to put off the potent addicting drugs until all other measures have been exhausted."[13]

Thus, only the very advanced or terminal stages of cancer or other diseases lay beyond the dangers of addiction. For the patient marginalized and separated from normal medical care not by depravity but by mortality, very high doses of opioids were often prescribed, sedating the person "into lethargy" until death.[14] A nurse-observer recorded the following incident in 1979, illustrating the contrast between normal standards of practice and the treatment of the dying: "Dr. Long told Annette to increase Mr. Piel's morphine to 60 mg. every two hours. Annette looked at Dr. Long and repeated the dose, and Dr. Long confirmed it." Annette said, "In nursing school . . . we were taught that 15 mg. was the maximum dose to be given at any one time. I know he's been on a gradually increasing dose, but I still remember that earlier learning." The patient was given the prescribed amount twice and died before the next scheduled dose.[15]

For nonterminal hospitalized patients, opioid prescriptions were handled very differently. A 1973 study of thirty-seven inpatients at Montefiore Hospital in the Bronx, for example, revealed that nurses and house staff consistently administered only 25 to 50 percent of the *prescribed* opioid dosages for pain, even though twenty-three of the patients complained that their pain was not well-controlled. The authors concluded, "For many physicians these drugs may have a special emotional significance that interferes with their rational use." A follow-up house staff survey and chart review of 110 patients at Mount Sinai Hospital in New York in 1983 found that the most common prescription was for 100 mg. of meperidine (Demerol), a minimally effective dosage.[16] Both of these studies reported that house staff were not well informed about appropriate opioid dosages but equated good practice with limited medication, even if pain relief was inadequate.

Nurse-educator Ada Jacox vividly described the pain medication "drama" on hospital wards in 1977: "It is still 'one-half to three-fourths of an hour early' before the four-hour interval between dosages expires . . . The battle goes on repeatedly with the patient complaining of pain . . . and the nurse insisting that the patient must wait. The patient becomes more irritable and anxious as his pain increases; meanwhile, the nurse becomes impatient and begins to believe that the patient . . . is 'addicted' to his medication . . . The drama is absurd."[17]

Marginalizing the addict had clearly not solved the problem of iatrogenic addiction or of chronic pain management. The search for an effective drug to prescribe for the treatment of severe long-term or chronic pain continued; one was needed that would fill the perceived niche between aspirin for moderate pain and morphine for very severe acute pain or terminal illness. Bayer and other German manufacturers had developed several new synthetic opiates, including oxycodone (1916), hydromorphone (1924), and hydrocodone (1925), although all showed addictive potential similar to morphine itself. The National Research Council (NRC) saw this as an opportunity to build the nation's pharmacological infrastructure and to end American dependence on domestic patent medicine enterprises and European manufacturers. Harvard pharmacologist Reid Hunt wrote, "A thorough study of the morphine molecule might show a possibility of separating the analgesic from the habit-forming property." Under the sponsorship of the NRC's Committee on Drug Addiction (created in 1929 and later incorporated into the National Institutes of Health), multiple opioid derivatives were investigated, including meperidine (Demerol), propoxyphene (Darvon), and methadone. These were variously effective for pain management, and methadone eventually became the treatment of choice for addiction, but none of the compounds proved to be free from risks of dependency or abuse.[18]

By the early 1980s, a new community of physicians and researchers had redefined severe chronic pain as a serious medical problem and readdressed the issue of the opioid prescription in its treatment. Two groups of researchers, one at St. Christopher's Hospice in London, the other at Memorial Sloan-Kettering Cancer Center in New York, had had extended opportunities to study opioid prescriptions for patients whose terminal diagnoses precluded the risk of addiction. As cancer treatments improved, these patients had longer prognoses, during which they often had to cope with chronic or recurrent pain. The founding of the International Association for the Study of Pain in 1973 brought these researchers together; the decision of the World Health Organization (WHO) to develop its Cancer Pain Programme in 1982 gave them their platform.[19]

Both research teams developed their opioid prescription practices with patient collaboration. Raymond Houde and nurse Ada Rogers conducted more than seventeen thousand analgesic studies on cancer patients at Sloan-Kettering from 1951 through the early 1970s; by varying the prescription and titrating the dosages, they treated each patient "as a whole human being."[20] Houde commented dryly on "the popular misconception that drugs with this capability enslave, demoralize, and lead the unwitting patient down the primrose path to addiction," asserting, "We have not seen any outward signs of elation in our cancer patients receiving on-demand narcotics."[21] He employed a crossover study design, in which each patient received a series of doses of an experimental analgesic, randomized against a comparison series of morphine or aspirin. "I have no way of knowing actually what these people are feeling," Houde explained. "So the only thing I can do is to have them serve as their own control."[22] Rogers collected the data and observed each patient closely, staying at the hospital at least ten hours every day. Their multiple comparison series constituted a detailed database of information about the relative strength and duration of pain relief and of side effects, such as nausea and respiratory depression, associated with the opioid variants. "We sought out a balance between a drug's good effects and its bad effects . . . The only way we could determine that, of course, was in relative terms."[23]

At St. Christopher's Hospice, physicians Cicely Saunders and Robert Twycross developed a protocol for "regular giving," writing prescriptions to ensure the patient a sufficient level of analgesia that she never had to sit in pain, anxiously watching the clock, but was free to spend time with family or pursue a favorite activity. Under this regime, they found that patients showed fewer signs of dependence than the textbooks predicted; they did not inevitably escalate to higher dosages and were sometimes even able to reduce the dose.[24] Their outcome measures were defined not in terms of analgesic characteristics, but patient well-being. "Your ultimate aim is not just to see the pain gone, but to see a patient free of pain doing something," Saunders commented. Twycross wrote that the patient given effective relief would "begin to live again . . . Freed from the day and nightmare of constant pain, his last weeks or months take on a new look."[25]

Working from very different perspectives, these two research teams demarginalized the dying cancer patients and normalized their roles in pain management—Houde and Rogers by making them essential research collaborators in differentiating drug properties, and Saunders and Twycross by allowing them to "live again," rather than succumb to pain or drug-induced coma. Their collaboration produced a set of WHO-endorsed prescription guidelines for cancer pain relief, which stressed dosage titrated to the patient's pain and given on a regular schedule.[26] The pain management community, not without internal debate, has since worked strenuously to apply this model to legitimate chronic opioid prescriptions for chronic noncancer pain, a serious problem affecting an estimated 9 percent of adult Americans, who report constant pain at levels that are "almost the worst one can possibly imagine."[27] The most recent set of expert practice guidelines for such prescriptions appeared in 2009.[28]

In one classic paper from 1981, pain management specialists Kathleen

Foley and Richard Kanner conducted a retrospective review of 103 patients seeking treatment at the Memorial Sloan-Kettering Pain Clinic over two years. They identified 45 cancer patients and 17 noncancer patients who received opioid prescriptions for extended periods. There was "no evidence of abuse" reported among the cancer patients; increased dosage in this group "was associated with rapidly progressive disease." Only 2 of the 17 noncancer patients, both of whom had "a long history of drug abuse behavior," exceeded their prescribed doses. The authors stated firmly, "*The dearth of clinical studies therefore offer limited support to the belief that chronic narcotic use for analgesia is associated with a high risk of addiction* . . . drug abuse and drug addiction should not be the primary concern of the prescribing physician." A few years later, another retrospective review of 38 noncancer patients who were prescribed opioid analgesics found evidence of dependence or excessive use for only 2, both with prior histories of substance abuse.[29]

Other studies mirrored these results, but with small cohorts. Many patients resisted taking opioids and stigmatized peers who did so. In 2000 Steven Passik and colleagues surveyed 52 cancer patients and 111 women with HIV/AIDS, asking them whether they had engaged in "aberrant" drug-taking behaviors, whether they would consider doing so, and whether they thought such behaviors were common. Patients were more likely to report past than present aberrant behavior, but they stated that such behavior might be justified to control pain. However, regardless of motivation, the cancer patients thought that 25 to 50 percent of their peers got "high" from their pain meds, and the women with AIDS estimated this to be true of 50 to 75 percent in their cohort. "Many patients believe that using illicit drugs for pain control . . . and getting 'high' from pain medications are common behaviors . . . Patients do not seem to generalize from their own behavior to that of pain patients in general."[30]

The cancer pain researchers had demarginalized the terminal cancer patient and created a legitimating model for long-term opioid prescriptions for pain. However, those who used opioids for chronic noncancer pain in this period still found themselves perceived as living prototypes of Kolb's irresponsible delinquent.

The Patient and the Opioid Prescription: Mary's Story

One such prototype was Mary, who suffered a sudden onset of severe rheumatoid arthritis in the early 1980s. She was then in her early twenties and had been married only a few years.[31] Her rheumatologist prescribed eight to ten tablets of Percocet (oxycodone with acetaminophen) a day, a dosage she often found insufficient. As she and her husband Tom told the author when interviewed at their home in 2001, Mary's life gradually became focused on getting the next prescription to control her pain: "I'd have to kind of get my courage up and say, 'I need another prescription.' I'd have to kind of spit it out at the end. Finally, one day, he said, 'I can't do this anymore. I'm afraid that you're addicted.'" Her doctor referred her to an inpatient pain clinic, where they taught her relaxation techniques, which she found only partially effective. When she "begged" for another prescription, he wrote a new scrip for six Lortab (hydrocodone with acetaminophen) daily. "So I'm on the magical six Lortab a day and trying to learn to live with the pain, because that only took care of part of it. It really wasn't enough."

Mary became a "doctor shopper," seeking new prescriptions from any doctor or dentist she saw, or visiting the Emergency Room to say her physician was out of town: "I would always tell them whatever problem I had was hurting a lot and could I get some pain medication for it. Wherever I could, I searched it out . . . I felt like I was a cheat and a thief and a liar . . . but I felt like, if I don't do this, I'm going to die, because it's so awful . . . They said I was supposed to learn to live with it, but I couldn't." Despite taking higher dosages, Mary did not experience "any major euphoria, or giddiness . . . or not being able to function at work." "But," she said, "it was a struggle. The pain is what disabled me, not the medication."

After a few years, what Mary described as "something great" happened. She was diagnosed with endometriosis, and her gynecologist gave her a second prescription for six Lortab. "I felt like I'd hit the gravy train." The pharmacists soon discovered the duplication and informed her physicians, and she "got busted." The rheumatologist was very angry, and the gynecologist agreed to take over the management of her pain medications. His prescription strategy was to have her pick up a fresh bottle of six tablets every day, an errand that became a difficult daily ordeal:

> I will never forget what it felt like walking into that drugstore every morning . . . You know how they usually have the pharmacy up on a little platform? I'd feel like I was walking back and looking up at them, like please God, please you people up there, will you give me my pain meds for the day? And I knew that everyone working there knew what was going on with me. I'd see them looking

> at me and kind of talking to each other and whispering. And by then, I felt like the whole town knew me and knew my problems. I think I must have been red flagged at all the pharmacies in the area . . . It was awful. And it really was me versus the world.

After Mary had had a hysterectomy, the gynecologist transferred her pain management to her psychiatrist, who began treating her for addiction. This development further reinforced Mary's image of herself as "a problem patient" that everyone despised. "Let's shuffle her around a bit, because we can't deal with her too long because we get so frustrated and so tired of her lying and cheating and such." She had several more surgeries during this period, necessitating prescriptions for stronger medications. These were entrusted to her husband Tom, or her mother, to dole out. "Everyone was really treating me like a child or like a criminal. And that's what I felt like."

Her relationship with her husband became increasingly strained and antagonistic. Tom recalled, "We had no lives . . . We were just coping day to day to day." He vividly recounted some of their conflicts over managing her prescriptions: "She'd find the pills where I hid them. I remember one time she found a bottle of Percocet that we had on hand as sort of backup. She emptied that out and filled the bottle with some aspirin . . . There were some fights about that . . . I was controlling Mary to some extent . . . She would lie to me. We said some awful things to each other."

The stigmatizing power of the opioid prescription had by then completely transformed Mary's life and self-image. The strategies employed by the physicians, of forecasting addiction, limiting her medications, and enforcing daily trips to the pharmacy, had severely damaged her self-esteem. She saw herself as dishonest, unreliable, and infantile, "a nasty little person." Her physicians were all staff members at a university-based medical center near her home, people whose authority and expertise she respected; and she internalized the definitions of her character and behavior spelled out by their prescription strategies. When her husband and her mother were made their agents and called on to enforce the prescriptions, her family relationships were damaged as well. The rest of the community appeared also to have united in condemning her behavior; she was left feeling isolated, "me versus the world."

At the age of thirty-seven, after suffering with her disease for thirteen years, Mary reached a breaking point in the mid-1990s. Her life had become such a struggle that she began to think seriously of suicide. "It had gotten to the point where there was nothing good there. All it was, was the fighting and the bad part." Her self-esteem was so low that she believed her family and friends would be better off if she were dead. Tom, who had borne with his wife's misery, dysfunction, and constant demands for medication during these years, was also close to giving up and was considering divorce.

After Mary told one doctor of "some suicidal thoughts," he wrote her a scrip for "this God-awful antidepressant that just knocked me out for like twelve hours and made me so doped up, I didn't know what end was up . . . I only took it for a couple of days and then threw it away. It's like he wasn't really considering the whole picture . . . let's give her a prescription and send her on her way." Yet none of her doctors were that casual about her opioid prescriptions.

At this nadir, through "divine intervention or just stupid luck," Mary scheduled an appointment with a local pain specialist: "I told him my whole ugly, nasty story about all the lies and the cheating and the stealing, and he just sort of sat there and nodded . . . He was talking to me like I was an adult, which was another thing I wasn't used to." The physician had her come in for several appointments and met with Tom as well. Then, said Mary, "He finally gave me the big test . . . He wrote me a prescription, which I filled on the way home . . . Tom's jaw hit the floor when he saw me walk in with a bottle of two hundred and fifty Percocet . . . And he finally just said, 'What do we have to lose, Mary? . . . If you've worked out something with him, go ahead.' Because he knew too. Basically, what this doctor had done was give me enough rope to hang myself."

But Mary kept to the schedule she and her new physician had worked out. She recorded her pain and dosages in a daily journal, and to her own surprise and pride, successfully managed her own prescription. "I proved to him and to my family and to myself that, yes, this is a pain problem . . . and we're dealing with that, and I'm an adult, and I'm managing it myself with the doctor." At the time of the interview, she had been able to reduce her dosage of Oxycontin (extended-release oxycodone) because she always had sufficient medication on hand and had lost her anxiety over whether or not the pain would become unbearable. If her pain flared up for any reason, she knew she could briefly increase the dose and would be able to go back to her usual level without problem: "The doctor trusts me . . . he's also given me back my self-esteem and the ability to control myself. I'm empowered."

Her psychiatrist, however, reacted negatively when Mary told her about the new prescription. "'You have no business having that quantity of pills . . .' I was just stunned." Even after the pain specialist wrote the psychiatrist a letter explaining his approach, she refused to cooperate and dropped Mary as a patient.

As the new century began, Mary and Tom felt that they had been given a new life. They made contacts with other pain patients and advocates and told their story at professional meetings. There they learned that Mary was not unique: "I've met grandmothers who talk about, 'Well, when the holidays were coming up, I knew I would be stressed out because all the family would always come to my house, and I'd have all the cooking and cleaning to do, and my pain levels would be up. So I'd have to start making appointments with my various doctors and go in and get my pain meds from these different doctors.' They were doing the same thing I was! . . . It is tragic and sad, but people have been forced to do that. It's just any way you can get it, you've got to do it."

Similar cases of drug-seeking and "doctor-shopping" behaviors caused widespread concern in the 1990s and early 2000s. Rush Limbaugh's much-publicized 2003 case, in which he obtained nineteen prescriptions for pain relievers from physicians in three different states in a five-month period, is only one notorious example.[32] Mary's case and those of others like her illustrate the way in which the opioid prescription could become a contested object between patients and physicians. The physician wary of the risk of addiction employed various strategies to simultaneously *grant* and *deny* patient access to opioids; he limited dosages, required extra patient effort to fill the prescription, and enlisted family members to monitor the patient. The patient countered by trying to extend access and to maximize her prescription options, by seeking prescriptions from other doctors, faking or exaggerating painful diagnoses, and hoarding supplies. The physician used his medical authority to support his actions and his use of the prescription to reinforce his own control and limit the patient's uncontrolled behavior. His actions may have seemed arbitrary and punitive to the patient but, accepting his authority, she nonetheless felt infantilized and stigmatized. As the writer of the opioid prescription, the physician became a harsh parent; the patient, an irresponsible child.

Mary's pain specialist overturned the rules of this contest and adopted a patient-centered model similar to that described by Houde and Twycross. He accepted his patient's reports of pain and transferred to her, through a liberal prescription, the right to manage her own access as a competent adult. The prescription itself was redefined as a treatment tool, which both parties could use cooperatively to manage the pain. Mary's story ended well, but the physician's prescription practices in this case could have exposed him to stigma and discredit. As the next case shows all too clearly, regulators and drug warriors had their own ways of interpreting prescription strategies.

The Pain Doctor and the Opioid Prescription: William Hurwitz's Story

In 2002 the Federal Drug Enforcement Administration (DEA) closed the practice of Dr. William Hurwitz of McLean, Virginia, seized his property, and charged him with one count of conspiracy to traffic in drugs, two counts of health care fraud, one count of racketeering, and fifty-eight counts of drug trafficking, including two counts involving serious bodily injury and two involving deaths. Dr. Hurwitz's indictment was based on his writing multiple high-volume opioid prescriptions for nineteen different patients seeking pain relief. His prescription practices had previously been censured by the Medical Board of the State of Virginia, but his fellow physicians on the board were inclined to believe he had been acting in good faith. After several of his patients were arrested for selling prescribed drugs, however, they were persuaded by prosecutors to tape-record conversations with Hurwitz in which he indicated his suspicions of their activities. This evidence led to his prosecution.[33]

The Hurwitz case was part of a deliberate policy developed by the DEA in the late 1990s and the 2000s, in response to heightened public concern over prescription pain reliever abuse. Hydrocodone was marketed in combination with aspirin under the trade name Lortabs and with acetaminophen as Vicodin. Vicodin was the most commonly prescribed narcotic pain reliever in 1999 and 2000, although its risks of abuse and dependence had been well documented. The Drug Abuse Warning Network (DAWN) reported an increase of 84 percent in emergency department reports of hydrocodone abuse or overdose between 1996 and 2000.[34] Vicodin became a recurring theme in popular culture: there were the reports of abuse by celebrities like Limbaugh, Matthew Perry, and the TV character Gregory House; and rock songs like "Feel Good Hit of the Summer" and "Vicodin Saturday Night" featured the drug.

But alarm over Vicodin paled in the wake of the Oxycontin panic. This potent extended-release formulation of oxycodone became a major drug of abuse after Purdue Pharma introduced it to the market in 1995, with a label warning not to crush the tablets, as this would cause rapid release. Street users immediately interpreted this to mean that crushing Oxycontin produced enough narcotic to create a satisfying rush when injected or snorted. Rates of abuse, diversion, overdose, and death reached alarming proportions by 1998, particularly in Appalachia.[35] DEA reviews of DAWN records found a 93.0 percent increase in oxycodone mentions by medical examiners between 1997 and 1998 and a 32.4 percent increase in oxycodone-linked emergency admissions from 1997 to 1999.[36] Both the Purdue formulation and its generic clones remain drugs of frequent abuse; between 2002 and 2006, the recreational use of Oxycontin by adolescents increased significantly, doubling among eighth-graders.[37]

The Oxycontin crisis was only part of a larger trend in prescription pain reliever abuse, particularly alarming among adolescents.[38] Rates of Vicodin abuse among adolescents remained at least 50 percent above those for Oxycontin in 2002–6, according to the National Institute on Drug Abuse's 2008 *Monitoring the Future* study.[39] The 2006 Partnership Attitude Tracking Study reported that approximately one-third of the teenagers surveyed believed that prescription medications were not addictive and were safer than illegal drugs.[40] Adolescents may obtain drugs through patterns of casual drug exchange, which are difficult for prescribers to detect. Nearly half (47.3%) of the teenagers studied in the 2005 *National Survey on Drug Use and Health* received prescription drugs for free from friends, and another 10 percent bought pills from friends or relatives. Only 21 percent obtained prescriptions from doctors. The remaining 22 percent reported a variety of avenues, including theft from friends or relatives, dealers, theft from medical sites, Internet buys, and fake prescriptions.[41]

The DEA's "Action Plan" for Oxycontin from the early 2000s identifies the major sources of drug diversion as "fraudulent prescriptions, doctor shopping, over-prescribing, and pharmacy theft."[42] Similarly, the agency's current summary on oxycodone states that the main street sources are "forged prescriptions, professional diversion through unscrupulous pharmacists, doctors, and dentists, 'doctor-shopping,' armed robberies, and night break-ins of pharmacies and nursing homes."[43] The DEA's recently updated "Fact Sheet" on hydrocodone, "the most frequently encountered dosage form in illicit traffic," describes additional "widespread diversion" sources such as "bogus call-in prescriptions, altered prescriptions, theft and illicit purchases from Internet sources."[44]

In March 2004 the George W. Bush administration made clear its intention to use prescription data to target doctor shoppers, overprescribers, and "unscrupulous" professionals. In a press release from the Office of National Drug Control Policy, drug czar John Walters, FDA commissioner Mark McClellan, DEA administrator Karen Tandy, and Surgeon General Richard Carmona described a coordinated strategy to increase the number of prescription drug monitoring programs and to employ data mining to track down physicians, pharmacies, and individuals who might be involved in prescription drug diversion.[45] This initiative was only one facet of the combined agencies' drug control activities. The drug enforcers' own evidence demonstrates that several factors other than physicians' prescribing practices contribute to the problem of prescription drug abuse. All the agencies have stressed that opioids are "legitimately used to treat moderate to severe pain."[46] The DEA reiterates that "doctors operating within bounds of accepted medical practice have nothing to fear" and has presented statistics showing that physicians facing sanctions from the agency constitute 1 percent or less of the total number of those registered to prescribe narcotics in any given year.[47]

However, the DEA's approach to Hurwitz's case described an alternate identity for pain management specialists, particularly those outside of major medical centers who prescribed more than minimal opioid dosages. High-volume prescription patterns, the Hurwitz case implied, were perceived by regulators as outside the bounds of professional practice *by definition*.[48] The prescriber in such cases, if not criminal, was stigmatized as either gullible or irresponsible—and therefore as someone who had lost a portion of his professional identity and authority. He had abandoned his role as the rational authority in the therapeutic encounter and, instead of using the prescription to monitor and control the patient's use of opioids, allowed the user to define the rules of drug dosage and duration.

The Hurwitz case first came to trial in late 2004. The issue of whether or not high-volume prescription patterns could be considered within the bounds of professional practice was central to the case, and expert witnesses from the pain management field testified on both sides of the question. The defense moved to admit into evidence an FAQ (Frequently Asked Questions and Answers) the DEA had posted on its Web site in August 2004. This document had been developed by a Working Group of pain management specialists in consultation with the DEA, which had reviewed multiple drafts. The FAQ stated, "The number of patients in a practice who receive opioids, the number of tablets prescribed for each patient, and the duration of therapy with these drugs do not by themselves indicate a problem, and they should not be used as the sole basis for an investigation."[49] This de-emphasis on prescription volume and duration reflected the views not just of the pain management community, but also of the U.S. Federation of State Medical Boards' Model Guidelines on pain practice, which "minimize the importance of historically suspicious factors such as prescribing quantity and frequency, and place them in the proper context." DEA administrator Asa Hutchinson had endorsed this very approach in 2002: "We may look at numbers as a possible indicator of suspicious activity, but in the absence of other information about diversion, quantity alone is not an indication of violation."[50]

On October 4, 2004, the DEA abruptly erased the FAQ from its Web site, describing it as "inaccurate" and not an "officially sanctioned" document. It also requested that the Working Group members remove it from their Web sites. The judge in the Hurwitz case ordered the FAQ document disqualified as evidence. As an apparent replacement, an "interim policy statement" (IPS) on "Dispensing of Controlled Substances for the Treatment of Pain" appeared in the *Federal Register* on November 16.[51] The IPS characterized the prescription of "an inordinately large quantity" of pills of a controlled substance and "large numbers of prescriptions" as the first and second among "certain recurring concomitance of condemned behavior."[52] The pain management community saw these actions, which appeared to equate prescription volume with deliberate criminal activity, as confusing and alarming. As Drs. Samuel Hassenbusch and Scott Fishman of the American Academy of Pain Medicine wrote to DEA administrator Tandy on December 22, the IPS "heightened suspicions that it may not be safe to treat pain aggressively." They predicted, "The substance and delivery of the IPS will have a chilling effect on pain care for millions of Americans."[53]

In the courtroom, meanwhile, the prosecutor repeatedly cited the *number* of pills in some of Hurwitz's prescriptions, without identifying the specific dosages. The leading exemplar was Patrick Snowden, a patient with a severely injured foot, who had apparently received prescriptions for sixteen hundred opioid pills in one day. When Snowden took his original prescription to a local pharmacy, which did not have the specified dosage in stock, the pharmacist called Hurwitz and requested two substitute prescriptions for pills of a lower dosage. While these three scrips *together* totaled sixteen hundred pills, there was no evidence that Snowden had ever exceeded his established prescribed daily dosage or that he resold or diverted any of the pills he brought home from the pharmacy on that occasion.[54]

The exaggerated figure, however, successfully conveyed the message that high-volume prescriptions could not be considered professional practice. The prosecutor argued that these showed Hurwitz not to have been acting "in good faith," and Judge Leonard Wexler accepted this contention, instructing the jurors not even to consider the question of "good faith" in weighing the evidence. The jury accordingly convicted the defendant on all fifty counts. The foreman told the *Washington Post* that, since a prescription for sixteen hundred pills was "beyond the bounds of all reason," he had been forced to conclude that Dr. Hurwitz was not a "legitimate" medical practitioner. Hurwitz was sentenced, on April 14, 2005, to twenty-five years in prison and a $2 million fine. The same day, Karen Tandy told a press conference that he was "no different from a cocaine or heroin dealer peddling poison on the street corner," an analogy supported by the sixteen-hundred-pill prescription; she famously held up a plastic bag to show the assembled reporters just how many pills that was.[55]

The leaders of the pain management community feared "collateral effects": that the stigmatization and criminalization of high-volume opioid prescriptions endangered the professional identity not only of Hurwitz, but of many of his colleagues.[56] They rallied behind him and helped to secure the pro bono services of two leading criminal lawyers in appealing his conviction. The U.S. Court of Appeals for the Fourth Circuit ruled on August 22, 2006, that the trial judge had been in error in not allowing evidence of "good faith" practice of medicine. That court overturned the 2005 conviction and ordered a new trial, which began on March 26, 2007.[57]

William Hurwitz had by then been in jail for two and a half years, but he had not been silent. His statement of his own case was published in *Pain Medicine* in 2005. He argued that drug enforcement tactics directed toward physicians threatened to undermine patient confidentiality and the physician's "traditional ethical commitments" to the patient. Moreover, they were ineffective in curbing illegal traffic in prescription drugs. "There are simply not enough 'bad' doctors" to account for the full extent of recreational use of prescription drugs. Further, although such usage, according to the National Survey of Drug Use and Health, had risen rapidly in the early 2000s, abuse and dependence levels remained much lower; "mere exposure to pain relievers among those using the medicine for non-medical purposes does not lead to abuse or dependency in the majority" of cases. Most illicit users, Hurwitz wrote, were "adolescents [who] typically obtain prescription drugs from peers, friends, or family members."[58] He and his supporters presented opioid prescription practice for pain management as both *justified* by traditional medical ethics and as *detached* from the real problem of illegal drug use.

In his second trial, Hurwitz enjoyed a stronger defense and a more sympathetic judge, Leonie Brinkema, who threw out twelve charges, including the bodily injury counts. Two expert witnesses, Drs. Russell Portenoy of Beth-Israel Medical Center in New York and James Campbell of Johns Hopkins, stated firmly that his prescription record was "within the bounds" of good professional practice. When a prosecution witness, Dr. Robin Hamill-Ruth of the University of Virginia, characterized the Hurwitz prescriptions as "illegal and immoral," the defense countered by presenting testimony from her former patient, Kathleen Lohrey, who suffered from migraine headaches. Hamill-Ruth had prescribed the anxiolytic drug Bu-Spar, the side effects of which include headaches, for Lohrey. The patient wrote of her experience, "This of course is not the first time that I have been treated as a 'nut' or a 'junkie.' "[59]

The prosecution also revived the issue of enormous pill counts. One witness, FBI agent Aaron Weeter, presented a detailed chart of some of the larger Hurwitz prescriptions. "Would you agree that . . . we can learn nothing very important from the pill count alone?" defense attorney Lawrence Robbins asked him. Weeter said that he was "not qualified" to answer that.[60]

Hurwitz himself testified for two days and again admitted that several of his "misbehaving patients" should have raised red flags when they reported losing prescriptions and requested premature refills. His justification was that he had been governed by his duty to believe and provide care for his patients. "Ultimately, pain is what the patient says it is."[61] In light of these admissions, the jury found him guilty on sixteen remaining counts of drug trafficking. Judge Brinkema, describing the bulk of his practice as legitimate and supported by "respectable medical literature and expertise," handed down a greatly reduced sentence of four years and nine months on July 13, 2007.[62]

The DEA meanwhile published a subsequent ruling on "Dispensing of Controlled Substances for the Treatment of Pain" in September 2006. This document repeats the same factors identified in the 2004 interim policy statement as evidence of "recurring patterns indicative of diversion and abuse" but emphasizes that "the entirety of circumstances must be considered."[63] A "Final Rule," in November of 2007, approved the issuing of multiple prescriptions, up to a ninety-day supply, "for a legitimate medical purpose . . . in the usual course of professional practice."[64] Although the DEA's Web site has continued to highlight the arrests of physicians who write high-volume prescriptions, the press releases generally identify other suspicious factors as well: that the prescriptions were issued over the Internet, without physical examination, and/or for nonexistent conditions.[65]

But the Hurwitz case is not an isolated episode. Other physicians who followed similar prescribing patterns received less support and suffered more drastic fates. Ronald McIver, a South Carolina physician, took an aggressive approach to pain treatment, working to reduce his patients' pain not to a moderate 5 on the standard 0–10 scale, but to a mild 2. To do this, he prescribed what the prosecution's chief expert described as "excessive" and "just extremely high" dosages. Some of his patients sold or abused the opioids he prescribed, and one, who suffered from congestive heart disease, died of an apparent overdose. McIver was convicted in April 2005 on nine counts of distribution and one of dispensing drugs that led to a death and sentenced to thirty years. His appeals were all rejected.[66]

The behaviors of both Hurwitz's prosecutors and his champions suggest that the cultural meanings of the opioid prescription in practice continue to overspill legal and medical guidelines. Physician misconduct and inappropriate prescribing were of course only one among multiple avenues of prescription drug diversion. But multiple and high-volume prescriptions served as measurable flags to the regulators that the physician had stepped outside the safer bounds of prescribing practice, signals that there might be a story to be unearthed. Was this doctor a criminal, a fool, or, as some might characterize William Hurwitz or Ronald McIver, a saint or a Don Quixote? However he was judged, there was a loss of legitimacy and professional identity. The DEA has stated that its ultimate standard for criminal misconduct is whether or not the physician has "broken the link" between the drug and a legitimate medical condition.[67] However, "broken links" are not readily identifiable without chart audits. High-volume prescriptions continue to be easier to track.

The pain management community has recognized the potential dangers of writing opioid prescriptions for a chronic pain patient who may be at risk for abuse or diversion; the professional literature is thick on this subject.[68] Hurwitz and his defenders chose to reestablish medical authority over opioid prescribing by characterizing the problem as a medical challenge to be met within professional practice but in a neutral zone apart from the drug wars.

But in defining the actual practices that medical professionals should use in managing patients with opioids, the pain community leaders have recently departed from the patient-centered guidelines described by earlier researchers Houde, Rogers, Saunders, and Twycross. Instead, they outline the kinds of provisions which so demoralized Mary, the arthritis patient: requiring extra effort by limiting prescriptions to one pharmacy and office visits to a single prescriber at specified intervals, limiting prescriptions to weekly or biweekly dosages, "enumerations of behaviors" that may terminate the patient's access, and, not least, "pill counts." Although "theoretically, opioids have no maximum or ceiling dose . . . there is little evidence to guide safe and effective prescribing at higher doses."[69] To avoid professional marginalization and loss of authority, the physician is urged to reassert parental authority and surveillance over the potentially misbehaving patient. The stigma of the chronic or high-volume opioid prescription, it seems, must attach to someone.

Conclusion

The Harrison Narcotic Act and the legal and regulatory actions that shaped its enforcement established in U.S. law a tension in medical practice that reflects the dual identity of the opioid drugs themselves. On the one hand, there is legitimate medical prescription for the relief of severe pain; on the other, there is illegitimate prescription for the management, or maintenance, of addiction.[70] Both practices are sited in the medical office in the hands of the registered prescriber, and only the parameters of her practice—the volume, dosage, and duration of the prescription, the attendant guidelines and records of the transaction—define the line between legitimate and illegitimate. The registration requirements created by the Harrison Act make those parameters highly visible to professional and regulatory observers,[71] but they do not define where the line actually falls.

From the 1930s through the 1970s, the professional consensus on where that line fell appeared to be quite clear. Legitimate pain prescriptions were written for the relief of short-term pain—postinjury, postoperative, or in the terminal stages of disease—and generally limited to textbook-defined dosages (15 mg.). There is some evidence that many patients may have found these limited dosages inadequate to control their pain. But higher dosages, or longer-duration prescriptions for chronic pain, crossed the line. The patient who asked for more medication risked the stigma of the addict, the "delinquent" who had no place in mainstream society. The prescriber, as the professional authority, controlled the pain prescription to separate the momentarily weak from the permanently depraved. Only those near death, already separated from normal life, could be allowed extended high-dosage opioids for analgesia.

In the 1970s and early 1980s, a new group of pain management specialists argued for a new set of pain prescription parameters and a shifting of the legitimacy line to allow for higher dosages. This goal was accomplished in the case of cancer pain, again for a group of patients who were already seen as set apart from society. However, for patients suffering from severe chronic noncancer pain and their physicians, the definition of new parameters offered both hope and frustration, relief and risk. The prescription and its definition of legitimate opioid use became contested and acquired new layers of meaning. When the once clear and bold line between legitimate and illegitimate use became blurred, the prescribing physician found himself having to define and defend his own identity in terms of where he drew the line in his own practice. The dilemma was only exacerbated when, in response to the specialists' call for more effective pain medications, new prescription opioids were developed and marketed and, by the late 1990s, widely abused. The Substance Abuse and Mental Health Services Administration reported in 2010 that some 35 million Americans had misused prescription opioids at some time in their lives and that treatment admissions for the problem had increased nearly 600 percent between 1998 and 2008.[72]

Where the physician had formerly used the limited-dosage opioid prescription to restrict his patient's pain relief within acceptable bounds, to underscore the line between addict and legitimate user, and thus to reinforce his professional role, he now had to find alternative ways of shaping his practice to accomplish these goals. Some doctors, of course, simply refused to prescribe opioids at all or continued to follow the old guidelines. Others, like Mary's several doctors, created rules and restrictions that reinforced their professional authority and their definitions of acceptable use by stigmatizing their patients. The patients were allowed higher dosages, but the new prescription practices defined them as illegitimate and irresponsible users who had to be monitored and controlled. Patients such as Mary often responded by trying to subvert the doctor's practices and developing alternative prescription practices to obtain the pain relief they needed.[73]

Still other physicians tried to negotiate prescription strategies that acknowledged patients as having legitimate need for analgesia and appropriate ability to use opioids responsibly, reflecting the approaches of Houde and Twycross. In some cases, as with Mary and the doctor she finally found to treat her, or with the patients described by Kathleen Foley and her colleagues, these strategies were successful for both patient and physician. However, physicians such as William Hurwitz, who liberalized their opioid prescription practices, ran the risk of attracting illegitimate users—patients who diverted the drugs for recreational use—and the surveillance of medical boards and drug enforcement agencies. The intention to provide medical relief for pain was not sufficient. If physicians were seen as not exercising their professional authority adequately to demarcate legitimate use, the stigma of irresponsibility and illegitimacy fell on them. By 2010 the pain management specialists had disseminated a new set of prescription practice guidelines to aid physicians in defining legitimate use and in reasserting control over the wayward patient.

The opioid prescription has become a delicate and often hard-to-wield tool, used not simply to provide relief for severe pain, but to carve in each case the fine line between rational and irrational, controlled and uncontrolled, right and wrong. Medical judgment alone is not a sufficient guarantee, intolerable pain not a sufficient justification. Opioid use, in American society, is a story of failure—to tolerate suffering, to resist temptation, to manage the patient. Stigma must be assigned, and the prescription itself determines where it will fall.

CHAPTER 9

Busted for Blockbusters

"Scrip Mills," Quaalude, and Prescribing Power in the 1970s

David Herzberg

It was the fall of 1976 and Martin Siegel had problems. Many of them. And as he explained to his friend Robert Rosenberg one evening, at one level or another all of them registered on a single scale: money. Money for alimony. Money for child support. Money to support the extravagant tastes of his new wife. And not least, money to support his drug habit. Siegel had been abusing drugs for decades, but his addiction to prescription narcotics such as Demerol and Talwin had taken a turn for the worse in the mid-1970s.

Martin Siegel was also a doctor. This made his problems better but also considerably worse. Better because even though his practice was foundering, he still earned close to one hundred thousand dollars a year—less than he was spending but a lifeline nonetheless. Worse because his prescription pad gave him such easy access to the drugs that were killing him. And worse because that prescription pad also helped him make friends with people like Robert Rosenberg—a drug friend and a small-time crook who supplied Siegel with "street" drugs like cocaine in return for prescription sedatives, stimulants, and narcotics.

Rosenberg had the perfect solution to Siegel's money problems: they should open a diet clinic. Why a diet clinic? Because diet clinics prescribed amphetamines, and people taking amphetamines had trouble sleeping at night. They needed sedatives—sedatives like the prescription drug methaqualone, popularly known as Quaalude and highly prized in America's latest drug counterculture. Rosenberg and a friend of his, an accountant named Gary Ellentuck, would supply the patients, run the business, and direct customers to a friendly pharmacist. Another friend, small-time mobster Herman "Woody" Witt, would bankroll the operation. All Siegel would have to do was put his name on the incorporating papers and sign the prescriptions, and the money would flow in.

That is how Dr. Martin Siegel became what a CBS News exposé later called New York City's "King of 'Ludes.'" Business boomed after the clinic opened in January 1977, and Siegel wrote thousands of prescriptions. In October they moved into a larger, nicer suite, and by March 1978 Witt was ready to take on a new doctor, a friend of Siegel's named Lawrence Glass. Glass soon made "partner" and was taking a 50 percent share of the profits. "If Siegel is the King," Rosenberg and Ellentuck kidded him as they watched the CBS exposé together, "you must be the Prince of 'Ludes!'"[1]

But Siegel's addiction was getting worse; Glass had actually been hired to take some of the burden off his increasingly erratic friend. His promotion to "partner" marked the moment when Siegel became entirely incapacitated. This advancement must have made Glass a little nervous, though, since he too was now addicted to Quaalude and cocaine. By the time the diet clinic closed in January 1979, Glass, like Siegel, had become a liability.[2]

If the clinic's physicians were faring poorly, however, the business as a whole was thriving. Already in late 1978, Rosenberg and Ellentuck had opened a franchise of the clinic in Chicago, and their profits rose further in 1979 when they abandoned diet clinics in favor of "stress clinics" that peddled Quaalude exclusively. Under the new model, Witt opened offices in Boston; Fort Lee, New Jersey; Los Angeles; and other cities. The flagship clinic—the Manhattan Center for Research into Stress & Pain Control—occupied a luxurious suite on Madison Avenue. With both Siegel and Glass sidelined, Witt hired a new front man: a relatively respectable physician from South Africa, Ronald Asherson, who had been the associate medical director for Pfizer in New York. To write the actual prescriptions, he employed dozens of physicians in part-time capacities.

But Asherson, like Siegel and Glass, was a drug addict. And while it had been slow in coming, state and federal authorities had begun to take an interest in the "King of 'Ludes'" and his colleagues. They started to close down the clinics one by one, until the entire operation was defunct in 1981. Everyone except the physicians then went on trial for drug trafficking and tax evasion, including Rosenberg, Ellentuck, Witt, the "friendly" pharmacists, and even receptionists and accountants. Siegel, Glass, and Asherson were drug-free and cooperated with the state to avoid prosecution. Siegel and Asherson had to fly in for the trial; they had fled the country to escape addiction and related troubles. Whatever profits had not already been spent on drugs were now being spent on lawyers, but to no avail: all defendants were convicted on multiple counts. It was the largest prosecution of its kind to date, a highly touted victory for the Drug Enforcement Administration (DEA).

What are we to make of this story? Siegel, Glass, and Asherson are not the kind of physicians who usually appear in histories of medicine. They were not therapeutic pioneers, they made no important discoveries, they wielded little political power, and they were not influential with their peers or patients. They were addicts, people at the margins of the profession selling their prescribing power to all comers. Nor were their "patients" the typical subjects of medical historiography. They were not diagnosed with an illness, and in any case they certainly were not receiving treatment for the most obvious illness they might have been suffering from (addiction). They were medical interlopers, visiting the medical system not as patients desiring therapy but as customers seeking convenient access to drugs.

And yet Quaalude clinics were less marginal than they first appear. The Manhattan Center alone employed dozens of physicians and saw thousands of "patients," and it was hardly the only clinic of its kind, even in New York City alone. Indeed, legal prescriptions were one of America's largest sources of abused drugs. Two out of every three drug abusers seeking treatment or visiting emergency rooms had been using prescription drugs, many of them supplied by the 250 million or more doses estimated to have been "diverted" annually from the licit to the illicit market. A 1979 study by the National Institute on Drug Abuse found that nonmedical prescription drug use was more widespread than all illegal drugs other than marijuana.[3] Scrip mills like the Manhattan Center were only one significant source of this illicit market, but as relatively formal and organized commercial outposts, they offer a window into a hidden world of disreputable and outright criminal prescribing. Such illicit prescribing has been persistent enough, and substantial enough, that it needs to be recognized as a structural—if undesirable—element of the medical system.

Incorporating illicit prescribing into the postwar medical system in this way adds a new dimension to recent scholarship on prescription medicines as commercial as well as therapeutic objects. Historians have explored how new drugs, sophisticated marketing, relatively weak regulation, and an intensifying consumer culture all contributed to a soaring use of prescription pharmaceuticals since midcentury.[4] But where does the act of prescribing itself fit into the story? For good reason prescribing has been analyzed as the object of external forces such as market research, marketing, patient demand, and so forth—it is a good place to gauge the impact of commerce on therapeutics. But as scrip mills demonstrate, prescribing itself can be a commercial act, not just a therapeutic act influenced by commerce. There were, in fact, vast quasi-licit markets for pharmaceuticals that depended on physicians' prescriptions just as much as the legitimate medical markets that grew alongside them. These need to be incorporated into any full accounting of the enormously profitable markets that came to be built around physicians' prescribing monopoly in the postwar era.[5]

Scrip mills also remind us that prescribing belongs in the history of addiction and drug abuse. In the extensive historiography of drugs in America, physician prescribing usually appears in the nineteenth and early twentieth century, after which attention turns to illegal "narcotics" and the social politics of antidrug campaigns. These social politics almost always involved the deployment of state power to police socially marginal populations, for example, Mexican immigrant laborers or inner-city African Americans.[6] An unintended result of this scholarly focus is to reinforce distinctions between drugs and medicines and to reinforce presumed connections between addiction and nonwhite or poor social groups. Quaalude clinics and other scrip mills present an alternate geography of drug abuse and addiction: one that was inhabited by relatively privileged people and professionals; one that was specifically tailored to benefit from the class, race, and gender biases of antidrug campaigns; and one that faced far less intrusive and punitive policing. Exploring the world of scrip mills helps us to crack open a much broader and more complete history of illicit drug use.

Finally, scrip mills also became an important site of government intervention into medical practice. David Rothman and others have examined how lawyers, bioethicists, and government regulators became "strangers at the bedside" of medical decision making as physicians lost professional autonomy in the 1960s and 1970s.[7] Pharmaceutical historians have added an important commercial dimension to this narrative by exploring the increasing influence of drug companies and drug marketers in medical education and practice. The scrip mill story shows that prescription drug crises also led to new kinds of physician oversight: expanded federal surveillance to distinguish between legitimate and illegitimate prescribing and a strong law enforcement system anchored in the DEA to police the divide. By allying with antidrug crusaders, federal drug reformers notched rare victories in the quest to impose new controls over what they saw as a pharmaceutical system run amok. These victories revise the more familiar tale of successful corporate and professional resistance to drug regulation in the postwar era.[8]

The Quaalude clinic story, then, provides an opportunity to examine physicians' prescribing power as it was entangled and redefined in the twentieth century's war on drugs. It reveals how prescribing could be a commercial as well as a therapeutic act; it highlights how prescribing remained central to American drug cultures long after the criminalization of narcotics; and it shows how prescribing created new openings for tighter federal scrutiny and policing of medical practitioners and their patients.

Context: The Licit Drug Wars

The Quaalude clinic trials hinged on a single central issue: what is the difference between "prescribing" and "drug trafficking"? As both defendants and prosecutors recognized, this was no simple question—and that in itself requires some explanation. After all, for the better part of a century, prescribing had been defined as the very antithesis of trafficking. Prescribing was noncommercial, while trafficking was profit-driven; prescribing was therapeutic, while trafficking fed abuse and addiction; prescribing was the exclusive hallmark of a respectable profession, while trafficking could be practiced by anyone; and, since the Harrison Narcotic Act of 1914, prescribing was only lightly regulated by the state, while trafficking faced criminal sanctions enforced by dedicated police agencies. And yet codifying and enforcing these seemingly clear and bright lines turned out to be quite difficult. Drug laws represented an early "stranger at the bedside" of medical decision making, but they were hampered by the need to respect physicians' autonomy and to accommodate high-volume commerce in prescription medications. This left the regulatory environment with enduring structural niches for illicit prescribing—an unwanted but seemingly ineradicable part of the medical system.

The first effort to clamp down on those structural niches emerged in the 1940s and 1950s, in response to medical and popular fear about abuse of medicines like barbiturate sedatives, amphetamine stimulants, and synthetic narcotics not covered by the Harrison Act. Like early antinarcotic campaigns, this early effort assumed that physicians were part of the solution, not part of the problem. As Nicolas Rasmussen notes in chapter 1, federal regulators limited such "dangerous drugs" to physicians' prescriptions in 1951. Unlike the Harrison Act, the new regime did not include a "good faith" caveat: all prescriptions were presumed to be legitimate medical decisions. And, as Food and Drug Administration (FDA) historian John Swann notes, prescribers received virtually no attention in the agency's midcentury campaign against illicit trafficking in barbiturates and amphetamines.[9]

The prescription-only regime did not bring an end to problems of abuse, however. Indeed, the problem only seemed to accelerate as new drug discoveries and an increasingly market-oriented pharmaceutical industry led to soaring use of prescription medications. By the mid-1950s a parade of hair-raising scare stories and popular outrage pushed Congress to hold hearings on the "menace" of barbiturates and amphetamines, but implacable industry and medical opposition successfully staved off further regulations.[10] In 1965, however, reformers finally notched a victory with the Drug Abuse Control Amendments. This law required physicians as well as pharmacists to keep records for certain classes of sedatives and stimulants; it allowed physicians to authorize a maximum of five refills on the same prescription; and it created a (mostly token) enforcement agency within the FDA.[11] But there were no criminal penalties, and even states that had clear authority to discipline physicians for noncriminal offenses rarely used it. Then too, putting a medicine on the list of controlled drugs was no easy task in the first place, since it often required a lengthy case-by-case battle against drug manufacturers and the medical profession.

Drug reformers gained a more important victory with the signature drug law of the era, the Comprehensive Drug Abuse Prevention and Control Act of 1970. This law established a much stronger enforcement agency, the DEA, to oversee a new Schedule of Controlled Substances that included both classic "street" drugs like heroin and very widely used prescription medicines like Valium. The drugs were organized into five classes depending on their medical value and abuse potential. To make, prescribe, or sell the drugs in this Schedule required a special license from the DEA. Record-keeping was tightened, and criminal penalties (including jail time) were imposed for infractions. For the "worst" drugs, those in Schedule I or II, the DEA could even impose production limits at the manufacturer level.[12]

These were significant new constraints on physicians' therapeutic autonomy, and they effectively created a second locus of federal control over pharmaceuticals outside the FDA. Housed in the Department of Justice instead of Health and Human Services and benefiting from the persistent cultural power of the "drug war," this alternate regulatory center enjoyed authority that pharmaceutical reformers elsewhere could only dream of (see, e.g., chapter 2, by Scott H. Podolsky). They could, for example, limit manufacturing to medical need as determined by experts, aggressively phase out dangerous drugs even if they had some medical value, and track physicians' prescribing habits and punish them for excessive prescribing.

Despite its unprecedented regulatory reach, however, the 1970 drug law still included important compromises forced by the political power of organized medicine and the pharmaceutical industry. The DEA had virtually no authority to reject a physician's application for a license, for example, or to revoke an existing one. That power lay with state medical boards, which, the DEA complained, rarely used it; according to one study commissioned by the agency, nearly 80 percent of state medical and pharmacy boards did not consider violations of federal drug laws grounds for taking action against a physician's license.[13] Even criminal cases against physicians were problematic, because, the DEA acknowledged, medical expertise "necessarily [gave doctors] wide latitude in judgment." Proving that a physician had criminal intent required lengthy and expensive investigations that had to produce near-perfect evidence before wary federal prosecutors would take up the case.[14]

As a result, the DEA denied or revoked only slightly more than fifty licenses to prescribe controlled substances in the 1970s.[15] And this did not even count another issue: whether or not a particular drug was on the Schedule of Controlled Substances in the first place. Placing a drug on the Schedule was no easy task, given resistance from manufacturers. The immensely popular minor tranquilizer Valium, for example, had been a target since the first 1965 law, but vigorous industry defense kept it off the Schedule until 1973—not accidentally, that was when its patent ran out.[16]

By the 1970s, then, the federal government had taken important and unprecedented steps toward acknowledging and regulating the gray areas between trafficking and prescribing. These steps marked the entrance of important new "strangers at the bedside" of medical decision making. But they were also shaped by a desire to accommodate physicians' traditional autonomy and the dictates of commercial markets for pharmaceuticals. Prescribers faced the lightest regulatory oversight of all the links in the chain of pharmaceutical distribution. Pharmaceutical companies continued to enjoy broad freedom to market their products to physicians and the public, helping to normalize the use of potentially abusable drugs. They were also able to delay the Scheduling of their most prized products long enough that their market niche—licit and illicit—could be well established. These compromises ensured the continued existence of illicit markets even as they constrained them.

Illicit Prescription Markets: The Example of Quaalude

Not surprisingly, prescription drug abuse survived and in many cases flourished even as the new regulatory regimes were being built. Perhaps the most prominent example was amphetamines, which, as Rasmussen has shown, were widely abused beginning as early as the 1940s by soldiers, jazz musicians, "beatniks," college students, avant-garde artists, and others, adding up to an astonishing 15 million Americans by the late 1960s. The FDA estimated that half of all amphetamine pills in the 1960s were dispensed without the benefit of a prescription. And, as Swann has noted, barbiturate abuse was so pervasive as early as the 1950s that the FDA devoted a majority of its enforcement budget to illicit pharmacy sales of the drugs. Barbiturates and tranquilizers continued to be nonmedical drugs of choice in later decades; in 1976 and 1977, for example, the federal Drug Abuse Warning Network (DAWN) reported that these drugs accounted for fully one-third of all drug-related cases seen in the nation's emergency rooms—more than twice as many as heroin and far more than cocaine, marijuana, and hallucinogens combined.[17]

Quaalude was a notable but not dominating part of this cornucopia of uppers, downers, and narcotics. In 1979, its fourteenth year on the U.S. market,

Quaalude was ranked by DAWN as the eleventh-most-common drug seen in emergency rooms, behind perennial leaders such as Valium and heroin but ahead of any single type of barbiturate.[18] Its notoriety and its complicated path through the changing regulatory system make it a useful case study that illustrates both the new powers and the continuing restraints federal authorities navigated as they sought to police the lines between trafficking and prescribing.

Methaqualone (as the generic chemical was named) was first synthesized in 1951 by Indian researchers looking for new analgesics. Recognized as a hypnotic four years later and unprotected by patent, the drug was developed commercially in Europe and Japan by the early 1960s. William H. Rorer Company introduced the drug to America as the sleeping pill Quaalude in 1965. The timing was no accident: as noted earlier, 1965 was the year that long-simmering public concern about addiction to barbiturates (the most widely used sleep medications) had been enshrined in the Drug Abuse Control Act, which imposed new limits on their use. Methaqualone was free of such restrictions. Despite worrying evidence from Germany and Japan, it was a new drug in the United States with no troubling track record. Rorer had not undertaken any tests for addiction potential, nor had the FDA requested any—in fact, addiction liability was not taken into consideration in the drug's approval at all.[19] The best federal regulators could do was to require a label that stated, "no cases of addiction have been reported; however, addiction potential has not been established."[20]

Rorer made the most of the situation, advertising its sleeping pill as a non-addictive alternative to the increasingly demonized barbiturates. A 1971 ad, for example, referred to the drug as "non-barbiturate Quaalude" and took pains to point out, again, that it was "chemically unrelated to barbiturates" and produced no "hangover" or "'drugged' after-effects in the morning."[21] One physician recalled receiving boxes of free Quaalude samples in the mail, with instructions suggesting they be "disperse[ed] to your patients as nonbarbiturate, nonaddictive, sleeping pills."[22] Rorer and a small number of competitor companies reaped decent, though perhaps not spectacular, rewards: $7 million, with more than 100 million doses prescribed in 1972.[23]

Not all of this use was legitimate. In fact, by all accounts, Quaalude quickly became popular with the youth counterculture (see fig. 9.1), among whom, reports indicated, it was called "the love drug" for its ability to ease social

Figure 9.1. "Captain Quaalude" T-shirts and other Quaalude-themed fashion wear (belt buckles, jewelry, etc.) provided evidence of the drug's sudden popularity among American youth. Reprinted as Exhibit No. 5, "Captain Quaalude T-Shirt," in U.S. Senate, Subcommittee to Investigate Juvenile Delinquency of the Committee on the Judiciary, *Methaqualone (Quaalude, Sopor) Traffic, Abuse, and Regulation*, 93rd Cong., 1st sess. (Washington, DC: Government Printing Office, 1973), 99.

inhibitions and prolong sex. Recreational users would ingest the drug and then purposefully stay awake through a period of drowsiness, ultimately reaching a pleasurable "high" that could be (dangerously) strengthened with the addition of alcohol.

The illicit market for Quaalude was not entirely separate from the drug's licit markets. At least some illicit users, for example, appear to have been attracted by Quaalude's reputation as nonaddictive. And even "street" sales usually hinged on brand names, with dealers and buyers alike focused on the assumed purity, reliability, and safety of pills stamped by a legitimate manufacturer and backed by FDA approval. Ironically, such assurances appear to have been as important in illicit circles as in the medical realm.[24]

Authorities' response to this first wave of abuse was minimal. The FDA revised its required warning label in 1970, advising that "psychological dependence occasionally occurs but physical dependence [has] rarely [been] reported."[25] But marketing campaigns for the drug continued unabated, and both use and abuse grew. Within a few years, a classic moral panic erupted, with medical observers like the *New England Journal of Medicine* and the *Journal of the American Medical Association* and popular media like the *Washington Post* and *Rolling Stone* describing a "silent but pervasive" Quaalude "epidemic" that was "all over the place and getting even bigger."[26] By 1973 Quaalude hearings were under way in the U.S. Senate, where one witness accused pharmaceutical companies of being "as much a pusher as the man on the corner who is selling a few bags of heroin."[27]

Fearing strict new regulations, drugmakers began to voluntarily implement some new restraints: Parke-Davis (which sold a competing brand of methaqualone) stopped sending free samples—Rorer eliminated only "unsolicited" samples—and heightened security at manufacturing plants.[28] It was to no avail, however. After relentless prodding from congressional antidrug crusaders and their allies, the newly created DEA placed methaqualone on Schedule II—the second-most-tightly-restricted category. The agency aggressively lowered quotas for domestic production, and medical use plummeted.[29]

Illicit use of the drug did not disappear, however. An illicit market for Quaalude had already been established—smaller, perhaps, than the market for amphetamines, barbiturates, and narcotics, but still substantial. This market remained closely linked to licit supply chains. Quaalude was difficult to make, so there were few bathtub-manufactured pills like those found in the early underground amphetamine trade.[30] There were some factory-produced counterfeit pills, but only a small minority. More common were supply chains drawing from legitimate markets: stolen prescription pads, misused free samples, stolen or fraudulently purchased pills (such as the six hundred thousand pills "diverted" from a Parke, Davis warehouse in Detroit in 1972), "doctor shopping" for legitimate prescriptions, and pharmacists' sales of the drug without prescriptions.[31]

Taken together, these routes provided a steady supply of Quaalude for illicit markets. The result was a profusion of outlets for Quaalude sales. Notable venues in major cities like New York were so-called juice bars, or night clubs that did not serve alcohol (and thus paid for no liquor license) but which levied a steep cover charge for patrons who preferred the Quaalude discreetly sold there.[32] In less glamorous settings, many thousands of pills were sold sans prescription by shady pharmacists, passed informally between friends, or—to judge by reports of pills seized in police busts—vended alongside heroin, marijuana, and cocaine by "street" dealers. And then there were "scrip doctors" like the notorious Max Jacobson, best known for his regular treatment of President John F. Kennedy and other luminaries with "vitamin shots" liberally dosed with amphetamines.[33]

Finally, of course, there were scrip mills like those run by Herman "Woody" Witt: entire clinics devoted to the sale of prescription drugs for nonmedical use. These represented only one retail option in the vibrant and diverse landscape of illicit prescription drug sales, and the scale and formality of their organization made them distinctive. And yet these qualities also make them a useful window into the shadowy world of illicit prescription markets. Scrip mills handled a large volume of sales and involved many traffickers and clients. They were also visible targets for state and federal authorities, whose investigations and prosecutions left long and valuable paper trails. Examining their well-documented rise and fall reveals how interactions between physicians, patients, the state, and pharmaceutical companies defined the highly contested act of prescribing.

The Prescribing Business

First, it will be helpful to get a sense of how scrip mills worked: how professional autonomy, pharmaceutical commerce, and drug-war biases enabled entrepreneurs like Herman Witt to formally incorporate large-scale drug-trafficking operations whose every transaction was reported, in triplicate, to state and federal authorities.

Scrip mill success hinged on mimicking the forms of medical therapeutics just enough to claim the special protections afforded to physicians' judgments about prescribing. Thus, when clients came to Witt's first diet clinic at 35 East Thirty-fifth Street in New York City, they acted and were treated much like patients: they reported their medical problems in writing and undressed for a physical exam; they were weighed, measured, and examined; and every three months they were given an order for a range of blood tests and an EKG.

Upon payment, they signed two additional forms acknowledging that they were receiving "potent" drugs and promising to follow the doctor's instructions carefully. The whole process was remarkably efficient; Glass and Siegel together saw between thirty-five and forty patients a day, and the clinic was only open between 3:00 and 7:00 p.m.[34]

This efficiency was possible because of the uniform therapy the clinic offered: 97 percent of visits (8,556 out of 8,673) ended with a prescription, and every patient got exactly the same prescription, for the same number of pills, to be taken in the same manner.[35] Insofar as possible, these prescriptions followed the letter (if not the spirit) of the *Physicians' Desk Reference* (*PDR*) guidelines. For example, when the Manhattan Center first opened in 1979, doctors were told to prescribe exactly fifty-three tablets per month: one-fourth tablet three times per day, and one tablet at night.[36] When the *PDR* changed its recommendations, clinic doctors changed their prescribing, too, dropping the monthly number to forty-five and adjusting their patient advice accordingly.[37] Patients were given two refills, which allowed them a questionable but not illegally long Quaalude regime of three months, after which they were required to come back to the clinic and be evaluated before receiving more. (Here they departed more from *PDR* standards, which strongly frowned upon long-term use.)

This prescribing uniformity was no accident, of course. It was policy, and several part-time physicians at the Manhattan Center had been fired for refusing to follow it. Dr. Peter Sarosi, for example, survived only two days on the job. The first day, he prescribed for seven of his nine patients; he was fired after repeating this subpar performance—seven out of eight patients—on day two. He had also made the curious (to his employers) choice of prescribing less than the appointed amount, saying that he wanted to "wean" patients off the drug. For this he had received a lecture from Sally Ungar, the receptionist, who told him flatly, "We prescribe Quaalude here for stress." Asherson fired him shortly afterward.[38]

Joseph DiBeneditto—a part-time doctor and DEA witness who worked at the clinic wearing a recording device—told the jury a similar tale. His first patient was a twenty-eight-year-old woman complaining of minor anxiety owing to demanding studies at the Fashion Institute of Technology. He recommended Valium instead of Quaalude, and she exploded at him. She had only come back to the clinic at all, she said, because she had been promised Quaalude. When he refused to budge, she stormed out. DiBeneditto was examining the next patient when he was interrupted by Ungar, who upbraided him and demanded a prescription. He refused. He made it through three more patients before Ungar returned and, claiming to act on orders from someone "high up," paid him for his two hours and fired him.[39] The cash-for-prescription regime was so strict that patients who did not receive Quaalude for whatever reason were not charged for their visit, even if they had been examined by a physician.[40]

Once they had their prescription, clinic clients faced another problem: many reputable drugstores would not stock Quaalude or any other Schedule II drug, and even if they did, they dispensed them with great caution.[41] So clients had to be steered toward cooperating pharmacies. Luckily for the clinics, Rosenberg had already lined up a friendly venue, Edlich's, where he had been taking his own illicit prescriptions for years. Rosenberg had spent enough money and time there to befriend Ben Rose, store manager and son of the owner. Rose agreed to accept the new stream of patients, and Rosenberg made sure that clinic clients were given a card for Edlich's on their way out.[42] The results were impressive: Edlich's went from filling a paltry thirty-five Quaalude prescriptions annually to filling more than one thousand the following year—over 90 percent of which were written by Siegel himself. By 1980 the pharmacy had become the single largest purveyor of Quaalude in New York State, dispensing nearly ten thousand prescriptions.[43]

The rigid and prolific prescribing regime was not the clinic's only suspicious element. Despite the range of seemingly irrelevant EKGs and blood tests (holdovers from the diet clinic days), for example, prospective patients were not given a drug screen. When Dr. Cary English interviewed for a part-time position and asked about this, he received a strange look and a pointed question: "Are you sure you want to work here?"[44] The clinics advertised by giving business cards to young women, who would distribute them at bars, discos, and restaurants. When the Chicago diet clinic closed, its staff contacted patients who had been rejected as too skinny and invited them to visit the new stress clinic.[45] Friends, family members, clinic staff, and investors like Witt, Rosenberg, and Ellentuck were often "treated" for free.[46] And, perhaps most disturbingly, Ellentuck and others sometimes invited friends to come to the clinic, dress up in a white lab coat, and pretend to be doctors, "examining" young women patients before forging their prescriptions.[47]

This is a damning list of warning signs, certainly enough to persuade a jury to convict. Yet Witt's clinics had operated in plain sight for three years and had sold tens of thousands of prescriptions before authorities shut them down. Part of their success came from their outward conformity to the professional standards that earned physicians some measure of protection amid American drug wars. It also came from their clientele, who were carefully screened to exploit the race and class politics of the drug wars.

Witnesses at the trial repeatedly described clinic patients as white, middle class, young adults, and evidence appears to back them up. Dr. Peter Sarosi, for example, saw nine patients on his first day; six were in their early thirties, three in their late twenties; five claimed to be in sales.[48] This homogeneity was the result of careful patient selection. To avoid association with the youthful drug "counterculture," for example, no one under twenty years old was permitted as a patient. Also excluded was anyone who *looked* like a person who had drug problems (a charge that carried clear racial overtones, given American stereotypes about addiction) and, more specifically, anyone who directly requested drugs.[49] Those left were largely white-collar, middle-class professionals who were able to maintain appearances. The offices, as one receptionist told the jury, had a "certain decorum," "very nice" and "very professional," like other doctors' offices.[50]

A crucial part of maintaining appearances was being able to make a credible claim to be suffering from an illness that merited treatment with Quaalude. In this endeavor, too, the class and racial background of the clinic's clients aided powerfully. Anxiety, insomnia, and "nerves" of all sorts had long been associated with the white-collar world. For nearly three decades, pharmaceutical companies had been waging massive marketing campaigns that turned this linkage into profits by persuading middle-class Americans (and their doctors) that their every problem was a genuine illness worthy of drug treatment. By the 1970s mood-altering drugs were among the most commonly prescribed medicines in the legitimate market, led by the 130 million prescriptions written each year for bestselling tranquilizer Valium.[51]

This sizable licit market, and its medical and cultural justifications, helped shield clinics against charges of abuse and addiction. During the trial the defense often described the clients—drug abusers and addicts all—as suffering from "symptoms" like jitteriness, fear of failure, and domestic strife that, in more stereotypical "junkies" (i.e., nonwhite or poor), would more likely be seen as predisposing character flaws or even drug withdrawal. Such politically charged ambiguities kept alive a space (however tenuous) for respectable, medical addiction.

Consider a typical instance of suspect prescribing: the "treatment" of clinic receptionist Jamie Kahn. Giving Quaalude to office staff was surely a red flag. And yet Kahn's medical records from the center show that she complained to the physician of anxiety and indicated that she had been prescribed Valium in the past. She checked off "yes" to a range of stress-related items on the self-history form, such as "Are you often anxious?" "Are you often irritable?" "Jumpy?" "Jittery?" Her biofeedback documents also described her as "clench[ing] or grind[ing] teeth." Moreover, Kahn admitted under cross-examination that she had not wanted to have a physical exam or a blood test but that clinic physicians had insisted.[52]

Also questionable was the treatment of Ellentuck's friend and co-worker Steven Greenhouse—the one who said he had been invited by Ellentuck to "play doctor" with female patients. Even here there was potentially mitigating information: at the time of his treatment, Greenhouse had spoken to clinic physicians about his obesity and the stress he suffered because of extreme marital strife.[53] At the Manhattan Center, even doubtful doctors described the patient population as—in the words of one part-timer—"young executives, pretty tense . . . They were telling me they have a tough job, tense, they suffer from anxiety because they were always afraid they could not perform, they either have to fulfill a quota of selling something or they have to perform in the job well enough, otherwise they might lose it."[54]

Prosecutors called these sham complaints. But the defense pointed out that psychological problems like stress and anxiety were much more difficult to assess than a broken bone.[55] And, what's more, the predominance of these ailments helped explain what, at first glimpse, appeared to be an absurdly high percentage of patients who received prescriptions for Quaalude. These were stress clinics, after all, where people went to be treated for stress. Clinic records and trial witnesses both agreed that patients of the Manhattan Center *did* have psychological problems. Would it be strange, the defense asked, if the vast majority of people going to an optometrist came out with eyeglasses?[56]

Such explanations hardly redeemed the diet and stress clinics, of course, and in the end jurors were not fooled. (As a prosecutor noted, would you expect everyone to leave an optometrist's office with a prescription for the exact same lenses?)[57] But the end had been a long time coming, and in the meantime, the clinics had flourished—not because they were ingeniously well run (they obviously were not), but because they were protected in some measure by physicians' prescribing autonomy, the openly commercial nature of legitimate prescription markets, and the ubiquity of drug prescribing for "respectable" white, middle-class Americans. Scrip mills inhabited a borderland between medicine and drug trafficking, sharing enough elements of legitimate prescribing to avoid easy identification as criminal enterprises.

Dr. Feelgoods?

As prescribers, physicians were central to the scrip-mill borderlands. The Manhattan Center alone, for example, employed almost fifty different physicians in its eighteen months of existence. Who were these doctors? Why did they work there? What did they think they were doing? Evidence from the clinics suggests that these doctors were not necessarily open criminals and medical pariahs—although some were. The clinics also attracted ordinary physicians cycling through temporary or part-time employment at the beginning of their career or during career transitions. This explains how they could at once be marginal, fringe operations while at the same time handling such large quantities of drugs and employing such a large number of physicians.

Siegel, the original physician, certainly knew what was going on at the clinics, and he largely fit the bill of a marginal or even criminal figure disconnected from mainstream medicine. He told the jury that he had been abusing drugs since he received his medical degree in 1959, beginning with Dexamyl, a barbiturate-amphetamine combination.[58] By the time he befriended Rosenberg, he had become a more committed addict, informally trading prescriptions for drugs and desperately on the lookout for new sources of income. New York State drug authorities were already investigating him for irregularities in his office narcotics records.[59] When Rosenberg spoke to him about opening a diet clinic, both men knew exactly what he was talking about. At the trial, when the prosecutor asked him what the purpose of the first diet clinic had been, Siegel was blunt: "To make money."[60]

Lawrence Glass's path to the clinics was different, and it hints at the porous nature of the boundaries separating licit from illicit medicine. A generation younger than Siegel, he had received his degree in 1973 and had befriended Siegel shortly thereafter during his internship. A youthful idealist, Glass went on to run a mobile crisis intervention unit in Eugene, Oregon, where he worked with young people struggling with a wide range of serious drug problems. From there he went to Guatemala with the Red Cross to do relief work after a major earthquake. When Glass came back to the United States in 1977, he was at loose ends and looking for work. Siegel offered him a place to stay and a job at his diet clinic.[61]

Glass was skeptical. Like many of his generation, he smoked marijuana; but unlike Siegel, he was no addict. Nor did his career thus far suggest that making money was his primary goal as a physician. He nonetheless agreed to test the waters, on the condition that he himself would write no prescriptions; he would just see patients and advise Siegel.[62]

The no-prescribing arrangement did not last long. As Siegel became increasingly unreliable, Glass soon started writing prescriptions (for a higher salary) and took on more responsibility at the clinic. He instituted a raft of new policies, starting with the rejection of "very, very thin people" as patients.[63] He imposed stricter limits on amphetamine prescribing, and he hired a psychologist. At one point he got cold feet and actually tried to stop prescribing amphetamines and Quaalude entirely, but that was too much for the clinic's investors. Without Quaalude, Ellentuck told him, there would be no practice at all.[64] At this point, it seems, Glass stopped resisting: he signed hundreds of prescriptions, began to use cocaine and Quaalude himself, and supplied his employers with drugs as his own addiction deepened.[65]

As Glass too became a liability, Witt decided to close up shop. Through his Chicago clinic, he had already seen that stress clinics were far more profitable than diet clinics anyway; shifting to that model in New York would solve two problems at once.

As medical director of the new Manhattan Center stress clinic, Witt hired Ronald Asherson, a South African physician who had received his degree from Capetown Medical School in 1957. In 1975, after several years as medical director of two large Swiss pharmaceutical companies for southern Africa, he moved to Manhattan to serve as associate medical director for Pfizer.[66] Once in New York City, Asherson began to spend time in one of the city's countless drug subcultures, and his career trajectory changed suddenly. He left the respectable job at Pfizer in 1976 and began a period of clinic-hopping, including a stint treating obese patients with protein shakes. Over the next few years he switched jobs every six or eight months. He also began supplying individual addicts with amphetamines, which might explain how he befriended Ben Rose (of Edlich's Pharmacy). It was Rose, of course, who told Asherson about the new "stress clinic."[67]

Like Glass, Asherson initially attempted to impose some semblance of real therapeutics at the clinic—perhaps because he was nervous about arrest, or perhaps because of his own embattled sense of medical ethics. He knew it was a scrip mill, of course, but he was skeptical about providing only one—precisely one—form of treatment. He told Witt that the Manhattan Stress Center should also provide biofeedback, meditation training, hypnosis, and other healing modalities. Witt thought that was fine, as long as nothing interfered with the Quaalude prescribing.[68] So Asherson hired Richard Kuhns, a psychologist, and bought some biofeedback equipment.[69] Kuhns seemed a safe choice: as a psychologist rather than a psychiatrist, he had no MD degree and thus had no legal authority to write prescriptions or, by implication, to interfere with the clinic's prescribing regime.

But the equipment worked for only six months, and Kuhns caused trouble after all by unexpectedly trying to reject some patients for treatment with Quaalude. Witt fired him and replaced him with an even less dangerous employee: "Dr." Edward Silverg, who had no degree whatsoever, medical or otherwise.[70] Asherson himself did not last very long at the clinic either. Under investigation for his own prescription writing, he fled the country briefly and was arrested for illegal sale of prescription narcotics upon his return in March 1980.[71]

Siegel, Glass, and Asherson had each taken a different path to the scrip mills, but their stories shared certain characteristics. Each had accepted the job at points of personal and professional transition. They were all connected to greater or lesser degrees with New York's drug subcultures: Siegel right from the start, Glass through Siegel, and Asherson through his social network. But they were also, at the same time, real physicians who had worked at ordinary practices and—at least in the case of Glass and Asherson—fought to maintain at least some connection to the legitimate world of medicine. And because of their cooperation with prosecutors, they did not lose their medical licenses; Glass was already re-employed as a physician by the time of the trial.[72]

The ambiguous lines between prescribing and trafficking were only further muddied by another population of physicians: the parade of part-timers. The Manhattan Center alone employed forty-five of them. Most had found the job through classified ads; successful candidates, Witt told Asherson, were foreign medical school graduates, doctors who "seemed to be in need of money," and people who did not ask many questions.[73] Physicians testifying at the trial claimed to have been doing a routine job search, not looking to join a scrip mill. Peter Sarosi described his position at the clinic as only one of several part-time jobs he held at the time; he also ran physical examinations for the United Parcel Service.[74]

If the part-timers were not committed traffickers looking for illicit work, why did they accept the job? Not all of them did. Emily Cole, for example, declined the job after her interview and immediately contacted the state Medical Association and Pharmacy Board.[75] Another doctor, Gregory Drezga (from Yugoslavia), was appalled to learn from "Dr." Silverg that no drug screens would be given despite the use of Quaalude. After talking to a friend about his doubts, he called New York State authorities, who connected him with investigators. He went to work at the clinic in November 1981 but did so as a government agent wearing a "wire."[76]

But not everyone walked away from the job, at least not right away. Freshly minted Mt. Sinai MD Cary English, for example, got plenty of warning signs (it was he who asked about drug screens and got the response "Are you sure you want to work here?"). But he still made it through a full day of Quaalude prescribing before his doubts got the better of him. He had never prescribed Quaalude before, so he went home and read a pharmacology textbook and the *PDR* and talked to co-workers at his other job (at Rikers Island hospital). He decided that he would not prescribe any more Quaalude. But he still went back to work the next day. His first patient demanded Quaalude. He refused. The second patient interrupted English's examination and told him she just wanted her prescription. Again, he refused. Silverg, whose job was at least in part to push reluctant doctors to prescribe with the program, reprimanded him: to work at the clinic, he said, meant to prescribe Quaalude. English quit.[77]

As we have already seen, Peter Sarosi, too, worked at the clinic for two full days before being fired. While we do not know what he thought about his job, it is revealing that on both days he refused to start new patients on Quaalude. Why? Perhaps he knew it was a scrip mill but did not understand how strict the rules were. But he, like English, may also have held on to the hope that the clinic was legitimate. After all, as mentioned earlier, the clinics stayed close to *PDR* guidelines aside from the intra-staff prescribing and drug trading.

It is difficult to tell how representative each of these witnesses was, but data introduced as evidence in the trial offer some clues. Sixteen of the forty-five doctors worked at the Manhattan Center for fewer than ten days, and eleven of these worked for fewer than five days. At least some of these must have been physicians who discovered the nature of the clinic only after working there and subsequently quit. Most of the rest (nineteen out of twenty-nine) worked for fewer than eight weeks—they did not settle into the job, but worked there briefly en route to some other position. For them, it was a professional way station. Only ten of the physicians worked at the center long enough that they must have knowingly embraced their role.[78]

Taken together, the four dozen Manhattan Center doctors present a surprising portrait of scrip doctors. While a few, like Siegel, were openly criminal and committed to drug trafficking, most of the others were fairly ordinary physicians who worked at the clinics temporarily and for practical reasons. They needed a job and were having trouble finding one, perhaps because their foreign medical degrees hurt their chances, or perhaps because, like Glass and Asherson, they were in a time of personal transition. Clinic procedures mimicked legitimate therapeutics just closely enough to have briefly fooled some of them, or, perhaps, to have allowed them to fool themselves. Most returned to being legitimate doctors afterward, and some of them went on to have long and respectable careers. Their experiences suggest, again, that scrip mills were a disreputable but still structural part of the medical system—a persistent and significant zone of drug trafficking distinct from "street" markets.

Policing Prescribers

Scrip mills flourished because of the way prescribing had come to be defined through negotiations between physicians, the state, patients, and pharmaceutical companies. Prescribing accommodated large-volume commercial flows; it was an arena of protected, autonomous professional judgment for physicians; and it was culturally associated with respectable populations seeking treatment rather than drug effects. Honoring these basic parameters made it very difficult to stamp out scrip mills like Witt's that by some measures did not look radically different from the rest of the market for pharmaceuticals. Witt's criminal trial, and the changes introduced to federal drug control regimes in its aftermath, help illustrate the challenges federal authorities faced in distinguishing licit from illicit prescribing and the ever more complex regulatory schemes required to police prescribing in a fundamentally commercial pharmaceutical system.

Defense lawyers were quick to point out the ways that Witt's clinics paralleled legitimate pharmaceutical markets. As they noted repeatedly, Quaalude was an FDA-approved drug, widely prescribed and marketed as less risky than other sedatives.[79] If the drug were truly so horrible, they said, why had the government not made it illegal? (Here, of course, they failed to mention the all-out industry campaign to fend off new drug regulations.)[80] When Sally Unger's lawyer argued, "The main problem [with Quaalude] is not a medical problem but a sociological concern that persons are abusing it," she echoed the exact arguments of pharmaceutical company experts and lobbyists defending troubled drugs on Capitol Hill.[81] Moreover, the defense also contended, the red tape surrounding Quaalude use made the prosecutors' scenario downright laughable. Were criminal doctors really so "idiotic" as to launch an illegal trafficking operation by "sending notice to two official agencies, one of the state and one of the federal government"?[82]

The defense's case was aided by the fact that even the most skeptical witnesses—the physicians who had refused to work at the clinics longer than a day or two—favored pharmacological treatment for anxiety and sleeplessness. They just preferred different medications. Emily Cole, for example, who had refused the job and immediately called Massachusetts authorities, admitted on the stand that she prescribed Valium even though Valium, too, was "euphoric," "over-used," and addictive.[83] Like other prosecution witnesses, she agreed that prescribing dangerous and addictive drugs was a legitimate, if unfortunate, part of medical practice. This attitude was not uncommon among American physicians in the 1970s, a decade in which vociferous condemnation of "overmedication" reduced the use of sedatives but still left them among the most widely prescribed medicines in the therapeutic armamentarium.

Part-time physician Gregory Drezga, too, found himself on difficult ground in condemning Quaalude. He acknowledged that the drug was useful for "situational insomnia" when combined with other treatment for two weeks or less. But when defense lawyers asked him whether he always referred patients to therapy every time he prescribed Dalmane (another sedative), he had to admit that he did not. Why was Quaalude different, then? Drezga's response is intelligent but also telling: "If there is a drug on the market which might be excellent medically but it has detrimental influence in the society, I think a prudent physician should not prescribe it if he can help it." He gave the example of heroin, the best painkiller but one that doctors had rightfully stopped prescribing because of its overall social impact.[84]

Drezga's reasoning was complex, nuanced, and startlingly open about the seemingly arbitrary lines that divided medicines from drugs. Illegal drugs, he suggested, were simply medicines that were currently being abused. Defense lawyers appeared to have a more compelling—or at least compellingly simple—line of reasoning. "We have here," one summed up, "the terrible problem of licensed physicians prescribing approved medications via licensed pharmacies by approved and licensed pharmacists . . . That is their crime."[85]

Compelling as it may have been, however, defense arguments did not win the day. Federal prosecutors won a conviction, as did New Jersey state prosecutors (who even convicted a few part-time doctors).[86] The DEA rightfully touted it as a significant victory and an indication that they were now taking "drug diversion" seriously as a national problem.

In part, the prosecutors' victory can be ascribed to the clinics' sloppiness. These were overtly fraudulent operations, rife with tell-tale signs such as the prescribing of drugs to staff, falsifying business papers, tax evasion, and a rigid prescribing regime that pushed beyond recommended limits. But drug enforcement officials could hardly count on such errors in their broader battle against illicit markets; only the largest, most formal operations were likely to suffer from them.

Perhaps the more important factor was the broad cultural backlash that had been brewing against prescription "wonder drugs" and against physicians' therapeutic autonomy more generally, in the 1960s and 1970s. As noted earlier, by the time Witt and his coconspirators sat in front of a jury, addiction scares had already toppled a host of former wonder drugs. Meanwhile physicians found their therapeutic decisions increasingly questioned—and often challenged—by lawyers, judges, bioethicists, and political activists of the "patients' rights movement." Prescription drugs may have continued to be used at shockingly high rates, but the cultural narratives that made sense of such widespread use had begun to shift. When Emily Cole was maneuvered into talking about Valium, her discomfort was in admitting that she prescribed the drug at all, not in claiming that it was addictive and harmful. The prevalence and legitimacy of these critiques helped make it easier for jurors to see the Quaalude clinics as shams.

Despite its success, however, the trial also revealed serious weaknesses in the campaign against illicit prescription markets. Convictions had required a lengthy and enormously expensive investigation—a suitable strategy for major, franchised operations like Witt's, but hardly affordable in most cases. Prosecutors had also depended on nuanced logic such as Drezga's that could rarely be relied on in the polemical landscape of American drug wars. This was the paradox: because illicit markets built on accepted elements of the postwar medical system—prescribing power, the commercial expansion of drug use, the medicalization of middle-class problems—they were difficult to stamp out without fundamentally changing the logic or rules of the broader system.

In fact, the Witt trial might be better understood as the exception that proved the rule of impotence in the face of thriving illicit prescription markets. Federal authorities, particularly the DEA, certainly treated it as such. Taking advantage of the cultural backlash against the medical system and the renewed "war against drugs" under President Ronald Reagan, the DEA pushed for and received real changes in pharmaceutical regulations. The agency increased pressure on state medical boards to mete out penalties to physician transgressors, for example, and won expanded authority to deny or revoke prescribing licenses in 1981. As already noted, the DEA revoked or denied the Controlled Substances licenses of only fifty physicians in the 1970s; the number rose to more than four hundred in the 1980s and to more than seven hundred in the 1990s.[87] The agency also implemented electronic compilation and tracking of prescriptions as well as diplomatic campaigns to persuade governments still allowing the manufacture of Quaalude to outlaw it.[88]

These were remarkably powerful new tools for overseeing medical practice, with an impact that reached far beyond the relatively limited number of willfully abusive "scrip doctors." The Schedule of Controlled Substances included many of the nation's most widely used prescription medicines and thus brought a vast range of common physician decisions under new federal scrutiny and regulation. This generated new and sophisticated streams of data about prescribing, distinct in both nature and purpose from the commercial prescription audits and other market-driven studies that shaped postwar pharmaceuticals.[89] And it provided a model—however warped by drug-war imperatives—for postwar drug reformers struggling (and usually failing) to establish real and significant limits on irrational prescribing in a profit-driven pharmaceutical system.

Conclusion

In the case of Quaalude, even these new tools were not enough. The DEA lowered the U.S. production quota to zero in 1982, when the only domestic manufacturer (Lemmon Pharmaceuticals) stopped making the drug. But in 1984 Congress still took the unusual step of passing a separate law to move Quaalude into the strictest category of controlled substances, Schedule I—

drugs of no therapeutic value—where it joined heroin, cocaine, LSD, and marijuana. Like heroin and cocaine before it. Quaalude was now strictly a "street" drug. For the drug's purveyors and for the authorities trying to police it, nuance and complexity were no longer a problem.

Quaalude's saga was short but revealing. In its scant two decades, the drug highlighted the continuing vitality and significance of illicit prescription markets—scrip mills and other less formal varieties that pervaded a postwar medical system awash in heavily advertised prescription medicines. These markets channeled pure and reliable drugs of known provenance to mostly white and affluent clients. Dating back to the origins of physicians' prescribing monopoly, such supply chains have been persistent and significant (if undesirable) elements of the medical system.

Recognizing these border zones helps illuminate several key aspects of physicians' prescribing power. It reminds us that physicians, too, were part of a postwar pharmaceutical system increasingly premised on the pursuit of profit as well as the delivery of therapy. Prescribing could be a commercial transaction, whether it occurred in a scrip mill or in a legitimate doctor's office after a patient threatened to go elsewhere if denied a prescription. This sort of prescribing was an important element of the postwar pharmaceutical boom. It was also a substantial arena of illicit drug trafficking—an alternate landscape of abuse and addiction whose exploration can add substance and detail to the open secret of white, middle-class drug use. And finally, such prescribing enabled an alliance between pharmaceutical reformers and anti-drug crusaders that produced expanded state surveillance and control over physician prescribing.

CHAPTER 10

The Afterlife of the Prescription

The Sciences of Therapeutic Surveillance

Jeremy A. Greene

The life of the prescription—if considered at all—is typically measured in hours, if not minutes. This ephemeral object, hastily scribbled on a wisp of paper or printed from a desktop computer, is handed from provider to patient and from patient to pharmacist and then forgotten by most parties once the medication is dispensed. But the prescription, once filled, is neither inert nor forgotten: each scrip leaves a residue of doctor, patient, diagnosis, and therapeutics for those interested in studying pharmaceuticals and their consumption. As Owen L. Wade of the British National Formulary noted with considerable optimism, by 1969 the computerization of claims data had transformed "such a mundane procedure" as the recording of a prescription sale into a technology providing penetrating insight into the nature of disease, drugs, and clinical care. "This is the satisfaction of the cook," Wade observed, "who finds old leftovers in the kitchen and creates a splendid dish."[1] Multitudes of individually forgotten prescriptions have, over the past half century, been assembled into vast systems of therapeutic surveillance that make dead scrips speak hidden truths about the otherwise opaque clinical realm.[2]

On one hand, the rise of systems of prescription surveillance can be viewed as part of a more general trend in the collection of health-related data over the twentieth century. These practices include the rise of national health surveys, the growth of actuarial sciences for life and health insurance, the collection of detailed morbidity and mortality statistics, hospital and physician practice censuses, and epidemiological surveillance programs for both infectious and noninfectious diseases.[3] Existing narratives of the rise of health surveillance often privilege the role of technology—especially the computerized database—in potentiating new uses for archived information. As in Wade's account, the sciences of surveillance thereby emerge as "serendipitous" consequences of data collection. Yet the history of the population sciences has taught us that health-related data are rarely collected without a reason, nor is their collection simply determined by the availability of information technology.[4] Exploring what was at stake in collecting prescription data—and who stood to benefit from its analysis—reveals a surprisingly complex narrative of private and public actors who shaped this therapeutic panopticon.

The dream of a database of drug prescribing long predated its construction. When Paul de Haen, a marketing consultant to the pharmaceutical industry and a prominent contributor to the new journal *Medical Marketing*, published one of the first textbooks of prescription drug marketing in 1949, he warned the field of pharmaceutical market research that it needed to develop more sophisticated tools for visualizing its current and future markets: "Market research in the pharmaceutical industry does not merely require the collection of certain statistical and commercial data . . . It covers a much broader field, and requires constant observation of the entire medical and pharmaceutical horizon, so that the physicians' prescription habits can be thoroughly studied. For this purpose special personnel should be trained."[5]

In the ensuing half century, these "special personnel" developed powerful new tools for studying the prescription as both object and action. But those interested in building an archive of prescriptions also found themselves in the company of strange bedfellows. Market researchers, sociologists of innovation and influence, health service researchers, academic physicians, and epidemiologists began independently to develop sampling techniques and technologies of surveillance to study doctor, patient, drug, and diagnosis through the records of old prescriptions. As they incorporated novel techniques from other social sciences and other industries, these researchers and entrepreneurs developed broad networks of personnel and surveillance instruments that enabled the "constant observation of the entire medical and pharmaceutical horizon" that de Haen envisioned.

While a promising literature has emerged on the history of databases and social sampling tools within the social sciences,[6] and historians of medicine have begun to attend to the roles of consumerism and the marketplace,[7] there has been very little scholarship to date on the vital interactions between pharmaceutical marketing research and the broader spheres of medical science and medical practice. This chapter traces the development of these parallel and interconnected sciences of the prescription as each held up the scrip as an imperfect window into medical practice and transformed this opaque object into a relational universe of data. The narrative begins with early prescription audits of the 1950s and concludes with the ambitious and largely unfunded proposals of the Congressional Joint Commission on Prescription Drug Use, which managed to set forth much of the paradigm of contemporary prescription surveillance by the publication of its report in 1980 but failed in its broader project of making the prescription database into a transparent and publicly accessible tool.

The Prescription as a Tool for Market Research

As noted elsewhere in this volume, the prescription was not required for sale of most "ethical" drugs in the first half of the twentieth century, and the system of federal laws mandating a prescription for the sale of pharmaceutical products was not fully established until 1951.[8] Consequently, pharmaceutical sales representatives in the first half of the twentieth century focused more of their effort on persuading pharmacies to stock their medications than on persuading physicians to prescribe them.[9] But in the late 1940s, as a robust pipeline of novel pharmaceutical agents led to an increasing market for brand-name prescription-only drugs, the act of prescribing had become a subject of increasing importance for the industry to understand.[10] Between 1947 and 1954, while over-the-counter drug sales remained relatively constant, prescription drug sales more than doubled in volume, only to double again before the end of the 1950s.[11]

Far from viewing the regulation of the prescription as a constraint, pharmaceutical manufacturers welcomed it as a central tool for ordering and rationalizing their market. As Richard Hull, director of marketing for Smith, Kline, and French, demonstrated at an industry seminar in 1958, the concentration of pharmaceutical marketing on prescribers instead of consumers created far more efficient channels for marketing and market research. Consider, for example, the utility of advertising in medical journals instead of popular magazines. A single ad placed in four medical journals cost less than three thousand 1958 dollars and provided "complete coverage" of the entire physician market. The same ad placed in the four most popular consumer magazines would cost in excess of eighty thousand dollars and would reach only a fraction of the potential consumers. In addition to being a more efficient target for promotion, Hull argued, physicians represented a highly visible and easily studied population, since physicians had to be licensed and because the American Medical Association maintained "what [had] been called the best professional directory service in the world."[12]

This directory provided a complete set of street addresses for the growing ranks of bell-ringing pharmaceutical "detail men" who represented an alternate and increasingly important direct channel between manufacturer and prescriber.[13] And yet even this well-circumscribed market of potential prescribers still contained more and less effective targets for promotion. Sales representatives in the field began to make note of physicians who were less responsive to sales tactics, gave them nicknames like "Dr. Snob," "Dr. Resistant," and "the Backslapper," and reported these observations to their managers, who sought to rationalize their activities.[14] "To call on every doctor on the block entails a great deal of waste motion at the company's expense," de Haen stated in his 1949 text, noting instead, "There are various methods of selecting the names of physicians who are most likely to prescribe a new product to a large group of patients." Arthur F. Peterson, a prolific writer of pharmaceutical sales training manuals in the 1940s and 1950s, advised sales representatives to keep logs of their visits in which they rated physicians on two axes: the first—coded A, B, C, or D—recorded the size of their practice, and the second—coded X, Y, or Z—recorded their receptivity to sales representatives and to novel medications. This data, reported to supervisors, formed a crude map of prescription density and marketing receptivity that enabled more strategic deployment of the sales force.[15]

Although the legal basis was uncertain, some enterprising detail men followed prescriptions from the physician to the pharmacy to investigate the impact of sales tactics on prescribing habits. Sales representatives at Sharp & Dohme in the late 1940s began to regularly ask local pharmacists about monthly rates of prescriptions for specific drug categories and which local physicians were big prescribers; at times pharmacists would allow them to peruse the books in which records of filled prescriptions were bound and kept for reporting to local and state pharmacy boards, in accordance with state laws and the guidelines of the American Pharmaceutical Association. "Just as our universe of physicians is clearly defined," noted one market researcher of the late 1950s, "so thanks to pharmacy laws and licensing provisions, is our universe of retail pharmacies . . . How fortunate Dr. Gallup would be if his sampling problems were as simple as ours."[16]

The mention of Gallup is apt. As the entrepreneurial detail men of the 1940s became the pharmaceutical marketing executives of the 1950s, they sought to expand their local data-gathering practices into a more centralized and reliable system of prescription surveillance. These efforts took place in a nation increasingly obsessed with surveying itself and representing its opinions and preferences using the tools of the social sciences.[17] One of the earliest applications of these new survey techniques to the forensic study of prescription was carried out by Raymond Gosselin, who at the time was pursuing a joint master's degree, in pharmacology at the Massachusetts College of Pharmacy and in business administration at Boston University. Gosselin's joint thesis, submitted in early 1950, came to be remembered as one of the first systematic and reproducible prescription drug audits.[18]

With the use of IBM punch cards and the tabulation team of the Boston University Statistical Laboratory, Gosselin charted a population-stratified random sample of all drugstores in Massachusetts. His research assistants opened hundreds of bound prescription ledgers and coded a random sample of scrips at each pharmacy. Shortly after his thesis was accepted by Boston University, Gosselin embarked on an entrepreneurial venture in nationwide prescription data sampling. R. A. Gosselin & Associates began to market its first prescription data product in the early 1950s: the National Prescription Audit (NPA), a subscription database that could segment the prescription drug market by local region and give quarter-to-quarter information on the performance of specific products and therapeutic categories. Gosselin's use of IBM punch cards produced a cumulative database of prescriptions that could answer questions with increasing sensitivity: by 1953 the original panel of 6,000 prescriptions had grown to 225,000, and by the end of the decade, industry executives declared the NPA sales figures to be more reliable than their own records.[19] The NPA was the leading product of R. A. Gosselin & Company until the company was sold to IMS Health in 1970, and it remains a key plank of pharmaceutical market research today.

For all its utility, the NPA depicted the prescription as a static record of sales and not as a clinical phenomenon in its full context. The NPA helped marketers visualize prescriptions at the point of purchase (the pharmacy), but its scope did not include the key site of prescription decision making (the clinic). To optimize prescriptions for their products, marketers needed to understand the prescription *as an action* within its proper context. One contemporary marketing executive noted, with regard to prescription audits, "[They] do not tell us *how* the drugs are being used. We still want to know what kinds of patients are being treated with a given compound, and in many cases what the physician expects to accomplish by its use . . . In other words we want to go beyond the prescription and learn about the patient for whom it is written."[20] To do so, pharmaceutical market researchers devised a novel data set that borrowed more from Nielsen television-viewing families than from Gallup polls. In 1956 a group of marketing executives from Smith, Kline, and French formed their own market research venture, Lea Associates, which offered as its flagship product the National Disease and Therapeutic Index (NDTI), a statistically representative panel of twelve hundred physicians who agreed to use case record diaries to record "basic diagnostic and therapeutic information about all patient contacts made by them during assigned two-day periods of their practices."[21] The NDTI accomplished for the act of prescribing what Nielsen did for the act of television watching: it turned a set of real-time prescriber decisions into a laboratory for drug market research. The therapeutic index allowed its users to link prescription audits to specific diagnoses and temporal changes in therapeutic practice, including off-label usage.[22] In effect, this tool created a "virtual clinic" into which marketers could peer to see how prescriptions, drugs, and diseases interacted inside doctor's offices.[23]

As the market for prescription data expanded beyond the pharmacy audit, companies such as Lea and Gosselin diversified their product lines with audits and index panels that mapped purchases by hospitals, responses to journal advertisements and direct mailings, and the actions of generalists, specialists, dentists, veterinarians, and pharmaceutical sales representatives. By 1968 Lea and Gosselin had merged into a single market research empire that offered eighteen major data products related to prescriptions and prescribing.[24] In addition to regular reports, these data empires ran customized analyses for a fee or sold raw data in the form of punch cards for subscribing firms to run on their own IBM machines.[25] In 1958, for example, Smith, Kline, and French purchased two hundred thousand IBM cards of prescription data from NDTI.[26] By the early 1960s, any pharmaceutical manufacturer or pharmaceutical marketing agency could be expected to run its own analysis of purchased data on in-house computers. Marketing teams joked about the displacement of human computers (which had typically been women) as mechanical computers became essential to the market research practices. Market research executives at Upjohn began to refer to their new IBM 1401 as "Miss Fourteen-Oh" and joked that she "dabbled in song and dance" as well, if crudely.[27]

The IBM punch card connected efforts of market researchers and purveyors of prescription data with the American prescribing physician in surprising ways. As I have argued elsewhere, the role of the American Medical Association (AMA) in compiling and distributing prescriber databases provides a unique case for examining the active role that the medical profession has taken in its own surveillance and objectification as a market of prescribers.[28] While the AMA did not provide prescription data as did the NDA or the NDTI, the organization was essential in promulgating a data product that became a veritable Rosetta Stone for translating prescription claims data into studies of the behavior of individual prescribers. In the mid-1940s, the AMA translated its directory of members into a complete computerized registry of all American physicians, using the same IBM punch card technology that Gosselin had used to create the initial prescription audit. In addition to its utility for state licensing boards, this database was understood to have value to market research in the pharmaceutical and medical device industries. Prospective clients were sent fliers promoting the completeness of the AMA records: "These cards," one pamphlet read, "are begun as soon as a student enters medical school and are kept for some time after a physician's death."[29] The compilation of prescriber data known as the Physician Masterfile remains a key source of revenue for the organization: AMA physician database products in 2005 provided $44.5 million in revenue, roughly 16 percent of the organization's total income.[30]

On a superficial level, marketers and market researchers first used this biographical data to weed out older and retired physicians and to subdivide the profession further into specialties relevant for the promotion of specific products.[31] For those interested in studying the social dynamics of prescribing and pharmaceutical promotion, the Masterfile became a crucial link connecting pharmacy data on prescription claims to the prescribing patterns of individual clinicians via a series of individual identifiers, including state licensing information, medical education information, national provider identifier codes, and U.S. Drug Enforcement Administration numbers for all U.S. physicians. By the end of the 1950s, pharmaceutical market researchers had tools for studying the act of prescribing with a precision unimaginable a decade earlier. One marketer crowed: "From the data we can learn how a given drug or class of drugs is being used. We can learn the relative frequency with which our sample of physicians sees or treats a given illness. We can learn how the illness is being treated. We can learn the sex and age distribution of patients. And in most cases we can learn the extent to which the condition is treated by specialists rather than by physicians in general practice. In fact, the possible types of tabulations and cross-tabulations are almost limitless."[32]

The Sociology of the Prescription

Whereas the technological basis for the prescription audit and diagnostic panel emerged first within the field of pharmaceutical market research, the theoretical basis for conceptualizing the prescription as a social action gained substance from the increasing interaction between industry and the applied social sciences in the 1950s and 1960s. In 1952 a sociologist at the University of Minnesota published a survey in the *Harvard Business Review* of the influence of industry sources on prescribing behavior, which showed that physicians consistently rated pharmaceutical sales representatives as the most important source of information on new drug products.[33] A robust literature of surveys of prescription behavior by other applied sociologists and scholars of marketing at business schools began to examine the demographic, geographic, economic, educational, and professional factors influencing the physician's prescribing of new medications.[34] These factors were measured to characterize the variation in the willingness of physicians to try new medications based on variables such as age, years in practice, size of practice, specialty, reading habits, and involvement with professional associations.[35]

But researchers felt that the self-reported basis of these studies clouded their ability to objectively study the dynamics of prescribing as a social action. In 1954 Ernst Dichter's Institution for Motivational Research (IMR) was hired by the Mass Motivation Study Committee of the Pharmaceutical Advertising Club to study the emotional and unconscious aspects of the prescribing decision. The principal tool of the IMR was the psychoanalytic "depth interview," in which the physician was "more apt to bare his real attitudes and motivations than he is in the conventional question-and-answer survey." Free-association games were supplemented by projective techniques, using storytelling cartoons and sentence completion exercises, which aimed to reveal the unconscious logic of prescribing with a minimum of self-conscious interference. "The prescriber, not the prescription, has to be considered," the study concluded; "in other words, it is the modern physician with his foibles, fears, hopes, and insecurities who is the theme of each ad and of each communication rather than the scientific and medical qualities of the drug."[36]

Such studies generated a vision of the prescriber as a dual entity—a conscious prescriber and an unconscious prescriber—which produced "a basic conflict aris[ing] from the physician's self image as a rational scientist . . . He expects himself to make up his own mind on the basis of objective evidence . . . And yet he finds himself confronted, like a housewife in a supermarket aisle, with a misery of choice which he tends generally to resolve by irrational and emotional factors."[37] But this research was expensive to conduct and required skilled personnel not available within the drug firm, so applied sociologists and market researchers began to converge on a more available source—the detail man—as a sampling device for prescriber research. Perhaps the most notable collaboration to this effect occurred between Columbia sociologists of influence and innovation and pharmaceutical market researchers in the 1950s and 1960s.[38] Through the mediation of the Bureau of Applied Social Research, a think tank associated with Columbia University and headed by the sociologist Robert Merton, pharmaceutical companies funded close analyses of the social forces influencing prescriber practices.

The bureau was, at the time, the nation's foremost sociological think tank; it specialized in crafting projects that simultaneously addressed concrete questions (of interest to its industrial funders) and abstract questions (of interest to its faculty members). An example of this research was a Pfizer-funded project researching "the epidemiology of a new drug," in which the Columbia sociologists guided Pfizer detail men in collecting data on the spread of usage of a novel Pfizer product in a set of well-defined local communities, using the sales representative as both vector and sampling device.[39] The research of the bureau emphasized the mapping of social relationships to model the influence of social networks on the prescribing behaviors and the spread of prescribing behaviors through the same social networks. For Pfizer, this project produced immediate feedback on marketing efforts; for the bureau, it produced empirical proof for the "social contagion" model of information transfer and diffusion of innovation through society. The book-length work based on the Pfizer studies, *Medical Innovation* (1966), remains to this day a key text in the sociology of influence, innovation, and information dissemination.[40]

The AMA also invested in the applied social study of the prescription in the early 1950s, beginning with a set of surveys and interview-based research projects by Ben Gaffin Inc., a Chicago-based public-opinion research firm. The Business Division of the *Journal of the American Medical Association* had become interested in studying effects of pharmaceutical promotion on physician education and prescribing behavior and contracted Gaffin to conduct a cross-sectional survey of five hundred physicians regarding their information on new drugs and their prescribing habits.[41] In 1955 the company was commissioned to perform a study of far larger magnitude that might use prescription data to provide more detail than could be found in surveys alone and generate a total understanding of all influences on prescription habits. Possibly the most detailed and intensive investigation ever performed into the factors influencing the physician's prescribing habits, the Fond du Lac Study staked out a relatively isolated county in Wisconsin and followed physicians, along with all possible contacts (other physicians, salespeople, pharmacists), across multiple locations (hospital, clinic, medical society, pharmacy) in the uptake of five new drug products.[42]

The granularity of detail of the Gaffin studies extended the contradictory image of the physician first developed in the Dichter study. Whereas Gaffin noted, "Practicing physicians in the U.S . . . tend to be idealistic, and . . . taken as a *group* they tend more to be motivated by helping mankind than do other groups, such as used car dealers," his report also stressed the "nonprofessional rationales" of prescribing behavior. He explained that a physician's act of prescribing was regularly influenced by nonprofessional concerns: "As a human being, he is comparatively quick or comparatively slow; he is comparatively hard-working or he is comparatively lazy; he is friendly or crabby; social or solitary; happy or unhappy. His morning contacts with his wife and children affect in a greater or lesser degree his attitude toward patients, toward co-workers, and toward detail men. His basic temperament, modified by his daily interpersonal relations, influence[s] all his actions and attitudes to some extent. His human-beingness is modified by his being a physician."[43]

The Automation of Pharmacovigilance

If the Fond du Lac Study represents the high-water mark in the collaboration between the organized medical profession, the pharmaceutical industry, and applied sociologists interested in studying influence and the dissemination of innovation, it carried significant costs for all parties. As the Eisenhower era of probusiness industrial-professional collaboration gave way to the more critical and consumer-protectionist politics of the Kennedy years, Senator Estes Kefauver (D-TN) initiated his well-publicized inquiry into the marketing and pricing of prescription drugs. On July 5 and 6, 1961, Hugh Hussey, the chairman of the AMA, was brought before Senator Kefauver's subcommittee and publicly confronted with the reports of Gaffin's research as evidence of marketing collusion between the AMA and the pharmaceutical industry. "We are not ashamed," Hussey maintained under questioning, "of what was found out by these surveys."[44] Nonetheless, the Fond du Lac Study was subsequently reclassified as confidential, the AMA was publicly ridiculed in the popular press, and a new generation of muckraking journalists used Fond du Lac as a starting point in documenting the collusion of marketing interests between the AMA and the pharmaceutical industry.[45]

Much as Kefauver's hearings were ultimately diverted from issues of marketing to issues of drug safety and efficacy, so too public discussions of drug prescription surveillance systems shifted from depictions of the dangers of drug marketing to the dangers of the drugs themselves.[46] Kefauver's initial bill, which focused on curbing the excesses of the marketing of branded drugs, was already doomed to die in committee when in 1962 a series of epidemiological reports initiated by the German anesthesiologist Widukind Lenz connected a recent increase in phocomelia, a birth defect that resulted in grossly visible limb deformities, with maternal use of the popular new antinausea medicine Contergan (thalidomide).[47] Images of thalidomide children became an international symbol of the failure of the medical profession and the regulatory state to protect vulnerable populations from the harmful effects of widely marketed new drugs. Contergan had been extensively marketed to physicians and consumers alike, and its premarket testing and postmarket promotion had emphasized its remarkably *nontoxic* safety profile by available standards of clinical pharmacology.[48] As Lenz's work was read internationally, his careful use of the correlative techniques of infectious disease epidemiology within the novel terrain of prescription drug use documented not only the unseen dangers of newly marketed drugs but also the need for a new discipline of pharmaceutical epidemiology to scour observational data for new drugs and seek adverse reactions that could clearly be associated with drug use in clinical practice.[49]

The recognition that the risks of new drugs might not be visible until they were consumed by large numbers of patients had been evident long before Lenz's epidemiology of thalidomide-associated phocomelia. Indeed, the history of federal drug regulation in the United States can be recounted as a succession of measures taken in response to dangers of drugs that became apparent only after widespread consumption by the general public.[50] However, the FDA in the 1960s still had very limited authority in the postmarket regulation of drugs. The agency had neither direct means to control physicians' prescriptions nor resources to gather data on prescribing of newly marketed drugs.

The AMA nominally maintained more influence in both arenas. In 1955 the AMA reconstituted its Council on Pharmacy and Chemistry into a new body, the Council on Drugs, and positioned itself as a national monitor of postmarketing drug experience. No single drug emphasized this shift more than Parke, Davis's Chloromycetin (chloramphenicol), discussed in chapter 2 by Scott H. Podolsky. Chloromycetin, a potent broad-spectrum antibiotic, was widely utilized after its release in 1951, only to become associated with a disturbing number of postmarketing drug-related adverse effects, including four hundred cases of aplastic anemia between 1953 and 1955.[51] The council swiftly established the Committee on Blood Dyscrasias to seek more reports on chloramphenicol's impact on the bone marrow; this committee was broadened in 1960 into the Committee on Adverse Reactions, which was tasked with compiling and reviewing all side effects from all drugs available on the American market. However, the AMA's approach depended entirely upon voluntary reporting, and the organization had neither authority nor adequate means to enforce data collection. Committee members complained loudly that the working of the system itself was doomed to failure; as one report noted, "physicians reported only a small fraction of all cases and the total number of patients receiving a drug was unknown."[52]

Although the 1962 Kefauver-Harris Drug Amendments included a proviso that manufacturers must establish records and make reports to FDA of "data relating to clinical experience and other data or information, received or otherwise obtained,"[53] for all new drugs, there was little budgetary support or regulatory authority to enforce this measure. By 1967 the agency had developed a protocol requiring manufacturers to report any published evidence of putative side effects of their marketed products. Any novel or unexpected adverse effect was to be reported to the agency within fifteen days; other relevant information regarding the safety or effectiveness of a drug was to be reported quarterly for the first year after approval, twice in the second year, and annually thereafter. This kind of information could become actionable only after years of case reports, and then only if one of the relatively few FDA staffers took active interest in pursuit of a specific question of drug harm.

The thalidomide disaster had, however, reawakened dormant interest in a collaboration between the FDA and the AMA (which had first flickered after the discovery of chloramphenicol toxicity) to monitor the safety profiles of all drugs on the market through the use of hospitals as surveillance tools. The two groups had utilized the new federal reporting system to launch a pilot program with five government-affiliated hospitals to screen for drug safety. This program was swiftly extended through the network of Public Health Service extramural grants to include numerous academic medical centers. It reached more than six hundred hospitals by 1964 and attracted the attention of academic physicians with interests in clinical epidemiology, such as Johns Hopkins University's Leighton Cluff, Harvard University's Thomas Chalmers, and Tufts University's Hershel Jick.[54]

The hospital offered a denominator of total patients exposed to drugs, against which to balance the numerator of reported drug-associated events, but it could not capture all patients *outside* the hospital taking the drug in question.[55] Leighton Cluff noted that an early validation system of reporting efforts at the Johns Hopkins Hospital had "proved completely unsatisfactory for detecting drug reactions." He explained, "During recent daily intensive surveillance of one hospital service, four times as many reactions were detected than had been reported on the cards from the entire hospital."[56] Would-be epidemiologists of adverse drug effects needed a way to circumvent the physician as a reporting device. "The importance of automatic data processing cannot be overemphasized," Cluff concluded, "for no other method allows reasonably complete, accurate, and efficient surveillance of drug usage."[57]

Cluff and his research team were supported by an emerging literature on the design of epidemiological computer systems that could create punch cards on a patient-by-patient basis in specially assigned inpatient drug monitoring units. Computerized drug monitoring involved the creation of three linked data sets for every drug received by every patient in a dedicated hospital ward.[58] D. J. Finney, an early theorist of computerized drug monitoring, expressed these data sets as a linked "P-D-E system," in which P(atient) population data would be systematically gathered within a set geographic or hospital catchment area, the D(rug) data would include records of all relevant prescriptions, and E(vent) collection would record all untoward reactions potentially attributable to the drugs prescribed.[59]

Proponents of drug monitoring imagined a linked system of inpatient surveillance wards circling the globe, which could act as pharmaco-vigilant sensors, detecting early signals of possible drug harms and providing descriptive data regarding their frequency, severity, and relative strength of association. Finney, for example, believed that the computerization of prescription data was transforming drug epidemiology from a reactive into a proactive field; while allowing that "much [was] due to Lenz for his discovery in 1961 [that thalidomide was associated with phocomelia]," he also boasted that "a *monitor* could have signaled a warning 1½–2 years earlier." Inpatient surveillance systems liberated pharmaco-epidemiology from the "weak link" of the reporting physician.[60]

This model was developed most extensively by the joint investigation of Dennis Slone, Hershel Jick, and Ivan Borda through the Boston Collaborative Drug Surveillance Program. Under the aegis of the clinical epidemiologist Thomas Chalmers, and with public and private support from both the U.S. Public Health Service and the Pharmaceutical Manufacturers Association, Slone, Jick, and Borda published a study in *Lancet* in 1966 that demonstrated the feasibility of implementing an automated hospital-based drug monitor system.[61] The study was based at the Lemuel Shattuck Hospital and added two new features to an inpatient medical ward to bypass the physician in the reporting of drug events. The first was a nurse hired expressly to serve as a human drug monitor, described as "a new member . . . added to the basic ward team whose primary role [was] the acquisition of accurate data."[62] The second was a nonhuman monitor: the automated computing system for processing data, whose database would be stocked through "a process of keypunching, storage on magnetic tape, and submission to computer programs" that would link hospital prescription data with patient demographics and clinical outcomes.[63]

Beginning in July 1966, Jick added a nurse monitor to the medical team of two wards in the Lemuel Shattuck Hospital to obtain all information about drugs given to all patients. Nurse monitors followed medical teams on rounds and sought to observe the *writing* of all prescriptions; they filled out a standardized "drug starting card" for every prescription made on every patient in the designated hospital ward and a similar "drug stopping card" for every drug discontinuation, along with demographic and clinical information related to all patients.[64] The Boston team became a model for an automated drug surveillance program that functioned "largely independent of clinical judgment in establishing a connection between a drug and an adverse event."[65] Early results showed that drug-related events were both more frequent and less severe than had previously been anticipated. More than one-third of patients on the Shattuck wards experienced at least one drug-associated adverse reaction during the first year of study.[66] By 1967 the Boston group had established a numerator-denominator approach for comparing drug usage between long-term and acute hospitals through a network of five hospitals in Boston.[67] By 1968 more than twenty-five hundred patients had been entered and discharged from the surveillance system; they had experienced more than twenty-six thousand monitored drug exposures, representing more than seven hundred individual drugs.[68] Commonly prescribed drugs, such as digoxin and heparin, could be reported in detail, yielding novel information related to their clinical pharmacology and their interactions with other drugs.[69] The system enabled the observation of not only obvious drug reactions (such as a rash) but also other clinical events (such as heart attacks or kidney failure) that could be associated with drugs only by careful epidemiological surveillance.

As the Boston Collaborative Drug Surveillance Program escalated its activities and exported its methods to other sites, a new series of drug scandals came to emphasize both the utility and the limitations of these forms of pharmacovigilance. Clioquinol, an anti-infective that had been in use since the 1930s, was found to be associated with subacute myelo-optic neuropathy in 1970, more than three decades after its initial introduction. An association between the synthetic estrogen diethylstilbestrol (DES) and a rare form of cervical clear cell adenoma was reported in 1971, with evidence of a twenty-year latency period between use of the drug and detection of the cancer.[70] The beta-blocker practolol became the focus of a broad scandal after it was associated with a potentially fatal inflammation of the skin and soft tissues (oculomucocutaneous syndrome) some five years after its broad release on the British market. These examples simultaneously underlined the necessity for drug surveillance units and underscored the impossibility of inpatient surveillance systems to capture drug-disease associations in which three decades or more might pass between drug exposure and adverse event. As Jick warned, in a systematic proposal for the theory and design of the emerging field of pharmacoepidemiology, the ability to study "drug-illness relations" required distinct methods depending on the time course and prevalence of prescription and adverse event. High-frequency events in high-prevalence diseases could be detected swiftly by case report, low-frequency events in high-prevalence diseases required careful active ongoing surveillance, and low-frequency events in low-prevalence diseases might simply never be adequately described.[71]

To address the growing problems of drug safety, prescription surveillance needed to extend outward: spatially, from the monitored wards of the hospital to the messier universe of outpatient care; temporally, from linkages visible in days or weeks of measurable hospital time to the longer stretches of months and years required to understand the impacts of chronic medication; and thematically, from the isolated connection of drug and disease to the study of all steps of diagnosis, prescription, adherence, consumption, and presentation that might occur in between. After all, to blame an adverse effect on a drug based solely on prescription and outcome records was to assume, first, that the drug was being prescribed correctly and, second, that the drug was being taken as prescribed. These assumptions were being undermined by another science of the prescription—the study of drug utilization—which used the prescription database to show how little was known of the rationality of drug prescription and drug consumption.

The Rational Prescription

As he delivered the 1975 Rho Chi lecture to the American Pharmaceutical Association in San Francisco, Edmund Pellegrino, a bioethicist and then chairman of the board of the Yale–New Haven Medical Center, argued for the urgent need for a more rational therapeutics, defined as "the parsimonious use of therapeutic agents only when they demonstrably alter the natural history of illness or demonstrably ameliorate symptoms, and do so with low enough toxicity and sufficient effectiveness to make their administration worthwhile."[72] The emphasis on rational therapeutics was not new: Pellegrino's talk echoed a long lineage of therapeutic reformers stretching back to Torald Sollmann's critique of the pharmaceutical marketplace when he was the founding chair of the AMA's Council on Pharmacy and Chemistry in the first decade of the twentieth century. Like many other therapeutic rationalists, Pellegrino urged that more attention needed to be paid to the cognitive, economic, and cultural forces leading physicians to prescribe drugs that were not necessary, too expensive, or too risky for patients. In response, he called for a new field of study capable of locating the systematic levers that could be pushed to encourage more rational prescribing:

> Education is not enough—nor is increasing the dose of clinical pharmacology. The therapeutic enthusiasm dominant in medicine and society is too pervasive to be surmounted by the average physician or pharmacist working alone. Education in drug use must be reinforced by a more systematic way to assure wise drugs usage—one which steers between the absolute autonomy of the physician on the one hand, and the evils of bureaucratic regulation on the other. Institutional mechanisms are needed to assure ongoing critical evaluation of drug use by individual physicians, by departments and by institutions . . . Drugs are among the most easily measured events in medical care as to kind, number, cost and side effects. They are high on the list in the current public demands for accountability and control of the quality of medical care.[73]

By the time of Pellegrino's speech in 1975, the study of drug utilization had developed from an early descriptive branch of health service research in the 1950s into an active field of policy research that sought mechanisms to effectively impose order on the heterogeneity of physician practice. The prescription, as research object, sampling tool, and site of intervention, was at the center of these efforts.

As Podolsky argues in chapter 2, the rhetoric of rational and irrational prescribing became particularly inflamed in the field of antibiotic overuse. The measurement of inappropriate prescribing in other therapeutic arenas proved even more challenging. One early foray into this work was conducted by Frank Furstenberg and a group of physicians associated with Baltimore City Hospital, which, like many state-run urban hospitals, provided a drug benefit to indigent populations.[74] Because the city of Baltimore paid for these medications, original copies of the prescriptions were sent to the City Health Department by pharmacies for payment. Furstenberg's group took a random sample of just over one thousand prescriptions filed by 159 physicians in 1950–51 and reviewed them with a physician, a pharmacist, and a biostatistician to assess the "appropriateness" of the prescription. Using the standards of the United States Pharmacopeia (USP) as an indicator of appropriate usage, more than one-third of all prescriptions were found to be "nonacceptable"; of the prescriptions by private physicians, less than half were for USP-indicated usages.[75] Charlotte Muller's parallel audits of prescriptions from New York City hospitals and hospital-associated outpatient clinics found that more than half lacked a clear indication for their use, were dosed incorrectly, or were prescribed against direct contraindications.[76]

The enactment of state-based Medicaid programs in 1966 broadened the scope of interest in public pharmaceutical provenance from hospital-by-hospital and city-by-city approaches to state and national levels of analysis. The expanded role of state and federal governments as pharmaceutical purchasers greatly increased the scale of public pharmacy audit data and created new possibilities for linking pharmacy audit data with hospital records and laboratory data. One of the first researchers to investigate Medicaid audits as a laboratory for examining drug utilization was the pioneering pharmacoepidemiologist Wayne Ray, who saw within these linkages an opportunity to study the epidemiology of prescription drug use. By the mid-1970s, a rather slim list of states had begun to tabulate their own Medicaid prescription payments on computer systems, along with rudimentary data regarding patent demographics and physician information. Working with a colleague who had begun a series of validation studies between Medicaid pharmacy claims and Medicaid outcome data, Ray argued that the sort of comprehensive surveillance that the hospital study afforded in the closed bounds of the drug monitoring ward could now be created in the outpatient setting by applying the health service approach of Furstenberg and Muller to the new electronic Medicaid data.[77]

The case of chloramphenicol provided a natural experiment. While Chloromycetin had been broadly marketed since 1949 for a wide variety of conditions, its arena of recommended use was severely circumscribed in the early 1950s after its association with aplastic anemia, bone marrow depression, and increased bacterial resistance (see chapter 2). Although by the early 1960s, the appropriate use of chloramphenicol was unequivocally limited to a handful of life-threatening infections, its sales remained high well into the 1970s.[78] This clear contrast between the limited use recommended for chloramphenicol by the medical literature and the evident popularity of this well-marketed drug made chloramphenicol usage a useful marker of irrational prescribing for drug utilization researchers such as Paul Stolley and Louis Lasagna at Johns Hopkins.[79]

Taking advantage of this "sensitive measure for inappropriate prescribing," Ray and his colleagues cataloged *all* chloramphenicol prescriptions dispensed to *all* ambulatory Medicaid patients in the state of Tennessee between July 1, 1973, and June 30, 1974, linked the prescriptions to diagnostic codes listed in physicians' notes, and characterized both the patients and the physicians, using links to Medicaid and state licensure boards.[80] Nearly half of all chloramphenicol use in adults and three-quarters of all use in children was associated with routine viral upper respiratory infections or otitis media, which usually required no antibiotic at all. Only one of 593 linked diagnoses (a case of rickettsial disease) demonstrated a clearly appropriate use of chloramphenicol.

Ray had demonstrated that prescription databases could be used to target the physicians responsible for irresponsible drug use. Fewer than 10 percent of the total prescribers accounted for more than half (55%) of the inappropriate chloramphenicol prescriptions. While nearly three-quarters of physicians used these drugs fewer than five times, a handful of "frequent prescribers" had issued nearly two hundred chloramphenicol prescriptions apiece. Ray was also able to use the prescription data to demonstrate demographic factors associated with frequency of inappropriate prescribing: rural physicians tended to prescribe chloramphenicol inappropriately more than urban physicians, and family practitioners and general practitioners used the drug more often than internists or pediatricians.

This use of Medicaid data to link outpatient prescriptions systematically with outpatient diagnostic codes served to link the safety concerns of pharmacovigilance with the health service questions of drug utilization studies. When the members of the Joint Commission on Prescription Drug Use held their first meeting in 1976, their chief motivation was to pursue this sort of linkage: to find a way to combine a database of clinical outcomes with a database of the use of prescription drugs so that both fields might inform each another.[81]

Focusing on the Prescription

The Joint Commission on Prescription Drug Use was formed in response to a press conference held by Senator Edward Kennedy on November 30, 1976, at which he announced that the new science of drug utilization studies had provided irrefutable evidence that prescription drugs were ill used in American society.[82] Kennedy called for Congress to work with the medical profession and the pharmaceutical industry to sponsor a public-private body of expertise whose explicit purpose would be to establish a postmarket surveillance system for prescription drugs.[83] As the commission noted in its final report, the purpose of systematic prescription surveillance was "not merely to learn 'something' about a drug but to glean information . . . useful in improving the rational use of drugs."[84]

The commission ran from 1976 until 1979 and issued its final report in the first month of 1980. Conceived as a public-private venture, it included representatives from the American Medical Association, the American Hospital Association, the American Pharmaceutical Association, the American Society for Clinical Pharmacology and Therapeutics, the American Society for Pharmacology and Experimental Therapeutics, and the American Society of Hospital Pharmacists. Also on the commission were the assistant secretary for health (in the Department of Health, Education, and Welfare), the president of the Institute of Medicine, and representatives from the fields of pharmaceutical market research, prescription sociology, adverse-event surveillance, and drug utilization research.

One of the first actions taken by the commission was a "fact finding trip" to several European nations to learn how countries with centralized national health systems, such as England and Sweden, had set up prescription surveillance systems. The Americans felt they had been outpaced by the efforts of European countries, especially the Scandinavians, who had long histories of centrally organized pharmacy records and more tightly controlled national formularies of allowable drugs.[85] Moreover, the World Health Organization had set up a regional European Drug Utilization Group in Oslo, which held a prominent conference on the overprescribing of prescription drugs in 1969 and then proceeded to develop methods of comparing utilization across drug classes and across national pharmacy standards. Ironically, even in countries such as Sweden, much of the prescription data came from the private sector.[86]

Returning to the United States, the commission worked to integrate these findings within the American context, in a way that would account for the social, epidemiological, marketing, and policy interests in the prescription as a source of data. Initially, the prospects for a harmonization of these four perspectives seemed quite rosy. At the first meeting, Howard L. Binkley, vice president for research and planning of the Pharmaceutical Manufacturers Association, provided a description and critique of presently available sources of data on trends in the prescribing and dispensing of prescription drugs, with an emphasis on how market research data found in the NPA and the NDTI could be linked to broader systems of private and public claims and outcomes data.[87] From the public sector, representatives from the FDA described their postmarket surveillance programs, and representatives from the Department of Health, Education, and Welfare described how the growth of Medicaid data might form a system of automated computerized clinical outcomes that would link effortlessly with prescription claims data and market research products. As more state Medicaid programs developed computerized Medicaid Management Information Systems, Medicaid data could provide a snapshot of outpatient pharmacy and clinical data "covered more broadly than by any other data base."[88] Given the 25 million lives covered by Medicaid in 1978, it was estimated that a cohort of ten thousand patients could be assembled in the first year of marketing for thirty (70%) of the new drugs introduced from 1973 to 1977:

> The following basic types of information would be available for each patient enrolled in the study: records of all prescription drug use during the study (regardless of where the claim originated); the diagnosis for which the study drug was prescribed (*if* the diagnosis can be inferred from a physician visit claim for a corresponding date); other diagnoses reported by a physician provider prior to prescribing the study drug and diagnoses reported for subsequent visits to the prescribing physician, referral physician or other physician; diagnoses recorded for hospital outpatient clinic visits, inpatient stays or non-hospital clinic visits; laboratory tests ordered (but not results); and patient demographic information. All transactions records for a specific patient bear a unique number assigned by his state Medicaid Agency; the number then serves to link multiple providers and to protect the patient's identity.[89]

By the second meeting, the commission had broken up into interdisciplinary teams organized thematically, with the liveliest attention received in Marcus Reidenberg's committee on "sources of available data, its quality, and limitations." This committee heard presentations from IMS Health (the corporate owner of proprietary prescription data sources such as NPA and NDTI), sociologists who had conducted physician panels and community surveys, the organizers of hospital prescription surveillance units, scholars using Medicaid databases, and representatives from health maintenance organizations (HMOs) who had developed prescription data sets. Early on, the subcommittee agreed that there was "a considerable amount of valuable data on the epidemiology and patterns of use of prescription drugs, and on patterns and incidence of disease and its treatment which, if made available in a proper format, could be a useful tool in knowing about and improving prescription drug therapy. The subcommittee noted that the major limitation of systems describing 'how drugs are used' was the lack of outcome data (i.e., what did the drug do?)."[90]

As the commission assessed its findings by 1979, it became clear that although several data sets existed, no individual data set—whether NDA, NDTI, hospital surveillance, or Medicaid data—contained enough information to deliver the "constant observation of the medical and pharmaceutical horizon" that de Haen had called for in 1949 to allow the full assessment of drug use in practice. The commission began to interview hybrid data sources that illustrated new links between the public and private nature of prescriber data sets. Fledgling HMOs such as Kaiser Permanente and the Group Health Cooperative of Puget Sound developed in-house proprietary databases that linked both prescription claim data and outcome data in the same place.[91] Exploratory work by Hershel Jick following the use of the blockbuster anti-ulcer drug Tagamet (cimetidine) in Puget Sound pharmacies suggested that this approach could be quite promising indeed. Another hybrid form was introduced by Noel Munson, a spokesman from Prescription Card Services (PCS), a private prescription data company that acted as a "fiscal intermediary" for public payment groups like Medicare and Medicaid and other groups that paid for prescription drugs. But these individual companies (e.g., PCS) appeared to code their data according to their own proprietary software.[92] Even within the Medicaid system, the promise of effortless data linkage remained a dream in the late 1970s, complicated by wide state-by-state discrepancies in patterns of coding, storing, and retrieving prescription data.[93]

The final report of the commission, issued in early 1980, illustrates the close interconnection among the four realms of interest in prescription research: marketing, social science, pharmaco-epidemiology, and health policy. By early 1978, IMS Health had been contracted jointly by the Joint Commission and the FDA to organize a full report on the possible availability of public and private sources for prescription data. IMS Health also drafted a full proposal for a new quasi-governmental entity—along the lines of the Institute of Medicine—to collect and link prescription data into a broadly accessible data universe that would encompass all aspects of prescription use, from diagnosis, to prescription, to purchase, to filling in the pharmacy, to use by patients, and ultimately to clinical outcomes of safety and efficacy.[94] From the outset, the Joint Commission had sought to design a system that would swiftly receive congressional funding. To that end, staffers from Kennedy's office emphasized the virtue of public-private collaboration as representing a practical and attainable goal, insofar as the interests of therapeutic rationalists could be reconciled with the interests of firms seeking to rationalize their marketing practices.[95]

If the 1980 publication of the Joint Commission report represented a high point of collaboration between market researchers, epidemiologists, therapeutic reformers, and sociologists in conceptualizing the prescription as an object of study and a building block of a system of therapeutic surveillance, it also represented a dream of collaborative work that soon dissipated. Like many other grand designs for federally sponsored health programs conceived in the later 1970s and proposed in the early 1980s, its speculative structures never materialized, its measures were left unfunded, and subsequent calls for a center for postmarketing surveillance were repeated, and unfunded, every few years for several decades. Only very recently, with the passage of the Food and Drug Administration Amendments Act of 2007, was a substantial public investment made in the construction of a linked public prescription database for pharmaco-epidemiological research, via the creation of the FDA's new automated pharmacovigilance program, the Sentinel System, still under construction in 2011.

Conclusion

The brief life, utopian vision, and tactical failure of the Joint Commission on Prescription Drug Use is a fitting place to end this narrative of the emergence, convergence, and discordance between the multiple sciences of the prescription that took shape in late-twentieth-century America. These fields shared a fascination with employing a piece of medical waste—the discarded prescription—in the construction of viewing structures that might allow many parties to peer into the otherwise obscure world of therapeutic practice. In this narrative, both the market and the state represent creative spaces for the generation of new techniques of prescription surveillance.

Ironically, in contrast to the common narrative of the American medical profession's long legacy of claimed autonomy and preference for isolationist self-regulation, this narrative also illustrates the active role of the profession—

from the organizational level of the AMA business offices down to the individual physician participating in marketing research—in leveraging the prescription as a means of creating a vast social-relational database in which the ideal of "private practice" might be made more public. Ed F. Linder, a pharmaceutical advertising consultant, noted in an article in the *New York State Journal of Medicine* in 1964:

> The physician is not merely a recipient of drug advertising. Very often he is an active participant in the grand design of the marketing effort for a drug. His clinical reports readily become sources of reference to be cited and quoted; results of scientific investigation find their way into promotional literature, sometimes even before publication of the original article. Even the least communicative practitioner indirectly makes his contribution to drug promotion as his prescription habits are carefully scrutinized by the searching eye of market analysis, and data so extracted serve to guide the conduct of campaigns.[96]

It is difficult to know to what extent the average physician practicing from the 1950s through the 1980s was conscious of the expanding network of surveillance technologies and market research professionals working to visualize his prescribing practices. While physicians participated in building individual pieces of this structure, most were likely unaware of the extent of the system of surveillance as a whole. Nonetheless, their participation was crucial in the transformation of the clinic into an open panopticon of prescribing.

The database of prescriptions is now both big business—witness the sale of the largest prescriber data firm, IMS Health, for $4 billion in November 2009—and a key topic in health policy and planning. Looking back from a twenty-first-century vantage point, where sales of pharmaceutical claim data have become key revenue streams for insurers and pharmacy chains and where electronic prescriptions are linked seamlessly to electronic medical records, the rise of a database of prescriptions may seem to have been inevitable. But the future of pharmaceutical data was anything but obvious to those actors trying to piece together early databases of prescriptions in the 1950s, 1960s, and 1970s. Pharmaceutical marketers assembled dynamic data sets to track the factors influencing prescribing habits by region, specialty, and the individual physician. Would-be therapeutic reformers and health service researchers used similar data sets to discern whether physicians prescribed "rationally" or were guided instead by "irrational" factors such as marketing, local practice, ignorance, greed, and haste. Applied sociologists tracked prescription rates to generate bold hypotheses regarding the social transmission of innovation. Epidemiologists and pharmacologists combed records of pharmacies, hospitals, health systems, and insurance plans to link patterns of drug prescribing to mortality, morbidity, and adverse effects.

Like so many other ambitious goals of health system planning in the late 1970s, the plan to combine these four streams of prescriber research into a national database of prescriptions never came to fruition. In its place has developed a robust marketplace of prescription data that intersects only obliquely with the broader public health functions of prescription surveillance. Nonetheless, the sciences of the prescription should not be understood to have withered in isolation. Much as the prescription itself serves as a tangible link between clinical practice and medical commerce, so too each of the nascent sciences of the prescription linked the clinical, commercial, and public health arenas. Neither fully public nor fully private, the prescription, once filled at the pharmacy, became a research tool emblematic of the curious hybrid of market, profession, and state that is American medicine.

Time Line of Federal Regulations and Rulings Related to the Prescription

1902 The Biologicals Control Act establishes the first sweeping federal regulation of pharmaceutical products.

1906 The Pure Food and Drugs Act (also referred to as the "Wiley Act") establishes what will become the FDA within the Bureau of Agriculture. It specifies that drug labels must accurately reflect their contents, especially for specific agents known to be dangerous.

1912 The Sherley Amendment prohibits the marketing of pharmaceuticals with therapeutic claims clearly intended to defraud the purchaser.

1914 The Harrison Narcotic Act restricts sales of narcotics exceeding a specified limit to sale by prescription, and it mandates strict maintenance of pharmaceutical ledgers by pharmacists and dispensing physicians.

1938 The Federal Food, Drug, and Cosmetics Act grants the FDA power to conduct factory inspections and demand proof of safety before a pharmaceutical company markets new drugs; liberalizes the definition of drug misbranding; and creates the category of nonnarcotic prescription-only drugs. The act is passed following a national uproar over the Elixir Sulfanilamide tragedy.

1948 The Supreme Court rules in *United States v. Sullivan* that the FDA has the power to define and restrict the sale of prescription-only drugs.

1951 The Durham-Humphrey Amendment codifies the federal law defining pharmaceutical products unsafe to use without expert guidance and formally restricts their sale to prescription by physicians, veterinarians, or dentists.

1962 The Kefauver-Harris Drug Amendments establish the phase I–III sequence of clinical trials required to demonstrate efficacy prior to new-drug approval. These amendments also become the basis for the retroactive Drug Efficacy Study of new drugs marketed between 1938 and 1962. The amendments are passed following widespread publicity of the risks of thalidomide.

1965 The Drug Abuse Control Amendments aim to restrict the abuse of depressants, amphetamines, and hallucinogens. Medicare and Medicaid establish a broad federal commitment to purchasing pharmaceuticals.

1968 The Bureau of Narcotics and Dangerous Drugs is formed, transferring responsibilities from the FDA to the Department of Justice.

1970 The Comprehensive Drug Abuse Prevention and Control Act recategorizes drugs of abuse in terms of comparative therapeutic value and addictive potential. The FDA requires the first patient package insert for oral contraceptives.

1972 The Over-the-Counter Drug Review process is initiated by the FDA.

1984 The Drug Price Competition and Patent Term Restoration Act (also referred to as the "Hatch-Waxman Act") provides additional patent protection to innovator companies and establishes a clear approval pathway for follow-on drugs.

1988 The Prescription Drug Marketing Act bans the resale of prescription drugs and mandates licensing of drug wholesalers by the states.

1997 The Food and Drug Administration Modernization Act regulates the marketing of drugs for unapproved indications.

2003 The Medicare Prescription Drug, Improvement, and Modernization Act (also referred to as the "prescription drug bill") establishes Medicare Part D, which commits Medicare to covering prescription drug costs within a tightly defined range.

2007 The Food and Drug Administration Amendments Act grants new powers to the FDA to monitor and regulate pharmaceutical products already on the market.

Notes

Introduction

The authors wish to acknowledge the thoughtful comments of Jean-Paul Gaudillière, Gregory Higby, Charles Rosenberg, John Swann, and the contributors to this volume.

1. W. E. Lockhart, "How to Write a Prescription in the Second Half of the Twentieth Century," *Southwestern Medicine* 3 (1963): 101–2.

2. Ibid., 101.

3. Kaiser Family Foundation, "Trends and Indicators in the Changing Health Care Marketplace," Exhibit 1.21, February 8, 2006, www.kff.org/insurance/7031/print-sec1.cfm, accessed December 1, 2010.

4. Michael Bartholow, "Top 200 Prescription Drugs of 2009," *Pharmacy Times*, May 2010, www.pharmacytimes.com/issue/pharmacy/2010/May2010/RxFocusTopDrugs-0510, accessed December 1, 2010; Bill Berkrot, "US Prescription Sales Hit $300 Billion in 2009," *Reuters*, U.S. ed., April 1, 2010, www.reuters.com/article/idUSN3122364020100401, accessed December 1, 2010.

5. Jeremy Greene, *Prescribing by Numbers: Drugs and the Definition of Disease* (Baltimore: Johns Hopkins University Press, 2007), 248–49.

6. In that eleven-year period, the number of prescriptions increased 71 percent, while the U.S. population grew by just 9 percent. Kaiser Family Foundation, "Prescription Drug Trends," May 2007, www.kff.org/rxdrugs/upload/3057_06.pdf, accessed December 1, 2010.

7. For studies working within the frameworks of biographies, life cycles, and trajectories of drugs, see, e.g., Godelieve Van Heteren, Marijke Gijswijt-Hofstra, and Tilli Tansey, eds., *Biographies of Remedies: Drugs, Medicines, and Contraceptives in Dutch and Anglo-America Healing Cultures*, vol. 66 of *Clio Medica*, Wellcome Series in the History of Medicine (Amsterdam: Editions Rodopi, 2002); Jean-Paul Gaudillière, ed., "Drug Trajectories," special issue of *Studies in History and Philosophy of Biological and Biomedical Sciences* 36 (December 2005); Andrea Tone and Elizabeth Siegel Watkins, *Medicating Modern America: Prescription Drugs in History* (New York: New York University Press, 2007).

8. See, e.g., Erwin Ackerknecht, *Therapeutics from the Primitives to the 20th Century* (New York: Hafner Press, 1973).

9. Glenn Sonnedecker, ed., *Kremers and Urdang's History of Pharmacy*, 4th ed. (Madison, WI: American Institute of the History of Pharmacy, 1986).

10. See, e.g., Gianna Pomata, *Contracting a Cure: Patients, Healers, and the Law in Early Modern Bologna* (Baltimore: Johns Hopkins University Press, 1998).

11. "Review of Dr. Otto Wall's *The Prescription: Therapeutically, Pharmaceutically, Grammatically, and Historically Considered*, Fourth Edition," *California State Journal of Medicine* (April 1918): 217.

12. Otto Wall, *The Prescription: Therapeutically, Pharmaceutically, Grammatically, and Historically Considered*, 4th ed. (St. Louis: C.V. Mosby, 1917), preface to the 1st edition, 9.

13. As one review of Wall's text noted, "the medical student who reads it carefully, will find added dignity in the scrap of paper called a prescription." "Review of Wall's *The Prescription*," 217.

14. Ibid., 241.

15. Wall, *Prescription*, 13.

16. The focus of this volume is restricted to Western medicine. For accounts of non-Western medical traditions, see, e.g., Paul Unschuld, *Medicine in China: A History of Ideas* (Berkeley: University of California Press, 1985); Dominik Wujastyk, *The Roots of Ayurveda* (New Delhi: Penguin, 1988); Bridie Andrews-Minehan and A. R. Cunningham, eds., *Western Medicine as Contested Knowledge* (Manchester, UK: Manchester University Press, 1997); Ted Kaptchuk, *The Web That Has No Weaver: Understanding Chinese Medicine* (Chicago: Contemporary Books, 2000); Volker Scheid, *Chinese Medicine in Contemporary China* (Durham, NC: Duke University Press, 2002); Shigehisa Kuriyama, *The Expressiveness of the Body and the Divergence of Greek and Chinese Medicine* (New York: Zone Books, 2002); Helaine Selin, ed., *Medicine across Cultures: History and Practice of Medicine in Non-Western Cultures* (Norwell, MA: Kluwer, 2003); Carla Nappi, *The Monkey and the Inkpot: Natural History and Its Transformations in Early Modern China* (Cambridge, MA: Harvard University Press, 2009).

17. Greg Higby, "Evolution of Pharmacy," in *Remington: The Science and Practice of Pharmacy*, ed. A. R. Gennaro (Baltimore: Lippincott, Williams & Wilkins, 2000), 11–13.

18. Wilbur L. Scoville, *The Art of Compounding: A Text Book for Students and a Reference Book for Pharmacists at the Prescription Counter*, 4th ed. (Philadelphia: P. Blakiston's, 1914), 14.

19. Bernard Fantus. *A Text Book on Prescription-Writing and Pharmacy with Practice in Prescription-Writing, Laboratory Exercises in Pharmacy, and a Reference List of the Official Drugs Especially Designed for Medical Students*, 2nd ed. (Chicago: Chicago Medical Book, 1905), 70.

20. Frederic Henry Gerrish, *Prescription Writing Designed for the Use of Medical Students Who Have Never Studied Latin*, 4th ed. (Portland, ME: Loring, Shortt & Harmon, 1888), 5.

21. Ibid., 6.

22. Scoville, *Art of Compounding*, 7.

23. Ibid., 6. Although full fluency in Latin might have been preferred, most compounding pharmacists had to "depend upon the unaided memory, or upon tables of technical terms" (6).

24. Jeremy A. Greene, "What's in a Name? Generic Drugs and the Persistence of the Pharmaceutical Brand," *Journal of the History of Medicine and Allied Sciences* (e-publication ahead of print, October 2010, http://jhmas.oxfordjournals.org/content/early/2010/09/21/jhmas.jrq049.abstract).

25. Gregory J. Higby, "The Continuing Evolution of American Pharmacy Practice," *Journal of the American Pharmaceutical Association* 42 (January–February 2002): 12.

26. Peter Temin, "The Origin of Compulsory Drug Prescriptions," *Journal of Law and Economics* 22 (1979): 91–105.

27. Federal Food, Drug, and Cosmetic Act of 1938, 502(f), 52 Stat. 1050–51 (1938), as cited in Temin, "Origin of Compulsory Drug Prescriptions," 96.

28. Statement of W. G. Campbell, Food, Drugs, and Cosmetics (S. 1944), in *Hearings before a Subcommittee of the Senate Committee on Commerce*, 73rd Cong., 2nd sess., 1934, as cited in Temin, "Origin of Compulsory Drug Prescriptions," 96.

29. Memorandum submitted by Winthrop Chemical Company, November 25, 1938, in *Hearings on Proposed Regulations Nov 17 and 18, 1938*, book 2, accession 88-52A-89, box 144, folder 603, National Record Center, Washington, DC, cited in Harry M. Marks, "Revisiting 'The Origins of Compulsory Prescriptions,'" *American Journal of Public Health* 85 (1995): 110.

30. Marks, "Revisiting," 111. See also Dominique Tobbell, "Allied against Reform: Pharmaceutical Industry–Academic Physician Relations in the United States, 1945–1970," *Bulletin of the History of Medicine* 82 (Winter 2008): 878–912.

31. Daniel P. Carpenter, *Reputation and Power: Organizational Image and Pharmaceutical Regulation at the FDA* (Princeton, NJ: Princeton University Press, 2010); John Swann, "FDA and the Practice of Pharmacy: Prescription Drug Regulation before the Durham-Humphrey Amendment of 1951," *Pharmacy in History* 36, no. 2 (1994): 55–70. Swann points out that as early as 1938, the FDA had taken the lead in proclaiming that sulfa drugs, aminopyrine, and cinchophen would be dangerous if labeled for "indiscriminate" use by the general public. By 1941 the FDA had identified more than twenty drugs or drug classes too dangerous to sell without a prescription, but it maintained that this list was merely a set of examples and insisted that manufacturers had ultimate responsibility for the labeling decision.

32. Swann, "FDA and the Practice of Pharmacy," 60. As far back as the 1880s, firms reported self-restricting their distribution of possibly harmful drugs to pharmacists who agreed to fill them only with prescriptions. Conversely, many drug manufacturers who produced items that were clearly safe for public consumption sought the increased professional veneer of the "prescription-only" status.

33. It is worth mentioning here that the 1938 and 1951 legislation did not entirely restrict authority of the prescription to physicians—both dentists and veterinarians were explicitly mentioned as well. Nonetheless, in nonoral forms of human medicine, the MD (and later the DO as well) was understood to wield the power of the

prescription pad. In most MD–non-MD clinical border zones (e.g., the line between psychologists and psychiatrists, or between optometrists and ophthalmologists), the power of the prescription remained a key distinction.

34. Thomas Maeder, *Adverse Reactions* (New York: William Morrow, 1994); P. D. Stolley and L. Lasagna, "Prescribing Patterns of Physicians," *Journal of Chronic Diseases* 22 (1969): 395–405; P. D. Stolley, M. H. Becker, and J. D. McEvilla, "Drug Prescribing and Use in an American Community," *Annals of Internal Medicine* 76 (1972): 537–40; Wayne Ray, "Prescribing of Chloramphenicol in Ambulatory Practice—An Epidemiological Study among Tennessee Medicaid Recipients," *Annals of Internal Medicine* 84, no. 3 (1976): 266.

35. Nora D. Volkow, "Testimony at Congressional Caucus on Prescription Drug Abuse," September 22, 2010, www.nida.nih.gov/Testimony/9–22–10Testimony.html #foot, accessed December 1, 2010.

36. Ibid.

37. Andrew Barry, "Pharmaceutical Matters: The Invention of Informed Materials," *Theory, Culture and Society* 22, no. 1 (2005): 51–69.

CHAPTER ONE: Goofball Panic

The author wishes to thank David Courtwright, Jeremy Greene, David Herzberg, Liz Watkins, and other colleagues and reviewers for comments on drafts, and to acknowledge funding from the Australian Research Council under grants DP0984694 and DP0449467.

1. Rita H. Kleeman, "Sleeping Pills Aren't Candy," *Saturday Evening Post*, February 24, 1944, 17, 85.

2. "Sister Aimee's Death Laid to Sleeping Pills," *Chicago Tribune*, October 14, 1944, sec. 1, p. 1; "Aimee's Death Blamed on Overdose of Capsules," *Los Angeles Times*, October 14, 1944, sec. 1, pp. 1, 4; "Lupe Velez Suicide over Love Tragedy," *Los Angeles Times*, December 15, 1944, sec. 1, pp. 1–3; "Lupe's Estate Worth $200,000; Burial Monday," *Chicago Tribune*, December 16, 1944, sec. 1, p. 9; "Raymond Lays Lupe's Death to 'Fake Marriage' Error," *Los Angeles Times*, December 17, 1944, A1; H. W., "Doctor's Pen Is More Potent Than Most Prescriptions," *Washington Post*, December 29, 1946, S6.

3. Peter Temin, "The Origin of Compulsory Drug Prescriptions," *Journal of Law and Economics* 22 (1979): 91–105; Harry Marks, "Revisiting 'The Origins of Compulsory Drug Prescriptions,'" *American Journal of Public Health* 85 (1995): 109–15. I argue that the clashing, insightful views of Temin and Marks are both flawed by omission. Both authors neglect the role of public concern over barbiturates, without which we cannot understand the fact that federal prescription-only controls were actually imposed, despite resistance from manufacturers and pharmacists.

4. Stanley Cohen, *Folk Devils and Moral Panics: The Creation of the Mods and Rockers* (Oxford: Martin Robertson, 1972); Jason Ditton, *Contrology: Beyond the New Criminology* (London: Macmillan, 1979). For a useful literature review, see Erich Goode and Nachman Ben-Yahuda, "Moral Panics: Culture, Politics, and Social Construction,"

Annual Review of Sociology 20 (1994): 149–71; David Garland, "On the Concept of Moral Panic," *Crime Media Culture* 4 (2008): 9–30.

5. David Musto, *American Disease: Origins of Narcotic Control* (New Haven, CT: Yale University Press, 1973); David Courtwright, *Dark Paradise: Opiate Addiction in America before 1940* (Cambridge, MA: Harvard University Press, 1982); Eve K. Sedgwick, "Epidemics of the Will," in *Tendencies* (Durham, NC: Duke University Press, 1993), 130–42. On other countries, see Virginia Berridge, "Morality and Medical Science: Concepts of Narcotic Addiction in Britain, 1820–1926," *Annals of Science* 36 (1979): 67–85; Desmond Manderson, *From Mr Sin to Mr Big: A History of Australian Drug Laws* (Melbourne: Oxford University Press, 1993); Catherine Carstairs, "Deporting 'Ah Sin' to Save the White Race: Moral Panic, Racialization, and the Extension of Canadian Drug Laws in the 1920s," *Canadian Bulletin of Medical History* 16 (1999): 65–88; Erving Goffman, *Stigma: Notes on the Management of Spoiled Identity* (New York: Simon and Schuster, 1963).

6. Musto, *American Disease;* Alfred Lindesmith, "The Drug Addict as a Psychopath," *American Sociological Review* 5 (1940): 914–20; Goffman, *Stigma*, 4; Mara L. Keire, "Dope Fiends and Degenerates: The Gendering of Addiction in the Early Twentieth Century," *Journal of Social History* 31 (1998): 809–22; Caroline Acker, "From All Purpose Anodyne to Marker of Deviance: Physicians' Attitudes towards Opiates in the US from 1890 to 1940," in *Drugs and Narcotics in History*, ed. Roy Porter and Mikulás Teich (Cambridge: Cambridge University Press, 1995), 114–32; Caroline Acker, *Creating the American Junkie: Addiction Research in the Classic Era of Narcotic Control* (Baltimore: Johns Hopkins University Press, 2002); Nancy Campbell, *Discovering Addiction: The Science and Politics of Substance Abuse Research* (Ann Arbor: University of Michigan Press, 2007); Nicolas Rasmussen, "Maurice Seevers, the Stimulants, and the Political Economy of Addiction in American Biomedicine," *Biosocieties* 5 (2010): 105–23. For an excellent early summary and source for the older literature, see Peter Conrad and Joseph Schneider, *Deviance and Medicalization: From Badness to Sickness* (St. Louis: Mosby, 1980), chap 5.

7. Howard Becker, *Outsiders: Studies in the Sociology of Social Deviance* (New York: Free Press, 1963), chaps. 7–8 and passim; David Courtwright, "Introduction: The Classic Era of Narcotic Control," in *Addicts Who Survived: An Oral History of Narcotics Use in America, 1923–1965*, ed. D. T. Courtwright, H. Joseph, and D. Des Jarlais (Knoxville: University of Tennessee Press, 1989), 1–44; Joseph Spillane, *Cocaine: From Medical Marvel to Modern Menace in the United States, 1884–1920* (Baltimore: Johns Hopkins University Press, 2000), chap. 6. On patent medicines, the FDA, and addictive drugs, see James Harvey Young, *Pure Food: Securing the Federal Food and Drugs Act of 1906* (Princeton, NJ: Princeton University Press, 1989); Rufus King, *The Drug Hang Up: America's Fifty Year Folly* (New York: Norton, 1972), chaps. 3–6.

8. Christopher Callahan and German Berrios, *Reinventing Depression: A History of the Treatment of Depression in Primary Care, 1940–2004* (Oxford: Oxford University Press, 2005), chaps. 3–4; "Lullaby Pill Peril," *Newsweek*, March 13, 1939, 36–37; W. E. Hambourger and Council on Pharmacy and Chemistry, "A Study of the Promiscuous

Use of Barbiturates. I. Their Use in Suicides," *Journal of the American Medical Association* 112 (1939): 1340–43; "A Study of the Promiscuous Use of Barbiturates. II. Analysis of Hospital Data," *Journal of the American Medical Association* 114 (1940): 2015–19; Bureau of Legal Medicine and Legislation, "Regulation of the Sale of Barbiturates by Statute," *Journal of the American Medical Association* 114 (1940): 2029–36; Editorial, "Barbital and Its Derivatives," *Journal of the American Medical Association* 114 (1940): 2020–21, quote on 2020; Charles O. Jackson, "Before the Drug Culture: Barbiturate/Amphetamine Use in American Society," *Clio Medica* 11 (1976): 47–58; John Swann, "FDA and the Practice of Pharmacy: Prescription Drug Regulation before the Durham-Humphrey Amendment of 1951," *Pharmacy in History* 36, no. 2 (1994): 55–70; Toine Pieters and Stephen Snelders, "From King Kong Pills to Mother's Little Helpers: Career Cycles of Two Families of Psychotropic Drugs: The Barbiturates and Benzodiazepines," *Canadian Bulletin of Medical History* 24 (2007): 93–112.

9. On medicine's headlong embrace of science, see Kenneth Ludmerer, *Learning to Heal: The Development of American Medical Education* (New York: Basic Books, 1985); John Harley Warner, *The Therapeutic Perspective: Medical Practice, Knowledge, and Identity in America, 1820–1885* (Cambridge, MA: Harvard University Press, 1986); John P. Swann, *Academic Scientists and the Pharmaceutical Industry: Cooperative Research in Twentieth-Century America* (Baltimore: Johns Hopkins University Press, 1988); Harry Marks, *The Progress of Experiment: Science and Therapeutic Reform in the United States, 1900–1990* (Cambridge, MA: Cambridge University Press, 1997); Nicolas Rasmussen, "The Moral Economy of the Drug Company–Medical Scientist Collaboration in Interwar America," *Social Studies of Science* 34 (2004): 161–85.

10. Federal Food, Drug, and Cosmetic Act of 1938, 75th Cong. Ch. 675 (52 US Statutes 1040), June 25, 1938. On the 1938 amendments, see James Harvey Young, *The Medical Messiahs: A Social History of Health Quackery in Twentieth-Century America* (Princeton, NJ: Princeton University Press, 1992), chap 8; Swann, "FDA and the Practice of Pharmacy"; Marks, "Revisiting"; see also Daniel Carpenter, *Reputation and Power: Organizational Image and Pharmaceutical Regulation at the FDA* (Princeton, NJ: Princeton University Press, 2010), chap. 2. Temin, "Origin of Compulsory Drug Prescriptions," 99, cites the "potent but dangerous" phrase (from the FDA's 1939 Annual Report). For FDA compulsion in amphetamine's classification, see Nicolas Rasmussen, *On Speed: The Many Lives of Amphetamine* (New York: New York University Press, 2008), chap. 4 and passim. For 1930s consumerism, see Arthur Kallett and F. J. Schlink, *100,000,000 Guinea Pigs: Dangers in Everyday Foods, Drugs, and Cosmetics* (New York: Grosset and Dunlap, 1933); Nancy Tomes, "Merchants of Health: Medicine and Consumer Culture in the United States 1900–1940," *Journal of American History* 88 (2001): 519–47; and Lizabeth Cohen, *A Consumer's Republic: The Politics of Mass Consumption in Postwar America* (New York: Vantage, 2003), chap. 1.

11. John Morton Blum, *V Was for Victory: Politics and American Culture during World War II* (New York: Harcourt, 1976); Marilyn Hegarty, *Victory Girls, Khaki-Wackies, and Patriotutes: The Regulation of Female Sexuality during World War II* (New York: New York University Press, 2008) ("The promiscuous girl" quote on 144); John Parascan-

dola, "Quarantining Women: Venereal Disease Rapid Treatment Centers in World War II America," *Bulletin of the History of Medicine* 83 (2009): 431–59.

12. James Gilbert, *A Cycle of Outrage: America's Reaction to the Juvenile Delinquent in the 1950s* (Oxford: Oxford University Press, 1988), chap. 2; Douglas Henry Daniels, "Los Angeles Zoot: Race 'Riot,' the Pachuco, and Black Music Culture," *Journal of African American History* 87 (Winter 2002): 98–118; Rasmussen, *On Speed*, chap. 4.

13. Alles to A. J. Affleck, December 29, 1944, with attached "Memorandum to Secretary and Members of the California State Board of Pharmacy" (1944) by "Baes and Dowdy, Inspectors"; Anon., December 15, 1944, "Hypnotic Drug Law"; Anon. [Alles?], undated, "Brief Presented to Mr Kraft" (handwritten); Alles to O. J. May, January 7, 1945; "Senate Passes Two Assembly Measures," *Pasadena Star News*, January 18, 1945, unpaginated clipping, all in box 12, folder "California Board of Pharmacy Benzedrine Legislation A.B. 285—1945," Gordon Alles Papers, California Institute of Technology archives.

14. Business Bulletin, *Wall Street Journal*, March 22, 1945, sec. 1, p. 1; "Sharp Rise in Deaths by Sleeping Pills Brings Warning by Dr Gonzales on Sales," *New York Times*, July 22, 1945, sec. 1, p. 25; "Sleeping Pill Curb Drawn for Albany," *New York Times*, September 10, 1945, sec. 1, p. 21; "New City Drive to Clamp Down on the Sale of Misbranded or Harmful Drugs Here," *New York Times*, September 10, 1946, sec. 1, p. 9; "Weinstein Warns of New Controls to Check Sales of Sleeping Pills," *New York Times*, October 15, 1946, sec. 1, p. 35. Also Swann, "FDA and the Practice of Pharmacy."

15. Davis, opinion, in United States v. Sullivan, No. 3688, 67 F.Supp. 192 (U.S. District Court for the middle district of Georgia, Columbus division), June 19, 1946 (quoted). See also United States v. Sullivan, Trading as Sullivan's Pharmacy, No. 121, 332 U.S. 689; 68 S. Ct. 331; 92 L. Ed. 297, 1948; "Food and Drug Unit Acts to Stop 'Self-Doctoring': Court Authorizes Prosecution for Sale of Drugs without Prescription," *Toledo Blade*, February 4, 1948, sec. 1, p. 3.

16. Temin, "Origin of Compulsory Drug Prescriptions"; see also Carpenter, *Reputation and Power*, chap 2 and passim.

17. Kleeman, "Sleeping Pills Aren't Candy," quote on 17.

18. Editorial, *Saint Petersburg Evening Independent*, March 8, 1945, sec. 1, p. 14; "Unprescribed Use of Sleeping Pills Brings Warning," *Washington Post*, September 5, 1945, sec 1, p. 2; Bess M. Wilson, "Addiction Peril," *Los Angeles Times*, April 24, 1949, C1; "Clubwomen Score Drug Act Failure," *Los Angeles Times*, November 29, 1949, B1; "Students' Use of Drug Told by P.T.A. Official," *Christian Science Monitor*, July 25, 1946, sec. 1, p. 15 ("America's Opium"); Editorial, "Education Badly Needed to Fight the Sleeping Pill Problem," *Schenectady Gazette*, January 28, 1947, sec. 1, p. 10 ("vitality").

19. Booton Herndon, "More Victims of the DEVIL'S CAPSULES," *American Weekly (Milwaukee Sentinel)*, December 28, 1947, 1; Edith Nourse Rogers statement, *Miscellaneous Bills: Hearings before the House Committee on Ways and Means*, 80th Cong., 1st sess., May 14–16, 1947, 69–70; Edith Nourse Rogers testimony in *Control of Narcotics, Marihuana, and Barbiturates: Hearings before a Subcommittee of the House Committee on Ways and Means*, 82nd Cong., 1st sess., on H. R. 3490 and H. R. 348, April 7, 14,

and 17, 1951, 3–8; King, *Drug Hang Up*, chap. 25; "Legislature to Remain in Session Three More Weeks," *Nashua (NH) Telegraph*, June 3, 1947, 4.

20. Paul Green, "The Barbiturate Time Bomb," *American Druggist*, September 1945, 76, 139–40, 144, 146.

21. "Druggists Protest Health Board's Plan to Tighten Up Sales of Sleeping Potions," *New York Times*, May 9, 1947, sec. 1, p. 23; "New Curb Placed on Sleeping Pills," *New York Times*, July 9, 1947, sec. 1, p. 25.

22. Robert Geiger, "US Wolfs Sleeping Pills by the Hundred Tons," *Washington Post*, January 26, 1947, B6; "Legislature to Remain in Session Three More Weeks," *Nashua (NH) Telegraph*, June 3, 1947, 4; "Druggists Warn of 'Hysteria,'" *New York Times*, October 20, 1947, sec. 1, p. 19.

23. Wallace Werble, "Waco Was a Barbiturate HOT SPOT," *Hygeia* 23 (June 1945): 432–33; "Academy Endorses Sleeping Pill Curb," *New York Times*, November 1, 1947, sec. 1, p. 8.

24. Geiger, "US Wolfs Sleeping Pills" (quoted); "Barbiturate Curb Backed by Doctors," *New York Times*, April 3, 1947, 26; Marks, "Revisiting"; Nicolas Rasmussen, "The Drug Industry and Clinical Research in Interwar America: Three Types of Physician Collaborator," *Bulletin of the History of Medicine* 79, no. 1 (2005): 50–80.

25. Theodore Klumpp, "Sleep and Sleeping Pills," *American Mercury*, October 1945, 457–62; see Tom Mahoney, *The Merchants of Life: An Account of the American Pharmaceutical Industry* (New York: Harper, 1959), chap. 14.

26. "Sleep Pill Curbs," *Business Week*, October 20, 1945, 36; "New York's Drive on Sleeping Pills Pays Off," *Los Angeles Times*, August 22, 1948, C15; Geiger, "US Wolfs Sleeping Pills" (quoted).

27. *Cong. Rec.* 97 (82nd Cong., 1st sess., House of Representatives), August 1, 1951, 9327 (Rep. O'Hara) and 9331 (Rep. Hale).

28. Young, *Medical Messiahs*, chap. 25; Marks, "Revisiting," 111; John P Swann, "Drug Abuse Control under FDA, 1938–1968," *Public Health Reports* 112 (1997): 83–86; Sam Fine, undated interview, www.fda.gov/AboutFDA/WhatWeDo/History/OralHistories/PerspectiveonFDAOralHistories/default.htm#ref, accessed August 31, 2010. See "Drive Launched on Sleeping Pills," *Christian Science Monitor*, March 12, 1946, sec. 1, p. 2; "Nab 6 Druggists in U.S. Drive on Sedative Sales," *Chicago Tribune*, January 19, 1950, A6; "20 Convictions under Food and Drug Act Listed," *Chicago Tribune*, July 22, 1950, B8; Associated Press, "Risky Drugs, Poison Foods Bother FDA," *Washington Post*, February 7, 1951, sec. 1, p. 6.

29. Rasmussen, "Maurice Seevers"; Arthur Tatum, Maurice Seevers, and Ken Collins, "Morphine Addiction and Its Physiological Interpretation Based on Experimental Evidence," *Journal of Pharmacology and Experimental Therapeutics* 36 (1929): 447–75; Maurice Seevers and Arthur Tatum, "Chronic Experimental Barbiturate Poisoning," *Journal of Pharmacology and Experimental Therapeutics* 42 (1931): 217–31; Arthur Tatum and Maurice Seevers, "Theories of Drug Addiction," *Physiological Reviews* 11 (1931): 107–21.

30. Harris Isbell, "Addiction to Barbiturates and the Barbiturate Abstinence Syn-

drome," *Annals of Internal Medicine* 33 (1950): 108–21; Harris Isbell, Sol Altschul, C. H. Kornetsky, A. J. Eisenman, H. G. Flanary, and H. F. Fraser, "Chronic Barbiturate Intoxication: An Experimental Study," *Archives of Neurology and Psychiatry* 64 (1950): 1–28; "Sleeping Pills Declared Worse Than Morphine," *Los Angeles Times*, April 21, 1950, sec. 1, p. 1; Richard Williams, "To Sleep: Perchance . . . ," *Life*, October 13, 1952, 105–18.

31. Chalmers Roberts, "Bill to End Leniency to Dope 'Repeaters' Gets Green Light; DC Judges Criticized," *Washington Post*, April 8, 1951, M1; "War on Narcotics Urged," *New York Times*, May 16, 1951, sec. 1, p. 42; Howard Rusk, "Headway Seen for Control of Teen-Age Narcotic Users," *New York Times*, May 20, 1951, sec. 1, p. 62; Harold Hinton, "3 Minors Recount Narcotic Scourge," *New York Times*, June 27, 1951, sec. 1, p. 19 ("leprosy" quote); King, *Drug Hang Up*, chap 25.

32. Public Law 215, An Act to Amend Sections 303(c) and 503(b) of the Federal Food, Drug, and Cosmetic Act, 82nd Cong., 1st sess. (65 *U.S. Statutes* 648, 1951); Young, *Medical Messiahs*, chap. 12; Temin, "Origin of Compulsory Drug Prescriptions"; Marks, "Revisiting"; Carpenter, *Reputation and Power.*

33. "House Unit Urges US to Control Barbiturates," *Toledo Blade*, August 5, 1951, sec. 1, p. 10; Editorial, "Control of Barbiturates," *Journal the American Medical Association* 148 (1952): 1126–27; King, *Drug Hang Up*, chap. 25; cf. Marks, "Revisiting."

34. These were the same forces urging the FDA to regulate the Benzedrine Inhaler at the same time. Charles O. Jackson, "The Amphetamine Inhaler: A Case Study of Medical Abuse," *Journal of the History of Medicine and Allied Sciences* 26 (1971): 187–96; Rasmussen, *On Speed*, chap. 4.

35. Eve Edstrom, "4 Witnesses Relate Use of Narcotics by Children; Appear at Odds with Anslinger Testimony on Addiction," *Washington Post*, November 25, 1953, sec. 1, p. 1 ("orgies"); King, *Drug Hang Up*, chaps. 25, 26 ("dissolution" quote on 268); testimony of George Larrick, in *Juvenile Delinquency (National, Federal and Youth-Serving Agencies): Hearings before the Subcommittee to Investigate Juvenile Delinquency of the Senate Committee on the Judiciary*, 83rd Cong., 1st sess., pursuant to S. Res. 89, November 19, 20, 23, 24, 1953, part 1, p. 284 ("deviates"); Larrick testimony, *Hearings of Subcommittee on Health, Senate Committee on Labor and Public Welfare*, 88th Cong., 2nd sess., August 3, 1964, 20; Dodd testimony, ibid., 48.

36. Richard Spong, "Cheap Drugs Menace," *Fredericksburg (VA) Free-Lance Star*, September 19, 1955, 6; "What to Do about Drug Addicts," *Hartford (CT) Courant*, September 26, 1955, 10; "US Cracks Down on 'Goofballs,'" *Schenectady Gazette*, October 26, 1955, 5; A. E. Hotchner, "A Growing Threat to America's Youth: Dope . . . Ten Cents a Shot," *Los Angeles Times*, August 28, 1955, sec. 1, pp. 8–9; Norma Lee Browning, "The Facts about 'THRILL PILLS,'" *Chicago Tribune*, March 10, 1957, D18, D42.

37. Gilbert, *Cycle of Outrage*. As with delinquency and the "goofball" panics, anxiety over Communist infiltration can ultimately be related to a crisis in masculinity and gender roles. The powerful "nuclear fear" of the era can likewise be linked partly to gender anxieties, if we allow that impotence in the face of instant annihilation challenged masculine identity. See K. A. Cuordileone, "'Politics in an Age of Anxiety':

Cold War Political Culture and the Crisis in American Masculinity," *Journal of American History* 87 (2000): 515–45; David K. Johnson, *The Lavender Scare: The Cold War Persecution of Gays and Lesbians in the Federal Government* (Chicago: University of Chicago Press, 2004); Paul Boyer, *By the Bomb's Early Light: American Thought and Culture at the Dawn of the Atomic Age* (New York: Pantheon, 1985); Spencer Weart, *Nuclear Fear: A History of Images* (Cambridge, MA: Harvard University Press, 1998).

38. On the 1965 Drug Abuse Control Amendments, see King, *Drug Hang Up*, chaps. 22, 26; Swann, "Drug Abuse Control"; Rasmussen, *On Speed*, chap. 6. The 1947 and 1951 barbiturate figures are from drug trade sources; see Isbell et al., "Chronic Barbiturate Intoxication"; "Pills Used for Thrill," *Berks County (PA) Reading Eagle*, March 22, 1953, 40; and Phyllis Rosenteur, "Sleeping Pill Boom Gives New Life to Mickey Finn," *Washington Post*, June 26, 1957, C4; Elijah Adams, "Barbiturates," *Scientific American*, January 1958, 60–64. Jackson, "Before the Drug Culture," cites FDA sources for describing barbiturate consumption as plateauing in the early 1950s. The New York Academy of Sciences Committee on Public Health endorsed the 1960 barbiturate consumption figure of 6 billion 1-grain, or 4.5 billion 1.5-grain doses annually; see *Misuse of Valuable Therapeutic Agents: Barbiturates, Tranquilizers, and Amphetamines: Hearings before the Committee on Interstate and Foreign Commerce*, House of Representatives, 89th Cong., 1st sess., on H. R. 2, January 27–February 10, 1965, 57–62. FDA's 1960 barbiturate figures supplied to Dodd are described by him in the same hearings at p. 46. On partial replacement of barbiturates by minor tranquilizers, see Mickey Smith, *Small Comfort: A History of the Minor Tranquilizers* (New York: Praeger, 1985); Andrea Tone, *The Age of Anxiety* (New York: Basic Books, 2008).

39. King, *Drug Hang Up*, chaps. 25–26 and passim, accuses FDA of "puffing" barbiturate and amphetamine figures in the period. FDA might be faulted for expressing barbiturate consumption in terms of 1.0 grain sedating doses instead of 1.5 grain sleeping doses, typically used by the drug industry, but figures were derived from drug industry sources. Amphetamine consumption estimates offered by FDA also square with figures derived from industry and other non-FDA sources. Nicolas Rasmussen, "America's First Amphetamine Epidemic, 1929–1971: A Quantitative and Qualitative Retrospective," *American Journal of Public Health* 98 (2008): 974–85; and unpublished data in possession of author.

40. King, *Drug Hang Up*, chaps. 22, 26; Rasmussen, *On Speed*, chaps. 6–7; Swann, "Drug Abuse Control"; John Ingersoll, February 1, 1972, in *Diet Pill (Amphetamines) Traffic, Abuse, and Regulation: Hearing before the Subcommittee to Investigate Juvenile Delinquency*, Senate Committee on the Judiciary, pursuant to S. Res. 32, sec. 12, 92nd Cong., 1st sess., February 7, 1972, 226–32; also see statement of M. Costello, ibid., 204–6.

41. Stephanie Fuller, "Parents to Battle Teen Drug Use," *Chicago Tribune*, April 2, 1970, A1. On LSD use as stirring a moral panic, and an object lesson in the irresolvable problems of "disproportionality" as a criterion of moral panic, see B. Cornwell and A. Linders, "The Myth of 'Moral Panic': An Alternative Account of LSD Prohibi-

tion," *Deviant Behavior* 23 (2002): 307–30; Erich Goode, "Moral Panics and Disproportionality: The Case of LSD Use in the Sixties," *Deviant Behavior* 29 (2008): 533–43. See Becker's keen observations on how the LSD drug panic could actually have produced "bad trips" among users: Howard Becker, "History, Culture and Subjective Experience: An Exploration of the Social Bases of Drug-Induced Experiences," *Journal of Health and Social Behavior* 8 (1967): 163–76; Jay Stevens, *Storming Heaven: LSD and the American Dream* (New York: Grove Press 1998), chap. 24 and passim.

42. Sue Avery, "Police Sergeant Warns about Peril of Drugs in the Home," *Los Angeles Times,* March 25, 1970, C10; "Drugs in Suburbia: Children of Affluence, Bored and Disillusioned, Turn to Pot and Pills," *Wall Street Journal,* November 12, 1970, sec. 1, p. 1; "Drugs in Suburbia: Alarmed Towns Mount Attacks on Use of Pot, Pills by Young People," *Wall Street Journal,* November 18, 1970, sec. 1, p. 1; Nancy Baltad, "Easy to Get Drugs, Students Admit," *Los Angeles Times,* July 29, 1971, SF5. On the growing perceived connection between "street drugs" and pharmaceuticals, see Susan Speaker, "From 'Happiness Pills' to 'National Nightmare': Changing Cultural Assessment of Minor Tranquilizers in America, 1955–1980," *Journal of the History of Medicine and Allied Sciences* 52 (1997): 338–76; Andrea Tone, "Tranquilizers on Trial: Psychopharmacology in the Age of Anxiety," in Tone and Watkins, *Medicating Modern America,* 156–80; David Herzberg, " 'The Pill You Love Can Turn on You': Feminism, Tranquilizers, and the Valium Panic of the 1970s," *American Quarterly* 5 (2006): 79–103; Rasmussen, *On Speed,* chap. 7; Rasmussen, "Maurice Seevers."

43. "Housewives among Top Users of 'Drugs,' " *Los Angeles Times,* June 25, 1971, sec. 1, p 17; Carl Chambers and Dodi Schultz, "Women and Drugs: A Startling Journal Survey, the Drugs Women Use," *Ladies Home Journal,* November 1971, 130; Carl Chambers and Dodi Schultz, "Housewives and the Drug Habit: What They Take—And Why," *Ladies Home Journal,* December 1971, 66.

44. "U.S. Is on 'Collective Trip' from Drugs, Agnew Says," *Washington Post,* June 24, 1970, A2; David T. Courtwright, "The Controlled Substances Act: How a "Big Tent" Reform Became a Punitive Drug Law," *Drug and Alcohol Dependence* 76 (2004): 9–15; King, *Drug Hang Up,* chap. 28.

45. Courtwright, "Controlled Substances Act"; Joseph F. Spillane, "Debating the Controlled Substances Act," *Drug and Alcohol Dependence* 76 (2004): 17–29; William MacAllister, "The Global Political Economy of Scheduling: The International-Historical Context of the Controlled Substances Act," *Drug and Alcohol Dependence* 76 (2004): 3–8. U.S. manufacturers reported producing 25 billion milligrams of amphetamine and methamphetamine base (before conversion to salts) in 1969, and based on 1971 prescribing, they were allowed to produce 2.5 billion milligrams in 1972. See statement of John Edwards, in *Diet Pill (Amphetamines) Traffic, Abuse, and Regulation,* 9–13; John Ingersoll, "Amphetamines and Methamphetamines, Aggregate Production Quotas," February 10, 1972, ibid., 171–73; James Graham, "Amphetamine Politics on Capitol Hill," *Society,* January 1972, 14 (DOI: 10.1007/BF02695852); Rasmussen, *On Speed,* chap 7; Rasmussen, "America's First Amphetamine Epidemic."

CHAPTER TWO: Pharmacological Restraints

1. Jerry Avorn and Daniel H. Solomon, "Cultural and Economic Factors That (Mis)Shape Antibiotic Use: The Nonpharmacologic Basis of Therapeutics," *Annals of Internal Medicine* 133 (2000): 128–35; Tanya Stivers, *Prescribing under Pressure: Parent-Physician Conversations and Antibiotics* (New York: Oxford University Press, 2007).

2. Conan MacDougall and Ron E. Polk, "Antimicrobial Stewardship Programs in Health Care Systems," *Clinical Microbiology Reviews* 18 (2005): 638–56.

3. Edward H. Kass and Katherine Murphey Hayes, "The Infectious Diseases Society of America: The First Twenty Five Meetings (1962–1987)," unpublished manuscript, box 28, folder 52, introduction, Edward Kass Papers (hereafter EKP), Harvard Medical Library at the Countway Medical Library, Boston. See also Kass and Hayes, "A History of the Infectious Diseases Society of America," *Reviews of Infectious Diseases* 10, suppl. 2 (1988): 1–6.

4. Wesley W. Spink, "The Use and Abuse of Chemotherapy," *Minnesota Medicine* 25 (1942): 988–90; Richard A. Kern, "Abuse of Sulfonamides in the Treatment of Acute Catarrhal Fever," *United States Naval Medical Bulletin* 44 (1945): 686–94; "Abuse of Penicillin," *New England Journal of Medicine* 233 (1945): 830–32; Maurice A. Schnitker, "Some Abuses of the Antibiotic Agents," *Ohio State Medical Journal* 43 (1947): 1138–40; Perrin H. Long, "The Use and Abuse of Chemotherapeutic and Antibiotic Agents," *New England Journal of Medicine* 237 (1947): 837–39; William S. Hoffman, "Penicillin: Its Use and Possible Abuse," *Journal of the American Dental Association* 34 (1947): 89–99.

5. See, e.g., "Use and Abuse of the Antibiotics," *Rocky Mountain Medical Journal* 49 (1952): 581–82; A. L. Tatum, "Misuse of Antibiotics," *Wisconsin Medical Journal* 51 (1952): 881; Carl A. Hartung, "Abuses in the Use of Antibiotics," *Journal of the Tennessee State Medical Association* 46 (1953): 403–6; John F. Waldo, "Antibiotics—Their Use and Abuse," *Rocky Mountain Medical Journal* 50 (1953): 879–82. See also James C. Whorton, "'Antibiotic Abandon': The Resurgence of Therapeutic Rationalism," in *The History of Antibiotics: A Symposium*, ed. John Parascandola (Madison, WI: American Institute of the History of Pharmacy, 1980), 125–36.

6. See, e.g., Louis Weinstein, "The Use and Misuse of Antibiotics," *Boston Medical Quarterly* 2 (1951): 97–98; W. H. Oatway Jr., "Delays in the Diagnosis of Tuberculosis from the Incautious Use of Antibiotics," *Arizona Medicine* 8 (1951): 25–28; Wendell H. Hall, "The Abuse and Misuse of Antibiotics," *Minnesota Medicine* 35 (1952): 629–30.

7. Scott H. Podolsky, "Antibiotics and the Social History of the Controlled Clinical Trial, 1950–1970," *Journal of the History of Medicine and Allied Sciences* 65 (2010): 327–67.

8. Harry F. Dowling, "Twixt the Cup and the Lip," *Journal of the American Medical Association* 165 (1957): 657–61; Harry M. Marks, "What Does Evidence Do? Histories of Therapeutic Research," in *Harmonizing Drugs: Standards in Twentieth-Century Pharmaceutical History*, ed. Christian Bonah, Christophe Masutti, Anne Rasmussen, and Jonathon Simon (Paris: Editions Glyphe, 2009), 88.

9. Morton Mintz, "FDA and Panalba: A Conflict of Commercial, Therapeutic Goals?" *Science* 165 (1969): 875–81; Milton Silverman and Philip R. Lee, *Pills, Profits, and Politics* (Berkeley: University of California Press, 1974), 107–37; Daniel P. Carpenter, *Reputation and Power: Organizational Image and Pharmaceutical Regulation at the FDA* (Princeton, NJ: Princeton University Press, 2010).

10. "The Drug Efficacy Study," *New England Journal of Medicine* 280 (1969): 1177–79 (unsigned editorial by Maxwell Finland, in Finland's collected reprints at the Countway Medical Library, Boston); Calvin Kunin, "Impact of Infections and Antibiotic Use on Medical Care," *Annals of Internal Medicine* 89 (1978): 717.

11. Kass and Hayes, "Infectious Diseases Society of America: First Twenty Five Meetings."

12. William B. McIlwaine, "The Use of Sulfapyridine and Sulfathiazole in General Practice," *Virginia Medical Monthly* 68 (1941): 410–11.

13. Maxwell Finland, "The Present Status of Antibiotics in Bacterial Infections," *Bulletin of the New York Academy of Medicine* 27 (1950): 214.

14. E. Jawetz and M. S. Marshall, "The Role of the Laboratory in Antibiotic Therapy," *Journal of Pediatrics* 37 (1950): 545.

15. See, e.g., Weinstein, "Use and Misuse of Antibiotics," 90–98.

16. Hall, "Abuse and Misuse of Antibiotics," 629.

17. Allen E. Hussar, "A Proposed Crusade for the Rational Use of Antibiotics," *Antibiotics Annual* (1954–55): 380.

18. Ernest Jawetz, "Patient, Doctor, Drug, and Bug," *Antibiotics Annual* (1957–58): 295.

19. Jeremy A. Greene and Scott H. Podolsky, "Keeping Modern in Medicine: Pharmaceutical Promotion and Physician Education in Postwar America," *Bulletin of the History of Medicine* 83 (2009): 339–46.

20. C. Henry Kempe, "A Rational Approach to Antibiotic Therapy of Childhood Infections," *Postgraduate Medicine* 24 (October 1958): 339–40.

21. John Lear, "Taking the Miracle out of the Miracle Drugs," *Saturday Review* 42 (January 3, 1959): 35–41; Richard Harris, *The Real Voice* (New York: Macmillan, 1964), 17–20.

22. William A. Nolen and Donald E. Dille, "Use and Abuse of Antibiotics in a Small Community," *New England Journal of Medicine* 257 (1957): 33–34. Nolen and Dille generously defined "definite" indications as "any temperature elevation of 1°F above normal and all the following, even though there was no temperature elevation: cellulitis; abscess; lymphangitis; prophylaxis in any wounds; prophylaxis in patients with history of rheumatic fever; croup (diathesis in children); history of recurrent otitis; laboratory evidence of cystitis; and any venereal disease."

23. Thomas Maeder, *Adverse Reactions* (New York: William Morrow, 1994), 343–47.

24. Paul D. Stolley and Louis Lasagna, "Prescribing Patterns of Physicians," *Journal of Chronic Diseases* 22 (1969): 396.

25. Paul D. Stolley, Marshall H. Becker, Joseph D. McEvilla, Louis Lasagna, Mar-

lane Gainor, and Lois M. Sloane, "Drug Prescribing and Use in an American Community," *Annals of Internal Medicine* 76 (1972): 538; *Advertising of Proprietary Remedies: Hearings before the Subcommittee on Monopoly*, Select Committee on Small Business, U.S. Senate, 92nd Cong., 2nd sess., part 3, 1972, 1169. Stolley reported at the Nelson hearings on advertising of proprietary medicines that, based on the same data set, chloramphenicol had been the 15th-most-prescribed antibiotic and that 3 of the 6 most-prescribed antibiotics (and 6 of the top 20) had been fixed-dose combination antibiotics.

26. See "The Internal Medical Audit," *Journal of the Oklahoma State Medical Association* 53 (1960): 805–7; "Minimum Use of Antibiotics," *Record* (of the Commission on Professional and Hospital Activities) 1 (July 30, 1962): 1; "Therapeutic Use of Antibiotics," *Record* 1 (August 22, 1962): 1; Robert S. Myers, Vergil N. Slee, and Richard P. Ament, "Antibiotic Study Shows Need for Therapy Audit in Hospitals," *Bulletin of the American College of Surgeons* 48 (1963): 61–63. For an earlier, rough overview of hospital antibiotic use at a community hospital in New England, stimulated by concerns regarding antibiotic resistance, see Wei-Ping Loh and Russell B. Street, "Study of the Use of Antimicrobial Agents in a Community Hospital," *New England Journal of Medicine* 251 (1954): 659–60.

27. Hobart A. Reimann and Joseph D'Ambola, "The Use and Cost of Antimicrobics in Hospitals," *Archives of Environmental Health* 13 (1966): 631–36.

28. William E. Scheckler and John V. Bennett, "Antibiotic Usage in Seven Community Hospitals," *Journal of the American Medical Association* 213 (1970): 264–67.

29. United States Task Force on Prescription Drugs, *Final Report* (Washington, DC: Government Printing Office, 1969), xxi; United States Task Force on Prescription Drugs, *The Drug Prescribers* (Washington, DC: Government Printing Office, 1968), 3. For an articulation of such a definition of "rational" therapeutics in congressional testimony, see Charles C. Edwards, in *Examination of the Pharmaceutical Industry, 1973–1974: Hearings before the Subcommittee on Health of the Committee on Labor and Public Welfare*, U.S. Senate, 93rd Cong., 1st sess., part 2, 1973, 566.

30. Andrew W. Roberts and James A. Visconti, "The Rational and Irrational Use of Systemic Antimicrobial Drugs," *American Journal of Hospital Pharmacy* 29 (1972): 828–34. They included fourteen categories of "irrational" prescribing, summarizing that "therapy was judged irrational . . . in two major cases: where antimicrobial therapy was warranted but the specific therapy used was inappropriate and when therapy was unnecessary and unwarranted."

31. For such cost data, see Henry E. Simmons and Paul D. Stolley, "This Is Medical Progress? Trends and Consequences of Antibiotic Use in the United States," *Journal of the American Medical Association* 227 (1974): 1024.

32. Richard A. Gleckman and Morton A. Madoff, "Environmental Pollution with Resistant Microbes," *New England Journal of Medicine* 281 (1969): 677; Maxwell Finland, "Changing Patterns of Susceptibility of Common Bacterial Pathogens to Antimicrobial Agents," *Annals of Internal Medicine* 76 (1972): 1009–36.

33. Maxwell Finland, "Prophylactic Antimicrobial Agents," *Proceedings of the In-*

ternational Conference on Nosocomial Infections, Center for Disease Control, August 3–6, 1970 (Chicago: American Hospital Association, 1971), 312–14.

34. Hussar, "Proposed Crusade," 381.

35. While Nelson's hearings did not provoke the type of sweeping legislative reform brought about by the Kefauver hearings, they were responsible for such smaller-scale changes as the mandate for patient package inserts for oral contraceptives (the first such patient-directed inserts). See Elizabeth Siegel Watkins, *On the Pill: A Social History of Oral Contraceptives, 1950–1970* (Baltimore: Johns Hopkins University Press, 1998), 103–28. Moreover, the very spotlight they shone upon particular antibiotics led to dramatic reductions in their usage. Chloramphenicol prescriptions in America were reduced from 3.3 million in 1965 to 1 million in 1970, never again crossing the 1 million mark in subsequent years; clindamycin usage was reduced from 6.1 million prescriptions in 1973 to 0.8 million prescriptions in 1977, consequent to concerns and hearings regarding its role in producing infectious colitis (what would today be considered *Clostridium dificile* colitis). See *Competitive Problems in the Drug Industry: Hearings before the Subcommittee on Monopoly, Select Committee on Small Business*, U.S. Senate, 1969, part 6 (90th Cong., 1st sess., 1967); part 27 (94th Cong., 1st sess., 1975); Marion J. Finkel, "Magnitude of Antibiotic Use," *Annals of Internal Medicine* 89 (1978): 791.

36. *Advertising of Proprietary Remedies*, part 3, 989.

37. Charles C. Edwards to Leighton Cluff, ibid., part 3, 1084.

38. Harry Dowling, ibid., part 3, 1104.

39. Paul D. Stolley, ibid., part 3, 1164–91.

40. Simmons and Stolley, "This Is Medical Progress?" 1023–28.

41. Mintz, "FDA and Panalba," 876.

42. William L. Hewitt, in *Competitive Problems in the Drug Industry*, 91st Cong., 1st sess., part 12, 5048. For similar claims made at the time by other expert witnesses, see ibid., 5003, 5009, 5020, 5071, 5087.

43. Calvin Kunin, in *Advertising of Proprietary Remedies*, part 3, 1214; Kunin, "A Brief Exposition of the Problem and Some Tentative Solutions," *Annals of Internal Medicine* 79 (1973): 557–58. For earlier invocations of "fear" as a motivation, see, e.g., Ryle A. Radke, "Use and Abuse of Antibiotics," *Medical Bulletin of the United States Army Far East* 2 (1954): 2.

44. See *Competitive Problems in the Drug Industry*, part 12, 1969, 5175–76.

45. Nea D'Amelio, "Are Family Doctors Prescribing Too Many Antibiotics?" *Medical Times* 102 (1974): 53–54. At the same time, 10,000 medical residents were polled. Only 2,358 responded, and of them, 65.7 percent felt that antibiotics were being overprescribed (and 91% thought they were being overprescribed in private practice), while 45.5 percent agreed that the average person didn't require antibiotics more than once every five to ten years. Ibid.

46. Michael Halberstam, *The Pills in Your Life* (New York: Grosset and Dunlap, 1972), 67–68.

47. See Kunin, in *Advertising of Proprietary Remedies*, part 3, 1211–12; Kunin, review of *The Pills in Your Life*, *American Society of Microbiology News* 38 (1972): 696, wherein Kunin concluded, "Shame on you Dr. Halberstam!"

48. D'Amelio, "Are Family Doctors Prescribing," 56. Similarly, a resident responded: "I prefer the occurrence of resistant strains rather than see people suffering from glomerulonephritis and a rheumatic heart disease. New antibiotics usually come up in due time and I think time itself really causes the formation of new strains." In Nea D'Amelio, "What Young Doctors Told Us about Antibiotic Overkill," *Medical Times* 102 (1974): 148. Regarding the history of such faith in "fighting resistance with technology," see Robert Bud, *Penicillin: Triumph and Tragedy* (Oxford: Oxford University Press, 2007), 116–39.

49. D'Amelio, "Are Family Doctors Prescribing," 57–59. Among the polled medical residents, this could take the form of a more generalized stab against the grasping aspirations of rational medicine itself: "With some residents it seems to be a matter of pride not to use antibiotics. They must feel that they get 'Brownie' points. Some would rather have both the patients and themselves dead in a ditch before they would use an antibiotic without getting back the culture and sensitivity." In D'Amelio, "What Young Doctors Told Us," 148.

50. D'Amelio, "Are Family Doctors Prescribing," 58.

51. Howard R. Seidenstein, "Trends of Antibiotic Use in the United States," letter to the editor, *Journal of the American Medical Association* 228 (1974): 1098–99.

52. Ibid., 1099. Of note, such proposed antibiotic guideline and review programs should also be understood against the backdrop of the Professional Standards Review Organizations (PSROs) that were initiated in the early 1970s to evaluate services reimbursed by Medicare; the antagonism to the antibiotic review programs may have likewise reflected antagonism to such wider efforts to evaluate services more broadly. On the PSRO program, see Charles C. Edwards, in *Examination of the Pharmaceutical Industry*, part 2, 575; Calvin M. Kunin and Herman Y. Efron, "Audits of Antimicrobial Usage: Guidelines for Peer Review," *Journal of the American Medical Association* 237 (1977): 1001–2; T. S. Jost, "Medicare Peer Review Organizations," *Quality Assurance in Health Care* 1 (1989): 235–48; Anita J. Bhatia, Sheila Blackstock, Rachel Nelson, and Terry S. Ng, "Evolution of Quality Review Programs for Medicare: Quality Assurance to Quality Improvement," *Health Care Financing Review* 22 (Fall 2000): 69–74.

53. Senator Gaylord Nelson, in *Competitive Problems in the Drug Industry*, part 6, 2408.

54. Mark Lepper, ibid., 2467.

55. Phillip Lee, in *Advertising of Proprietary Remedies*, part 3, 1041–42, 1124.

56. Ibid., 1132.

57. Ibid., 1131, 1155.

58. Ibid., 1132.

59. Ibid., 1125–26, 1171–72. For discussion about the "cost-benefit ratio" of such measures, see ibid., 1172–74.

60. David M. Greeley to Maxwell Finland, April 10, 1957, box 5, folder 45; June M.

Cardullo to Finland, May 25, 1957, box 5, folder 45; M. J. Leitner to Finland, October 25, 1957; Finland to Leitner, October 28, 1957, both in box 5, folder 47, all in Maxwell Finland Papers (hereafter MFP), Harvard Medical Library at the Countway Medical Library, Boston; Jawetz, "Patient, Doctor," 292.

61. John E. McGowan Jr. and Maxwell Finland, "Usage of Antibiotics in a General Hospital: Effect of Requiring Justification," *Journal of Infectious Diseases* 130 (1974): 165–68.

62. L. E. Burney, "Staphylococcal Disease: A National Problem," in *Proceedings of the National Conference on Hospital-Acquired Staphylococcal Disease* (Atlanta: U.S. Department of Health, Education, and Welfare, 1958), 7.

63. John McGowan noted in 1979 that, while "the JCAH [today the Joint Commission] requires that hospitals receiving accreditation review usage of antibiotics, the Commission does not mandate any specific means for implementing this procedure." In McGowan, "Continuing Education for Improving Antibiotic Usage," *Quality Review Bulletin* 5 (1979): 32.

64. William A. Craig, Stephen J. Uman, William R. Shaw, Vadakepat Ramgopal, and Lloyd L. Eagan, "Hospital Use of Antimicrobial Drugs: Survey at 19 Hospitals and Results of Antimicrobial Control Program," *Annals of Internal Medicine* 89 (1978): 795.

65. Of course, as articulated by McGowan and Finland, the restrictive program not only could serve "as a deterrent against abuse of certain antibiotics, but has also provided continuing education of the hospital staff (and the consultants) in the proper use of those agents in the management of their patients." In McGowan and Finland, "Usage of Antibiotics," 165.

66. Kempe, "Rational Approach," 339.

67. Henry Simmons, in *Advertising of Proprietary Remedies*, part 3, 1052–53.

68. James Visconti, in *Examination of the Pharmaceutical Industry*, part 1, 1973, 619.

69. Ibid., 622.

70. Calvin Kunin, in *Advertising of Proprietary Remedies*, part 3, 1220. Kunin actually used the term in this instance to contrast such an effort with the more "glamorous" war on cancer. He was in fact critical of the potential for waste in such early-1970s categorical programs. See Kunin, "Impact of Infections," 717.

71. See, e.g., Leighton Cluff, in *Advertising of Proprietary Remedies*, part 3, 1199.

72. "'SK-Line' Antibiotics," in *Examination of the Pharmaceutical Industry*, part 1, 1182.

73. Calvin Kunin, "In Comment [on "This Is Medical Progress?"]," *Journal of the American Medical Association* 227 (1974): 1031–32; see also Kunin, "Clinical Investigators and the Pharmaceutical Industry," *Annals of Internal Medicine* 89 (1978): 842–45.

74. Charles V. Sanders, "Memorial: Jay P. Sanford, 1928–1996," *Transactions of the American Clinical and Climatological Society* 112 (2001): 44–46.

75. Kunin and Efron, "Audits of Antimicrobial Usage," 1001–2.

76. Stuart Levy, *The Antibiotic Paradox: How Miracle Drugs Are Destroying the Miracle* (New York: Plenum Press, 1992); Alliance for the Prudent Use of Antibiotics, www.tufts.edu/med/apua/About_us/about.html, accessed March 1, 2010.

77. Greene and Podolsky, "Keeping Modern in Medicine," 363–75.

78. Stephen R. Jones, Joel Barks, Turner Bratton, Everett McRee, Jeff Pannell, Victor A. Yanchick, Richard Browne, and James W. Smith, "The Effect of an Educational Program upon Hospital Antibiotic Use," *American Journal of the Medical Sciences* 273 (1977): 79–85.

79. *Advertising of Proprietary Remedies*, part 3, 1042.

80. Ibid., 1082–83.

81. John C. Ballin to Edward H. Kass, January 12, 1973; Leighton E. Cluff to Kass, February 8, 1973, both in box 25, folder 36, EKP. Thirty-seven participants attended the meeting. Since Edward Kass was unable to attend the meeting, George Gee Jackson chaired the meeting. In "Minutes of the Conference on Antibiotic Utilization [March 12, 1973] accompanying John C. Ballin to Participants," April 3, 1973, box 25, folder 36, EKP.

82. "Minutes of the Conference on Antibiotic Utilization," EKP.

83. Ibid.

84. Ibid.

85. Stuart Mudd, "Staphylococcal Infections in the Hospital and Community," *Journal of the American Medical Association* 166 (1958): 1177.

86. Ibid., 1178.

87. Simmons and Stolley, "This Is Medical Progress?" 1027.

88. John C. Ballin, Michael H. M. Dykes, Joseph B. Jerome, Mary E. Kosman, John R. Lewis, and Donald O. Schiffman, "In Comment," *Journal of the American Medical Association* 227 (1974): 1029–30. In the actual minutes of the meeting, sent to participants by Ballin himself, appears the line that when the participants were asked, in speaking for themselves and not their organizations, about antibiotic overuse, "there was a unanimous vote that there appears to be an inappropriate use of antibiotics and a massive overuse." Five of the six authors appear to have been present at the meeting. In "Minutes of the Conference on Antibiotic Utilization," box 25, folder 36, EKP.

89. Ballin et al., "In Comment," 1030. They continued: "Scientific knowledge increases rapidly and arises from a consensus of the scientific community. It cannot be laid down by the long, drawn out, and difficult-to-reverse process of regulatory fiat, however enlightened the latter may be." At the same time, the *Journal of the American Medical Association* rejected John McGowan and Maxwell Finland's paper "Usage of Antibiotics in a General Hospital: Effect of Requiring Justification," which was ultimately published in the *Journal of Infectious Diseases*. See Maxwell Finland to John McGowan, November 1, 1973, box 6, folder 30, MFP.

90. Kass and Hayes, "Infectious Diseases Society of America: The First Twenty Five Meetings," 1969–1978 (i), 32.

91. Merle A. Sande, "The Need for Controlled Clinical Studies in Antimicrobial Therapy," *Annals of Internal Medicine* 89 (1978): 858.

92. Harold C. Neu, "The Crisis in Antibiotic Resistance," *Science* 257 (1992): 1064; Bud, *Penicillin*, 192–212; Robert Bud, "From Germophobia to the Carefree Life and Back Again: The Lifecycle of the Antibiotic Brand," in *Medicating Modern America: Prescription Drugs in History*, ed. Andrea Tone and Elizabeth Siegel Watkins (New York: New York University Press, 2007), 32–36.

93. Ralph Gonzales, John F. Steiner, and Merle A. Sande, "Antibiotic Prescribing for Adults with Colds, Upper Respiratory Tract Infections, and Bronchitis by Ambulatory Care Physicians," *Journal of the American Medical Association* 278 (1997): 901–4.

94. Benjamin Schwartz, David M. Bell, and James M. Hughes, "Preventing the Emergence of Antimicrobial Resistance: A Call for Action by Clinicians, Public Health Officials, and Patients," *Journal of the American Medical Association* 278 (1997): 944–45.

95. Donald A. Goldmann, Robert A. Weinstein, Richard P. Wenzel, Ofelia C. Tablan, Richard J. Duma, Robert P. Gaynes, James Schlosser, and William J. Martone, for the Workshop to Prevent and Control the Emergence and Spread of Antimicrobial-Resistant Microorganisms in Hospitals, "Strategies to Prevent and Control the Emergence and Spread of Antimicrobial-Resistant Microorganisms in Hospitals: A Challenge to Hospital Leadership," *Journal of the American Medical Association* 275 (1996): 235.

96. John E. McGowan Jr., "Do Intensive Hospital Antibiotic Control Programs Prevent the Spread of Antibiotic Resistance?" *Infection Control and Hospital Epidemiology* 15 (1994): 478–83. This article presented the consensus statement developed by the participants in a 1994 workshop on antimicrobial resistance in hospitals, sponsored by the Center for Disease Control and the National Foundation for Infectious Diseases.

97. Timothy H. Dellit, Robert C. Owens, John E. McGowan Jr., Dale N. Gerding, Robert A. Weinstein, John P. Burke, W. Charles Huskins, et al., "Infectious Diseases Society of America and the Society for Healthcare Epidemiology of America Guidelines for Developing an Institutional Program to Enhance Antimicrobial Stewardship," *Clinical Infectious Diseases* 44 (2007): 159–77. More than three decades previously, at the 1973 AMA conference, William Kirby had stated that two options were available for approaching apparent antibiotic overuse: "1. It can be studied for 50 years—this is not too practical. 2. From a practical point of view, proper usage can be incorporated into a peer review mechanism." In "Minutes of the Conference on Antibiotic Utilization [March 12, 1973] accompanying John C. Ballin to Participants," April 3, 1973, box 25, folder 36, EKP.

98. Tamar F. Barlam and Margarita Divall, "Antibiotic-Stewardship Practices at Top Academic Centers throughout the United States and at Hospitals throughout Massachusetts," *Infection Control and Hospital Epidemiology* 27 (2006): 695, 700. Among academic hospitals, the figure was 89 percent.

99. Ibid., 697. "As one program director noted, it is 'unpleasant' to be involved with antibiotic oversight." Ibid.

100. See, e.g., Stivers, *Prescribing under Pressure.*

101. Harry M. Marks, "Making Risks Visible: The Science and Politics of Adverse Drug Reactions," in *Ways of Regulating: Therapeutic Agents between Plants, Shops and Consulting Rooms,* ed. Jean Paul Gaudillière and Volker Hess, Preprint 363 (Berlin: Max Plank Institut für Wissenschaftsgeschichte, 2009).

CHAPTER THREE: "Eroding the Physician's Control of Therapy"

Thank you to Elizabeth Watkins and Jeremy Greene for their insightful comments on an earlier version of this chapter. The chapter title is drawn from an editorial written in 1973 by a member of the American Medical Association's Department of Drugs: Donald O. Schiffman, "Editorial: Eroding the Physician's Control of Therapy," *Journal of the American Medical Association* 225, no. 2 (1973): 552.

1. Elliot J. Margolis, "Prescription Perspective," *Physician's Management,* September 1966, 13.

2. John Swann, "FDA and the Practice of Pharmacy: Prescription Drug Regulation before the Durham-Humphrey Amendment of 1951," *Pharmacy in History* 36 (1994): 55–70; Harry Marks, "Revisiting 'The Origins of Compulsory Drug Prescriptions,'" *American Journal of Public Health* 85, no. 1 (1995): 109–16.

3. Peter Temin, *Taking Your Medicine: Drug Regulation in the United States* (Cambridge, MA: Harvard University Press, 1980), 51–58; Swann, "FDA and the Practice of Pharmacy"; Marks, "Revisiting." For discussion of the economic and cultural implications of branding in the drug industry, see Robert Bud, *Penicillin: Triumph and Tragedy* (Oxford: Oxford University Press, 2007); Jeremy A. Greene, "No Such Thing as a Generic Drug? The Persistence of the Brand in American Pharmaceuticals," paper presented at the History of Science Society Annual Meeting, Pittsburgh, November 7, 2008.

4. See, e.g., *FDC Reports,* January 24, 1953, W6–W7.

5. "Michigan Bill Would Permit Using 1 USP, NF Item for Another," *American Druggist* 152 (February 18, 1952): 16.

6. Ibid.; *FDC Reports,* April 4, 1953, W8–W10.

7. "State Conventions Act on Substitution as Makers Charge Practice Is Epidemic," *American Druggist* 126 (July 2, 1952): 5–6.

8. *FDC Reports,* December 5, 1953, W4.

9. "Abbott Wins First in Series of Suits to Stop Substitution: Execs Pilot Drive," *American Druggist* 127 (May 11, 1953): 5; "Abbott Wins Second Substitution Case: Substitution Bill Signed," *American Druggist* 128 (August 2, 1953): 9; "Klumpp: Manufacturers May Go to Court to Halt Substitution," *American Druggist* 125 (May 26, 1952): 22; "American Druggist Survey Is the First to Explore All Aspects of Substitution," *American Druggist* 128 (July 6, 1953): 5–16, quote on 15.

10. The firms that established the NPC were Abbott; Ciba; Hoffman-LaRoche;

Lederle; McNeil Laboratories; William S. Merrell; Pfizer; G. D. Searle; Smith, Kline, and French; Squibb; Upjohn; and Winthrop-Stearns. *FDC Reports*, December 5, 1953, W3–W4; Theodore G. Klumpp, "E Pluribus Unum: An Address Delivered at the Mid-Winter Luncheon of the Drug, Chemical and Allied Trades Section of the New York Board of Trade, Inc., New York City, January 26, 1954," RG 4, subgroup 3, series 4, folder 04-00-03-004-0001, Sterling Drug Inc. Collection, Archives Center, National Museum of American History, Smithsonian Institution, Washington, DC (hereafter, Sterling Drug Collection).

11. *FDC Reports*, December 17, 1956, 13–14. See also "Stewart: You Must Prod the Rx Boards," *American Druggist* 130 (July 19, 1954): 13–14.

12. Klumpp, "E Pluribus Unum."

13. Ibid.

14. *FDC Reports*, December 17, 1956, 13–14; "86% of Manufacturers Say Substitution on Their Products Is Negligible Today," *American Druggist* 136 (July 15, 1957): 5–6.

15. Neil J. Facchinetti and W. Michael Dickson, "Access to Generic Drugs in the 1950s: The Politics of a Social Problem," *American Journal of Public Health* 72, no. 5 (1982): 468–75.

16. U.S. Department of Health, Education, and Welfare, *Health United States 1975*, DHEW Publication No. (HRA) 76-1232 (Washington, DC: Government Printing Office, 1976), 11–16.

17. The average annual price index for hospital fees and physician, dentist, and other professional fees (so-called medical care services) increased from 49.2 in 1950 to 74.9 in 1960 (correlating to an average annual percentage change of 4.2% between 1950 and 1955 and of 4.4% between 1955 and 1960). In contrast, the consumer price index for prescription drugs over the same period increased from 92.6 in 1950 to 115.3 in 1960, corresponding to an average annual percentage change for prescription drug prices of 1.9% between 1950 and 1955 and 2.0% between 1955 and 1960. Ibid., 72–73. All annual price index values are calculated from constant dollars.

18. "Brand-Name Drugs Called Big Expense," *New York Times*, May 14, 1958, 33; "The Not-So-Long Arm of the Law," *Consumer Reports*, October 1958, 503–9; "The High Price of Rx Drugs," *Consumer Reports*, November 1958, 597–99; *FDC Reports*, October 27, 1958, 14.

19. For more detailed scholarly analyses of the Kefauver hearings, see Richard Harris, *The Real Voice* (New York: Macmillan, 1964); Richard E. McFadyen, "Estes Kefauver and the Drug Industry" (PhD diss., Emory University, 1973); Robert Bud, "Antibiotics, Big Business, and Consumers: The Context of Government Investigations into the Postwar American Drug Industry," *Technology and Culture* 46 (2005): 329–49; Daniel Scroop, "A Faded Passion? Estes Kefauver and the Senate Subcommittee on Antitrust and Monopoly," *Business and Economic History On-Line* 5 (2007): 1–17; Dominique A. Tobbell, "'Who's Winning the Human Race?' Cold War as Pharmaceutical Political Strategy," *Journal of the History of Medicine and Allied Sciences* 64, no. 4 (2009): 429–73.

20. "Statement of Ethel Percy Andrus," in *Administered Prices in the Drug Industry: Hearings before the U.S. Senate Committee on the Judiciary, Subcommittee on Antitrust and Monopoly* (Washington, DC: Government Printing Office, 1959–62), December 11, 1959, 8262 ("monthly costs"), 8272 ("the only way"), 11701.

21. Theodore G. Klumpp, "Medical Progress and the Pharmaceutical Industry," commencement address, University of Chattanooga, Chattanooga, Tennessee, June 6, 1960, pp. 10–11, RG 4, subgroup 3, series 4, folder 04-00-03-004-0002, Sterling Drug Collection.

22. "Testimony of Dr. Austin Smith," in *Administered Prices in the Drug Industry*, April 20, 1960, 10895–96; "Testimony of Eugene N. Beesley," December 8, 1961, reprinted in *Drug Industry Antitrust Act: Hearings before the Antitrust Subcommittee (Subcommittee No. 5) of the Committee on the Judiciary House of Representatives*, 87th Cong., 2nd sess., May 24, 1962, 696.

23. "Testimony of Newall Stewart," in *Administered Prices in the Drug Industry*, May 13, 1960, 11701.

24. *FDC Reports*, July 3, 1961, 19.

25. Ibid.; *FDC Reports*, December 4, 1961, 18.

26. *FDC Reports*, December 5, 1960, 10–14.

27. Marks, "Revisiting"; Dominique A. Tobbell, "Allied against Reform: Pharmaceutical Industry–Academic Physician Relations in the United States, 1945–1970," *Bulletin of the History of Medicine* 82, no. 4 (2008): 878–912; Tobbell, "'Who's Winning the Human Race?'"

28. This was actually a continuation of the price-fixing allegations first made by the FTC against the tetracycline manufacturers in 1958. Samuel Mines, *Pfizer . . . an Informal History* (New York: Pfizer, 1978), 198–212.

29. *Competitive Problems in the Drug Industry: Hearings before the Subcommittee on Monopoly, Select Committee on Small Business*, U.S. Senate, 1967–77 (hereafter cited as *Nelson Hearings*).

30. *FDC Reports*, January 16, 1967, 3–5; "Social Security Act Amendments of 1967," *Congressional Record* 113, no. 24 (November 16, 1967): 32821–44; "Social Security Act Amendments of 1967," *Congressional Record* 113, no. 24 (November 21, 1967): 33518–28; Representative Multer, "A Bill to Amend the Federal Food, Drug and Cosmetic Act," *Congressional Record* 113, no. 13 (June 22, 1967): 16951–52.

31. See also Tobbell, "Allied against Reform."

32. D. G. Chapman, R. Crisafio, and J. A. Campbell, "The Relation between in Vitro Disintegration Time of Sugar-Coated Tablets and Physiological Availability of Riboflavin," *Journal of the American Pharmaceutical Association* 43, no. 5 (May 1954): 297–304; D. G. Chapman, R. Crisafio, and J. A. Campbell, "The Relation between in Vitro Disintegration Time of Sugar-Coated Tablets and Physiological Availability of Sodium p-aminosalicylate," *Journal of the American Pharmaceutical Association* 45, no. 6 (June 1956): 374–78; D. G. Chapman, L. G. Chatten, and J. A. Campbell, "Physiological Availability of Drugs in Tablets," *Canadian Medical Association Journal* 76 (1957): 102–5, quotation on 103; A. B. Morrison, D. G. Chapman, and J. A. Campbell,

"Further Studies on the Relation between in Vitro Disintegration Time of Tablets and Urinary Excretion Rates of Riboflavin," *Journal of the American Pharmaceutical Association* 48, no. 11 (1959): 634–37; A. B. Morrison and J. A. Campbell, "The Relationship between Physiological Availability of Salicylates and Riboflavin and in Vitro Disintegration of Enteric Coated Tablets," *Journal of the American Pharmaceutical Association* 49, no. 7 (1960): 473–78; A. B. Morrison and J. A. Campbell, "Physiologic Availability of Riboflavin and Thiamine in "Chewable" Vitamin Products," *American Journal of Clinical Nutrition* 10, no. 3 (1962): 212–16; A. B. Morrison, C. B. Perusse, and J. A. Campbell, "The Relationship between in Vitro Disintegration Time and in Vivo Release of Vitamins from a Triple-Dose Spaced-Release Preparation," *Journal of Pharmaceutical Sciences* 51, no. 7 (1962): 623–26; Denys Cook, "History of Bioavailability Testing at the Food and Drug Directorate," *Revue Canadienne de Biologie* 32, suppl. (1973): 157–62.

33. Lloyd Miller, "Physiological Availability and Homogeneity in U.S.P. Dosage Forms," Circular 30, U.S.P. Committee of Revision, September 28, 1961, p. 108, B82, F 11, USP Records, Wisconsin State Historical Society, cited in Daniel P. Carpenter and Dominique A. Tobbell, "Bioequivalence: The Regulatory Career of a Pharmaceutical Concept," *Bulletin of the History of Medicine* 85, no. 1 (2011): 93–131.

34. Carpenter and Tobbell, "Bioequivalence"; "Statement of C. Joseph Stetler, President, PMA," *Nelson Hearings*, November 16, 1967, 1367–1400, quotation on 1367.

35. "Statement of Stetler," 1367.

36. "Statement submitted by Alfred Gilman to Senator Nelson," *Congressional Record* 113, no. 19 (September 11, 1967): 25146–48.

37. Dale G. Friend to Duke C. Trexler, February 17, 1969, folder "Membership: Comments on Topics of Final Report 1969," Drug Efficacy Study Collection, Series 2, DES Panels, National Academies Archives.

38. Edward F. Skinner, "Generic Prescribing," *Journal of the American Medical Association* 198, no. 7 (1966): 792–97, quote on 793. See also Charles A. Ragan, "Editorial: Are We Headed for a Dark Age of Nondiscovery in Therapeutics?" *Journal of the American Medical Association* 202, no. 12 (1967): 1099–1100.

39. For further examples of the medical profession's concern with preserving the physicians' autonomy amid mounting government regulation, see chapter 2, by Podolsky, on professional and congressional debates concerning physicians' antibiotic prescription practices; see also Harry Marks, "Making Risks Visible: The Science and Politics of Adverse Drug Reactions," in *Ways of Regulating: Therapeutic Agents between Plants, Shops, and Consulting Rooms*, ed. Jean Paul Gaudillière and Volker Hess (Berlin: Max-Planck-Institute, 2009), 105–22.

40. Donald Janson, "AMA Says Brand-Name Drugs Do Not Always Cost More," *New York Times*, May 26, 1967, 24; AMA, "Generic Prescribing Doesn't Guarantee Lower Drug Costs, Chicago Survey Shows," news release, May 26, 1967, box 32, folder 32-15, John Adriani Papers, National Library of Medicine, Bethesda, MD (hereafter Adriani Papers).

41. For a thorough analysis of the FDA and its approach to prescription drug

regulation after World War II, see Daniel P. Carpenter, *Reputation and Power: Organizational Image and Pharmaceutical Regulation at the FDA* (Princeton, NJ: Princeton University Press, 2010).

42. John Adriani to Senator Russell Long, January 29, 1967, box 40, folder 40-2, Adriani Papers.

43. Adriani to Jean Weston, April 12, 1966, box 34, folder 34-3, Adriani Papers.

44. "Social Security Act Amendments of 1967," *Congressional Record* 113, no. 24 (November 16, 1967): 32821–44; "Social Security Act Amendments of 1967," *Congressional Record* 113, no. 24 (November 21, 1967): 33518–28; Multer, "A Bill to Amend the Federal Food, Drug and Cosmetic Act"; Senator Joseph Montoya, "Legislation to Deal with the Catastrophic Prescription Drug Expense of the Aged," *Congressional Record* 114, no. 2 (1968): 2220–23; Senator Russell Long, "H.R. 17550—Social Security Amendments of 1970—Amendment No. 929," *Congressional Record* 116, no. 24 (September 21, 1970): 32837–41; Senator Joseph Montoya, "Social Security Amendments of 1970—Amendment No. 1113," *Congressional Record* 116, no. 31 (December 15, 1970): 41580–82.

45. Adriani to Senator Russell Long, January 29, 1967, box 40, folder 40-2, Adriani Papers.

46. Adriani to Weston, May 18, 1967, box 34, folder 34-3, Adriani Papers.

47. Ibid.; Adriani to Weston, April 2, 1969, box 34, folder 34-3, Adriani Papers.

48. *FDC Reports*, March 6, 1967, 12.

49. For health care expenditure statistics, see U.S. Department of HEW, *Health United States 1975*. In 1973 personal health care expenditures for "drugs and sundries" were estimated at 11 percent of personal expenditures for all health care services (see p. 19). For average annual percentage change for consumer prices for medical care services, see pp. 70–71.

50. Pharmaceutical Manufacturers Association, *Drugs Anonymous?* (Washington, DC: PMA, 1967), box 1964, Parke-Davis Collection, Archives Center, National Museum of American History, Smithsonian Institution, Washington, DC. For information on *Compulsory Generic Prescribing—A Peril to the Health Care System*, see Senator Gaylord Nelson, "Drug Quality Standards," *Congressional Record* 113, no. 5 (March 7, 1967): 5630–31.

51. Senator Morse, "Drug Industry Seeks to Defeat Senator Nelson," *Congressional Record* 114, no. 21 (September 24, 1968): 27950–51.

52. See, e.g., Adriani's public assertion, "The drug industry *was* involved in this matter," because of his putative opposition to brand-name drugs. John Adriani, "Statement Released Monday, September 2, 1969," box 37, folder 37-1, Adriani Papers.

53. For a detailed study of the AMA's role in U.S. health policy, see Frank D. Campion, *The AMA and U.S. Health Policy since 1940* (Chicago: Chicago Free Press, 1984). For a briefer discussion on the lobbying efforts of the AMA, see Colin Gordon, *Dead on Arrival: The Politics of Health Care in Twentieth Century America* (Princeton, NJ: Princeton University Press, 2003), 234–36.

54. Elizabeth Siegel Watkins, *On the Pill: A Social History of Oral Contraceptives,*

1950–1970 (Baltimore: Johns Hopkins University Press, 1998); Helen Marieskind, "The Women's Health Movement," *International Journal of Health Services* 5 no. 2 (1975): 217–23; Sheryl Burt Ruzek, *The Women's Health Movement: Feminist Alternatives to Medical Control* (New York: Praeger, 1978); Mary K. Zimmerman, "The Women's Health Movement: A Critique of Medical Enterprise and the Position of Women," in *Analyzing Gender: A Handbook of Social Science Research*, ed. Beth B. Hess and Myra Marx Ferree (Newbury Park, CA: Sage, 1987), 442–72; Carol S. Weisman, *Women's Healthcare: Activist Traditions and Institutional Change* (Baltimore: Johns Hopkins University Press, 1998); Wendy Kline, "The Making of Our Bodies, Ourselves: Re-thinking Women's Health and Second Wave Feminism," in *Feminist Coalitions: Historical Perspectives on Second-Wave Feminism in the United States*, ed. Stephanie Gilmore (Urbana: University of Illinois Press, 2008); and Wendy P. Kline, *Bodies of Knowledge: Sexuality, Reproduction, and Women's Health in the Second Wave* (Chicago: University of Chicago Press, 2010).

55. On this point see, e.g., Paul Starr, *The Social Transformation of American Medicine: The Rise of a Sovereign Profession and the Making of a Vast Industry* (New York: Basic Books, 1982), 388–93; David J. Rothman, *Strangers at the Bedside: A History of How Law and Bioethics Transformed Medical Decision Making* (New York: Basic Books, 1991), 190–221; Bud, *Penicillin*, 140–62.

56. Carpenter and Tobbell, "Bioequivalence."

57. For details on the deal Senator Kefauver made with Tennessee pharmacists to exclude pharmacy practices from his congressional investigation, see Harris, *Real Voice*, 41.

58. William S. Apple, "'Pharmacy's Lib,'" *Journal of the American Pharmaceutical Association* 11, n.s., no. 10 (1971): 528–33, quotations on 528–29.

59. "National Pharmaceutical Counsel," *FDC Reports*, December 5, 1953, W3–W4; D. G. Baird, "Why Parke, Davis Is Advertising to Build Prestige for Pharmacists," *Sales Management*, September 15, 1946, reprint in NL box 31, folder NLd 1946, Parke-Davis Collection.

60. Apple, "'Pharmacy's Lib,'" 529–33.

61. *FDC Reports*, December 19, 1973.

62. "Editorial: Substitution of Drugs," *Journal of the American Medical Association* 212, no. 8 (1970): 1369.

63. Weston to Adriani, July 16, 1970; Adriani to Weston, August 10, 1970, both in box 34, folder 34-3, Adriani Papers; Council on Drugs, "Resolution 58—Maintenance of anti-substitution laws; resolution 63—substitution of prescription drugs" (and related materials), addendum to meeting minutes of council on drugs, May 7–8, 1971, box 28, folder 28-2, Adriani Papers.

64. D. N. Goldstein, "Save Antisubstitution," *Wisconsin Medical Journal* 70 (1971): 26A–27A, quotation on 27A; Vesa Manninen, John Melin, and Gottfried Härtel, "Serum-digoxin Concentrations during Treatment with Different Preparations," *Lancet* 2, no. 7730 (1971): 934–35; D. C. Blair, R. W. Barnes, E. L. Wildner, and W. J. Murray, "Biological Availability of Oxytetracycline HCl Capsules," *Journal of the American*

Medical Association 215 (1971): 251–54; J. G. Wagner, P. G. Welling, K. P. Lee, and J. E. Walker, "*In Vivo* and *in Vitro* Availability of Commercial Warfarin Tablets," *Journal of Pharmaceutical Sciences* 60 (1971): 666–67; J. Lindenbaum, M. H. Mellow, M. O. Blackstone, and V. P. Butler Jr., "Variation in Biologic Availability of Digoxin from Four Preparations," *New England Journal of Medicine* 285 (1971): 1344–47; M. H. Barr, L. M. Gerbracht, K. Letcher, M. Plaut, and N. Strahl, "Assessment of the Biological Availability of Tetracycline Products in Man," *Clinical Pharmacology and Therapeutics* 13 (1972): 97–108; and C. Macleod, H. Rabin, J. Ruedy, M. Caron, D. P. Zarowny, and R. O. Davies, "Comparative Bioavailability of Three Brands of Ampicillin," *Canadian Medical Association Journal* 107, no. 3 (1972): 203–9.

65. For a discussion of the regulatory and policy implications of this, see Drug Bioequivalence Study Panel, U.S. Office of Technology Assessment, *Drug Bioequivalence* (Washington, DC: OTA, 1974).

66. Milton M. Perloff, "Anti-Substitution Law Repeal—Pro and Con: The Case against Repeal," *Philadelphia Medicine* 67, no. 3 (1971): 83–90, quotation on 90; "Editorial: Drug Substitution—How to Turn Order into Chaos," *Journal of the American Medical Association* 217, no. 6 (1971): 817–18.

67. "2 Generic-Drug Bills Offered," *New York Times*, February 23, 1975, NJ56.

68. "Bill to Allow Drug Substitutions Advances in Assembly," *Los Angeles Times*, March 6, 1975, A29; "Generic Drug Bill Wins Senate Approval, 30-7," *Los Angeles Times*, September 12, 1975.

69. Martin Waldron, "Generic Drugs: The Fight Continues," *New York Times*, October 3, 1976, 316; Martin Waldron, "Assembly Approves Generic-Drug Prescriptions," *New York Times*, February 15, 1977, 67. For accusations against industry and physician lobbying tactics in Illinois, see William Griffin, "Drug Firms Accused of 'Dirty' Tactics against Bill," *Chicago Tribune*, November 15, 1977, B12.

70. "States Moving to Allow Generic Substitution," *Internist* 20, no. 4 (1979): 12–13.

71. Donald O. Schiffman, "On Therapeutic Equivalence and the Antisubstitution Laws," *Journal of the American Medical Association* 223, no. 5 (1973): 552–53, quotation on 552.

72. "Generic Substitution Becomes Law," *Missouri Medicine* 75, no. 9 (1978): 467; Sherry L. Hall, "Attorney General Limits Generic Drug Substitution," *Michigan Medicine* 74, no. 9 (1975): 148–49; "Bill to Allow Drug Substitutions Advances in Assembly," *Los Angeles Times*, March 6, 1975, A29.

CHAPTER FOUR: Deciphering the Prescription

I would like to thank Jillian Cunningham for research assistance, Robert Day for access to Jere Goyan's unpublished speeches and for insights into the origins of clinical pharmacy at UCSF, and Jeremy Greene for comments on the first draft of this manuscript.

1. Dan Kushner, "Are You Set to Become a Punching Bag between the Physician and the Patient?" *American Druggist* 182 (December 1980): 8, 42.

2. Even though direct-to-consumer advertising of prescription drugs did not begin until the 1980s, the pharmaceutical companies have a much longer history of promoting their products directly to consumers through a variety of informal and indirect marketing techniques, including ghostwriting popular articles, organizing public relations events, and crafting implicit advertisements. See David Herzberg and Jeremy A. Greene, "Hidden in Plain Sight: The Popular Promotion of Prescription Drugs in the 20th Century," *American Journal of Public Health* 100 (2010): 793–803.

3. See, e.g., John Parascandola, "The Pharmacist and VD Control in World War II," *Pharmacy in History* 47 (2005): 62–68; and Robert A. Buerki, "The Public Image of the American Pharmacist in the Popular Press," *Pharmacy in History* 38 (1996): 62–78. For a good overview of the history and historiography of pharmacy, see Gregory J. Higby, "Evolution of Pharmacy," in *Remington's the Science and Practice of Pharmacy*, 21st ed., ed. David B. Troy (Philadelphia: Lippincott, Williams & Wilkins, 2006), 7–19, especially the comprehensive bibliography at the end of the chapter.

4. "Notes on the Package Insert," *Journal of the American Medical Association* 207 (February 17, 1969): 1335–38.

5. For a thorough analysis of the pill patient package insert, see Elizabeth Siegel Watkins, *On the Pill: A Social History of Oral Contraceptives, 1950–1970* (Baltimore: Johns Hopkins University Press, 1998), 120–31; or Elizabeth Siegel Watkins, "Expanding Consumer Information: The Origin of the Patient Package Insert," *Advancing the Consumer Interest* 10 (Spring 1998): 20–26.

6. Watkins, *On the Pill*, 127.

7. Ibid., 121–22.

8. Ibid., 122–26.

9. Elizabeth Siegel Watkins, "'Doctor, Are You Trying to Kill Me?': Ambivalence about the Patient Package Insert for Estrogen," *Bulletin of the History of Medicine* 76 (Spring 2002): 89.

10. Watkins, *On the Pill*, 126–27.

11. Ruth R. Faden and Tom L. Beauchamp, *A History and Theory of Informed Consent* (New York: Oxford University Press, 1986), 94.

12. Barbara Resnick Troetel, "Three-Part Disharmony: The Transformation of the Food and Drug Administration in the 1970s" (PhD diss., City University of New York, 1996), 294–95.

13. *Federal Register* 40 (November 7, 1975): 52075.

14. Troetel, "Three-Part Disharmony," 309.

15. Letter from American Medical Association, March 8, 1975, FDA Docket no. 75P-0277, Records—FDA, quoted in Troetel, "Three-Part Disharmony," 313.

16. Letter from American Pharmaceutical Association, March 8, 1976, and letter from Pharmaceutical Manufacturers Association, March 3, 1976, FDA Docket no. 75P-0277, Records—FDA, Rockville, MD, cited in Troetel, "Three-Part Disharmony," 311–13.

17. Pierre S. Del Prato, "The Community Pharmacist's Viewpoint," Drug Information Association / American Medical Association / Food and Drug Administration /

Pharmaceutical Manufacturers Association Joint Symposium, *Drug Information Journal* 11, Special Supplement (January 1977): 15S–17S.

18. Ibid., 16S.

19. Mary Jo Reilly, "The Hospital Pharmacist's Viewpoint," Drug Information Association / American Medical Association / Food and Drug Administration / Pharmaceutical Manufacturers Association Joint Symposium, *Drug Information Journal* 11, Special Supplement (January 1977): 21S-25S.

20. Ibid., 23S.

21. *Federal Register* 41 (September 29, 1976): 43108–9. For a thorough analysis of the estrogen patient package insert, see Elizabeth Siegel Watkins, *The Estrogen Elixir: A History of Hormone Replacement Therapy in America* (Baltimore: Johns Hopkins University Press, 2007), 132–47; or Watkins, "'Doctor, Are You Trying to Kill Me?'" 84–104.

22. Watkins, *Estrogen Elixir*, 137.

23. Letter 0315, Docket no. 76N-0381, Records—FDA, Rockville, MD. The author of this letter also submitted it to *Pharmacy Times*, where it appeared in the "Letters from our Readers" section in the February 1978 issue, p. 18.

24. Letter 0341, Docket no. 76N-0381, Records—FDA.

25. Dennis B. Worthen, ed., *A Road Map to a Profession's Future: The Millis Study Commission on Pharmacy* (Amsterdam: Gordon and Breach Science, 1999), appendix B, 270–72. Of course, pharmacy has a much longer history. For a brief overview, see David L. Cowen, "Pharmacists and Physicians: An Uneasy Relationship," *Pharmacy in History* 34 (1992): 3–16. The definitive history of pharmacy is Glenn Sonnedecker, *Kremer and Urdang's History of Pharmacy* (Philadelphia: Lippincott, 1976).

26. Lawrence C. Weaver, Allen I. White, and Dennis B. Worthen, "Prelude to the Commission: The Intervening Years," in Worthen, *Road Map*, 51. See also E. C. Elliott, *The General Report of the Pharmaceutical Survey, 1946–49* (Washington, DC: American Council on Education, 1950).

27. Michael C. Shannon, "Transitions: Changing Emphases in Pharmacy Education, 1946 to 1976," in Worthen, *Road Map*, 9–11.

28. Weaver, White, and Worthen, "Prelude to the Commission," 54.

29. Ibid., 52–53.

30. Ibid., 47–48. See also Dichter Institute for Motivational Research Inc., *Communicating the Value of Comprehensive Pharmaceutical Services to the Consumer* (Washington, DC: American Pharmaceutical Association, 1973); and Buerki, "Public Image," 71.

31. Gregory J. Higby, "The Continuing Evolution of American Pharmacy Practice, 1952–2002," *Journal of the American Pharmaceutical Association* 42 (January–February 2002): 13. See also Bob Day, "The Beginning of Something Magnificent," *UCSF Pharmacy Alumni Association Newsletter*, Fall 2006, 3–4. For a different recollection, see Russell R. Miller, "History of Clinical Pharmacy and Clinical Pharmacology," *Journal of Clinical Pharmacology* 21 (1981): 195–97.

32. "A History of UCSF," http://history.library.ucsf.edu/1959.html, accessed December 3, 2009.

33. Weaver, White, and Worthen, "Prelude to the Commission," 28.

34. Higby, "Continuing Evolution," 12.

35. H. F. DeBoest, *Proceedings of the Third Professional Pharmacy Seminar* (Indianapolis: Lilly Research Laboratories, 1969), 16, quoted in Weaver, White, and Worthen, "Prelude to the Commission," 40–41.

36. Quoted in Weaver, White, and Worthen, "Prelude to the Commission," 45.

37. *Pharmacists for the Future: The Report of the Study Commission on Pharmacy* (Bethesda, MD: American Association of Colleges of Pharmacy, 1975), reprinted in Worthen, *Road Map*, 147, 149, 150.

38. Worthen, *Road Map*, 195, 175.

39. Ibid., 217–18.

40. Ibid., 195.

41. Higby, "Continuing Evolution," 12, 13. See also William E. Fassett, "Ethics, Law, and the Emergence of Pharmacists' Responsibility for Patient Care," *Annals of Pharmacotherapy* 41 (July 2007), 1264.

42. William A. Zellmer, "Patient Package Inserts," *American Journal of Hospital Pharmacy* 33 (June 1976): 535; Trisha Gorman, "Patient Package Inserts: Will They Replace the Pharmacist's Advice?" *Drug Topics*, April 1, 1977, 53, 71.

43. Gorman, "Patient Package Inserts," 51.

44. Lawrence Fleckenstein, "Attitudes toward the Patient Package Insert—A Survey of Physicians and Pharmacists," *Drug Information Journal* 11 (March 1977): 22–29.

45. *Federal Register* 42 (1977), 37642; 43 (1978), 47198. For the number of estrogen prescriptions, see Dianne L. Kennedy, Carlene Baum, and Mary B. Forbes, "Noncontraceptive Estrogens and Progestins: Use Patterns over Time," *Obstetrics and Gynecology* 65 (March 1985): 442; for the number of pharmacists, see *Pharmacists for the Future: The Report of the Study Commission on Pharmacy* (Bethesda, MD: American Association of Colleges of Pharmacy, 1975), reprinted in Worthen, *Road Map*, 154.

46. *Federal Register* 44 (1979): 40019–20.

47. Troetel, "Three-Part Disharmony," 328 (quote), 329.

48. Richard P. Penna, "Patient Information—The Time Is Now," 5; JSW, "Where Do We Stand with PPIs?" 12; JSW, "Status of Commercial PPIs in Question," 15; Richard P. Penna, "A Hard Look at FDA's Patient Labeling Project," 17, all in *American Pharmacy* NS19 (November 1979).

49. George F. Archambault, "I'm Worried—Aren't You?" 836; John A. Owen, "Palpable Spleen," 871, both in *Hospital Formulary* (October 1979); Irving Rubin, "Product-by-Product PPIs: Not in the Public Interest!" *Pharmacy Times*, September 1979, 29; "PP vs. PPI: Which Will Win?" *Pharmacy Times*, November 1979, 25; "Allocate More Time for Voluntary Patient Education!" *Pharmacy Times*, February 1980, 29.

50. For a contemporary analysis of the possible influence of the PPI on the phar-

macist's liability, see Craig Harman Walker, "The Patient Package Insert and Pharmacist Liability," *Tulsa Law Journal* 14 (1978–79): 590–614.

51. Diane Smalley, "Additional Comments on PPIs," *Journal of the American Pharmaceutical Association* NS17 (March 1977): 132; Jerrund Wilkerson, "Patient Information—Amen," *American Pharmacy* NS20 (January 1980): 4; Jere Goyan, "Pharmaceutical Education: A Ticket to Professional Survival or Extinction (Revisited)," First Annual Kenneth L. Waters Lecture, University of Georgia, College of Pharmacy, May 18, 1982. I am indebted to Professor Robert L. Day for providing me the text of this speech.

52. Louis A. Morris, "Rationale for Patient Package Inserts," *American Journal of Hospital Pharmacy* 35 (February 1978): 183.

53. *Informing Patients about Drugs: Summary Report on Alternative Designs for Prescription Drug Leaflets* (Santa Monica, CA: Rand Corporation, 1981), 2–4.

54. Pharmaceutical Manufacturers Association v. Food and Drug Administration, Civil No. 77-291, U.S. (D. Del.), 484 F. Supp. 1179 (1980).

55. *Federal Register* 45 (1980): 60754. The drug classes were ampicillins, benzodiazepines, cimetidine, clofibrate, digoxin, methoxsalen, propoxyphene, phenytoin, thiazides, and warfarin.

56. Kushner, "Are You Set to Become a Punching Bag," 8, 42; "Mandatory PPI's: RPh's Still Don't Want Them, but Get Set for FDA's Limited Program," *American Druggist* 182 (November 1980): 17; William A. Zellmer, "Thinking Positive about PPIs," *American Journal of Hospital Pharmacy* 37 (November 1980): 1487.

57. Executive Order No. 12291, *Federal Register* 46 (February 19, 1981): 13193; 46 (April 28, 1981): 23739; 47 (February 17, 1982): 7200–7201; 47 (September 7, 1982): 39147. See also Troetel, "Three-Part Disharmony," 333–35. Patient package inserts remained a requirement for oral contraceptives, estrogens, progestins, and isoproterenol.

58. Stuart L. Nightingale, "Written Patient Information on Prescription Drugs," *International Journal of Technology Assessment in Health Care* 11 (1995): 402–3; G. William Crist, "Written Patient Information: The PHARMEX Approach," *American Pharmacy* NS21 (July 1981): 51.

59. William H. Shrank and Jerry Avorn, "Educating Patients about Their Medications: The Potential and Limitations of Written Drug Information," *Health Affairs* 26 (May–June 2007): 731–40.

60. Nightingale, "Written Patient Information," 403–5.

61. Louis A. Morris, Ann Myers, Paul Gibbs, and Chang Lao, "Estrogen PPIs: An FDA Survey," *American Pharmacy* NS20 (June 1980): 22–26.

62. Elizabeth R. Kaczmarek, Jesse E. Stewart, and Richard A. Hutchinson, "PPIs: Community Pharmacists' Compliance," *Drug Intelligence and Clinical Pharmacy* 15 (February 1981): 117–19, quote on 119.

63. Morris, Myers, Gibbs, and Lao, "Estrogen PPIs," 25–26.

64. See, e.g., Jerome A. Halperin, "Equity for the Pharmacist: Recognition with Responsibility," *Drug Intelligence and Clinical Pharmacy* 14 (July–August 1980): 489–92; Donald C. Brodie, "Pharmacy's Societal Purpose," *American Journal of Hospital Phar-*

macy 38 (December 1981): 1893–96; Richard P. Penna, "Pharmacy: A Profession in Transition or a Transitory Profession?" *American Journal of Hospital Pharmacy* 44 (September 1987): 2053–59.

CHAPTER FIVE: The Right to Write

This research was supported in part by the Robert Wood Johnson Foundation Investigator in Health Policy Program and an ANA/ANF/AAN Distinguished Nurse Scholar Fellowship at the Institute of Medicine.

1. Laura Raines, "The Right to Write," www.ajc.com/hotjobs/content/hotjobs/careercenter/pulse/2006/0506_pulse0601.html, accessed February 2, 2010. The "right to write" denotes, here, the legal right to prescribe gained through legislation, as in this case through the state's Nurse Practice Act.

2. Staff, "MAG News from the Capital," *Medical Association of Georgia Journal* 95, no. 2 (2006): 23.

3. Arlene Keeling, *Nursing and the Privilege of Prescription, 1893–2000* (Columbus: Ohio State University Press, 2007), 16–19, reference to "de facto diagnosis" on 19.

4. Ibid., 16–19.

5. Lillian Wald, *The House on Henry Street* (New York: Holt, 1915), 31.

6. Keeling, *Nursing and the Privilege of Prescription;* Vanda Summers, "Saddle-Bag and Log Cabin Technic: The Frontier Nursing Service, Inc.," *American Journal of Nursing* 38, no. 11 (1938): 1183–88.

7. Leonard I. Stein, "The Doctor-Nurse Game," *Archives of General Psychiatry* 16, no. 6 (June 1, 1967): 699–703.

8. Julie Fairman and Joan Lynaugh, *Critical Care Nursing: A History* (Philadelphia: University of Pennsylvania Press, 1998), 70–92.

9. Ibid., 74.

10. C.L., telephone interview by Julie Fairman, 1987.

11. Fairman and Lynaugh, *Critical Care Nursing,* 73–74.

12. Henry Silver, Loretta C. Ford, and Susan Stearly, "A Program to Increase Health Care for Children: The Pediatric Nurse Practitioner Program," *Pediatrics* 39, no. 5 (May 1967): 756–60; Charles E. Lewis and Barbara A. Resnick, "Nurse Clinics and Progressive Ambulatory Care," *New England Journal of Medicine* 277, no. 23 (December 7, 1964): 1236–41.

13. Deidre Wicks, *Nurses and Doctors at Work: Rethinking Professional Boundaries* (London: Taylor & Francis, 1998).

14. Fairman and Lynaugh, *Critical Care Nursing,* 70–92. This point is also made by the following authors: Stein, "Doctor-Nurse Game," 699–703; Rose Coser, *Life on the Ward* (East Lansing: Michigan State University Press, 1962); Rose Coser, Howard Becker, B. Greer, Everett C. Hughes, and Anselm Strauss, *Boys in White: Student Cultures in Medical Schools* (Chicago: University of Chicago Press, 1961).

15. Andrew Abbott, *The System of Professions: An Essay on the Division of Expert Labor* (Chicago: University of Chicago Press, 1988).

16. Julie Fairman, "Alternate Visions: The Nurse-Technology Relationship in the Context of the History of Technology," *Nursing History Review* 6, no. 1 (1998): 129–46; Julie Fairman, "Watchful Vigilance: Nursing Care, Technology, and the Development of Intensive Care Units," *Nursing Research* 4 (1992): 56–60. Thanks to Cynthia Connolly for suggesting the AIDS units as places where physicians took on care that was typically in nursing's realm, such as emotional support and family assessment.

17. Barbara Bates, interview by Julie Fairman, November 13, 1997.

18. Julie Fairman, "Delegated by Default or Negotiated by Need: Physicians, Nurse Practitioners, and the Process of Clinical Thinking," *Medical Humanities Review* 13, no. 1 (1999): 38–58.

19. Julie Fairman, *Making Room in the Clinic: Nurse Practitioners and the Evolution of Modern Health Care* (New Brunswick, NJ: Rutgers University Press, 2008).

20. This description is a composite of cases found in the Board of Medicine and Board of Nursing minutes, Pennsylvania Department of Professional and Occupational Affairs, 1965–80. See Fairman, *Making Room in the Clinic.*

21. Barbara Safriet, "Health Care Dollars and Regulatory Sense: The Role of Advanced Practice Nursing," *Yale Journal on Regulation* 9, no. 2 (1992): 417–88.

22. See, e.g., Kathi Gannon, "Pharmacy Has Mixed Reaction to Nurse Practitioner Prescribing," *Drug Topics* 134, no. 1 (1990): 15.

23. See various dates in 1978 in the Board of Medicine minutes, Pennsylvania Department of Professional and Occupational Affairs, files, 1965–80, Harrisburg, PA.

24. Deborah Ann Sampson, "Determinants and Determination: Negotiating Nurse Practitioner Prescribing Legislation in New Hampshire, 1973–1985" (PhD diss., University of Pennsylvania, 2006), *Dissertations available from ProQuest,* paper AAI3225536, January 1, 2006, http://repository.upenn.edu/dissertations/AAI3225536.

25. Safriet, "Health Care Dollars."

26. Patricia Donovan, "Medical Societies vs. Nurse Practitioners," *Family Planning Perspectives* 15, no. 4 (1983): 166–71.

27. William Biel, "Professional Nursing: Sermchief v. Gonzales: The Missouri Supreme Court Interprets the Nursing Practice Act," *University of Missouri at Kansas City Law Review* 53 (1984): 98–108, http://heinonline.org, accessed January 14, 2010.

28. Staff, Missouri Community Health Corp., "Issues in Question," Jefferson City, 1983.

29. "Consumer Access and Barriers to Primary Care," *The Center to Champion Nursing in America,* http://championnursing.org/aprnmap, accessed August 15, 2011. The eighteen states included Alaska, Arizona, Colorado, Hawaii, Idaho, Iowa, Maine, Maryland, Montana, New Hampshire, New Mexico, North Dakota, Oregon, Rhode Island, Utah, Vermont, Washington, and Wyoming.

30. Safriet, "Health Care Dollars," 419.

31. Peter Temin, "The Origin of Compulsory Drug Prescription," *Journal of Law and Economics* 22 (1979): 91–105.

32. Milton Freudenheim, "Doctors Battle Nurses over Domain in Care," *New York Times*, June 4, 1983, 1.

33. Elizabeth Hadley, "Nurses and Prescriptive Authority: A Legal and Economic Analysis," *American Journal of Law and Medicine* 15 (1989): 245–99.

34. Jean Whalen, "'A Necessity in the Nursing World': The Chicago Nurses Professional Registry, 1913–1950," *Nursing History Review* 13 (2005): 49–75.

35. "ANA Board Approves a Definition of Nursing Practice," *American Journal of Nursing* 55 (1955): 1474.

36. Lyndia Flanagan, comp., *One Strong Voice. The Story of the American Nurses' Association* (Kansas City, MO: ANA, 1976), 218; Dorothy V. Moses, "Report of the Committee on Standards of Geriatric Nursing Practice," *Proceedings of the 46th Convention of the ANA*, May 13–17, 1968 (New York: ANA, 1968), 72.

37. "ANA Board Approves a Definition of Nursing Practice," 1474.

38. Safriet, "Health Care Dollars."

39. Harry M. Marks, "Revisiting 'The Origins of Compulsory Drug Prescriptions,'" *American Journal of Public Health* 85, no. 1 (January 1995): 109–51.

40. Fairman, *Making Room in the Clinic*, 28–30.

41. See, e.g., Peter Richman, Gregory Garra, Barnet Eskin, Ashraf Nashed, and Ronald Cody, "Oral Antibiotic Use without Consulting a Physician: A Survey of ED Patients," *American Journal of Emergency Medicine* 19, no. 1 (2001): 57–60.

42. Frances Hughes, Sean Clark, Deborah Sampson, Eileen Sullivan Marx, and Julie A. Fairman, "Nurse Practitioner Research: An Historical Analysis," in *Nurses, Nurse Practitioners*, ed. M. Mezey and D. O. McGivern, 5th ed. (New York: Springer, 2010).

CHAPTER SIX: The Best Prescription for Women's Health

I am grateful to the many people who helped an idea transform into this chapter. In particular, I'd like to thank Liz Watkins and Jeremy Greene for imagining that my work might fit into this volume; Carol Downer of the Federation of Feminist Health Centers and Eileen Schnitger and Shauna Heckert of the Chico Feminist Women's Health Center for giving me access to their files; Carol Downer, Eileen Schnitger, Brenda Hanson-Smith, Gena Pennington, Suzanne Willow, Carol Ervin, Lorraine Carolan, Susan Riesel, Susan Anderson, and Dana Gallagher for agreeing to be interviewed and for corresponding with me until I got the details right; Anna Piechowski and Jocelyn Bosley for transcribing the interviews; Katie Robinson for research assistance and help with the map; and Lisa Saywell for organizing my transcripts and enhancing the map.

1. September 20, 1972, search warrant, Carol Downer files, Los Angeles. Many of the unpublished sources used in this chapter are still in the hands of their owners. I have noted the location of the original.

2. Feminist Women's Health Center, December 6, 1972, press release, Carol Downer files.

3. Ibid.

4. Bonnie Lefkowitz, *Community Health Centers: A Movement and the People Who Made It Happen* (New Brunswick, NJ: Rutgers University Press, 2007).

5. The idea of midlevel health care workers was frequently marketed to physicians as a way to rid themselves of the mundane tasks and thus make their own practices more interesting. See, e.g., Russell N. DeJong, "Use of Women's Health Care Specialists in a Women's Health Clinic," *Journal of Reproductive Medicine* 26 (1981): 287; Len Hughes Andrus and Mary Fenley, "Assistants to Primary Physicians in California," *Western Journal of Medicine* 122 (1975): 81; and Donald R. Ostergard, John E. Gunning, and John R. Marshall, "Training and Function of a Women's Health-Care Specialist, a Physician's Assistant, or Nurse Practitioner in Obstetrics and Gynecology," *American Journal of Obstetrics and Gynecology* 121 (1975): 1036.

6. For more on the history of advanced-practice nurses, see Julie Fairman, *Making Room in the Clinic: Nurse Practitioners and the Evolution of Modern Health Care* (New Brunswick, NJ: Rutgers University Press, 2008); chapter 5 in this volume; and Arlene W. Keeling, *Nursing and the Privilege of Prescription, 1893–2000* (Columbus: Ohio State University Press, 2007). Dennis Moriarty and David Hecomvich, *Physician Manpower: An Approach to Estimation of Need in California* (Sacramento, CA: Comprehensive Health Planning Program, Health Quality Systems, State Department of Health, 1973); American Medical Association, Council on Health Manpower, *Expanding the Supply of Health Services in the 1970s: Report of the National Congress on Health Manpower* (Chicago: American Medical Association, [1970?]); *Report of the National Advisory Commission on Health Manpower*, vol. 1 (Washington, DC: Government Printing Office, November 1967). As policymakers wrung their hands about the crisis in medical manpower, many experts suggested that the problem was not a shortage of physicians but rather an oversupply in specialty practices and in particular geographic areas.

7. Although it is a crude marker, it is noteworthy that Lyndon Johnson was the last Democratic presidential candidate to carry a majority of the vote in Butte (where Chico is located) and Glenn (contiguous to Butte) counties.

8. Douglas Shuit, "Bucolic Battleground," *Los Angeles Times*, April 28, 1985, A3.

9. The college was Humboldt State College, now Humboldt State University. The southern part of Humboldt County has been called the "heartland of high-grade marijuana farming in California." David Samuels, "Dr. Kush: How Medical Marijuana Is Transforming the Weed Industry," *New Yorker*, July 28, 2008.

10. For more on the women's health movement, see Leslie J. Reagan, "Crossing the Border for Abortions: California Activists, Mexican Clinics, and the Creation of a Feminist Health Agency in the 1960s," *Feminist Studies* 26 (2000): 323–48; Kathy Davis, *The Making of "Our Bodies, Ourselves": How Feminism Travels across Borders* (Durham, NC: Duke University Press, 2007); Laura Kaplan, *The Story of Jane: The Legendary Underground Feminist Abortion Service* (New York: Pantheon Press, 1995) (Maginnis quote on 22); Sandra Morgen, *Into Our Own Hands: The Women's Health Movement in the United States, 1969–1990* (New Brunswick, NJ: Rutgers University

Press, 2002); Wendy Kline, *Bodies of Knowledge: Sexuality, Reproduction, and Women's Health in the Second Wave* (Chicago: University of Chicago Press, 2010).

11. West Coast Sisters, "Self-Help Clinic"; and "Self-Help Clinic part II," n.d., Chico Feminist Health Center files, Chico, CA.

12. Carol Downer, "Women Professionals in the Feminist Health Movement," 3, n.d., Carol Downer files.

13. Feminist Women's Health Center, "Well Woman Health Care in Woman Controlled Clinics," 1976, Chico FWHC files.

14. Eileen Schnitger to author, November 24, 2010, e-mail. See also FWHC, "Well Woman Health Care."

15. For information about hostility from the local community, see "Chico Physician Cover-up" binder, Chico FWHC files.

16. "Chico Witnesses," in ibid. Women at the Chico clinic speculated that McDowell's difficulties with the local hospital emerged as a result of his association with the center and his suspended privileges at an Oregon hospital.

17. Chico FWHC interview, September 9, 1975, 29, 64, Carol Downer files. (The interview included Dido Hasper and Janice Turrini, founders of the Chico Center. Lorraine Rothman and Lynne Randall appear to be asking most of the questions. The extant transcript does not generally note who asked the questions or who answered them.)

18. Brenda Hanson-Smith, interview by the author, March 19, 2008. Watson came to Chico from Oakland only twice a month, every other Saturday.

19. Ibid.; the "Chico Physician Cover-up" binder confirms that Lorenz reviewed Hanson's charts.

20. The 1974 Nurse Practice Act gave health facilities the freedom to decide on the ground the "policies and protocols" appropriate and acceptable for nursing as long as they did not contradict any guidelines established by the Board of Medical Quality and the Board of Registered Nursing. For the history of prescription authority of nurses in California, see Elizabeth Harrison Hadley, "Nurses and Prescriptive Authority: A Legal and Economic Analysis," *American Journal of Law and Medicine* 15 (1989), 271–78.

21. Hanson-Smith interview.

22. Chico Feminist Women's Health Center, "Participatory Clinic," n.d., Detroit Feminist Women's Health Center Collection, Walter P. Reuther Library, Wayne State University, Detroit, MI.

23. See chapter 7, by Heather Munro Prescott. For FWHC suspicion of IUDs and oral contraceptives, see Federation of Feminist Women's Health Centers, *A New View of a Woman's Body* (New York: Simon and Schuster, 1981), 105–20. For birth control counseling in the clinic, see FWHC, "Well Woman Health Care."

24. Chico FWHC interview.

25. Hanson-Smith interview.

26. Ibid. I have no corroboration of Lorenz's quote.

27. Hanson-Smith interview; Mabel W. Daley (Office of Family Planning) to Diane (Dido) Hasper, December 16, 1975, Chico FWHC files.

28. Hilary Abramson, "Feminists vs. the State," *Sacramento Bee Magazine,* May 1, 1988, 8, citing a March 11, 1975, memo from Butte-Glenn Medical Society Directors' Meeting. Rebecca Chalker, "Chico Feminist Women's Health Center Files Second Major Anti-Trust Suit against ob/gyns," n.d., "Anti-trust Internal Documents" folder, Chico FWHC files. Hans Freistadt, a semiretired ob/gyn in Oroville, twenty-six miles north of Chico, agreed to provide backup in December 1975. "Norcal Appeal" binder, Chico FWHC files.

29. Hanson-Smith interview.

30. Ibid.

31. "Chronology," "Physician Cover-up" binder.

32. Chico FWHC interview.

33. Shauna Heckert and Diane Hasper to Barbara Avid [*sic*], May 20, 1977, Chico FWHC files.

34. Ibid.

35. See, e.g., documentation of an investigation prompted by the lieutenant governor's office, "Mike Curb" file, Carol Downer files.

36. Josette Mondanaro and Lyn Headley to Philip Weiler, memo re Feminist Health Centers, May 15, 1979, Carol Downer files.

37. Mondanaro to Carol Downer, May 16, 1979, "Briefing Notebook," Chico FWHC files.

38. For about a year on and off between 1980 and 1981, Mike Curb was acting governor as Jerry Brown was frequently outside the state campaigning for the presidency.

39. "Mike Curb," file.

40. Philip Weiler, chief deputy director of the Department of Health Services, to Jeannie Clayton, April 7, 1981, "Briefing Notebook," Chico FWHC files.

41. Dido Hasper, Chico FWHC, to Barbara Aved, Office of Family Planning, July 30, 1981, Chico FWHC files.

42. Aved, OFP, to Thora Delay, Chico FWHC, July 28, 1981; Delay to Aved, August 14, 1981; Aved to "Family Planning Providers," October 8, 1981, Chico FWHC files.

43. *North Coast Weekly Journal,* www.northcoastjournal.com/091406/cover0914.html, accessed October 4, 2006.

44. Gena Pennington, interview by the author, October 9, 2006. There are some discrepancies about the years physicians started to work in the clinic and when Pennington started to work there. Although she may have begun as a volunteer, at some point she became a paid part-time employee.

45. Ibid.; Suzanne Willow, interview by the author, December 30, 2009.

46. *Report of the National Advisory Commission,* 2.

47. Dick Howard and R. Douglas Roederer, *Health Manpower Licensing: California's Demonstration Projects* (Lexington, KY: Council of State Governments, April 1973), 2.

48. Ibid., 15.

49. Ostergard et al., "Training and Function," 1030.

50. A similar program, the Gynecorps Training Program, was developed in 1972 by the Department of Obstetrics and Gynecology at the University of Washington School of Medicine. See Richard M. Briggs, Barbara S. Schneidman, Eleanor N. Thorson, and Ronald D. Deisher, "Education and Integration of Midlevel Health-Care Practitioners in Obstetrics and Gynecology: Experience of a Training Program in Washington State," *American Journal of Obstetrics and Gynecology* 132 (1978): 68–77.

51. Ibid., 68; Ostergard et al., "Training and Function," 1036.

52. Briggs et al., "Education and Integration," 76.

53. Howard and Roederer, *Health Manpower Licensing,* 4.

54. Ibid., 3.

55. Willow interview.

56. Henshell died in May 2006. Information on her part of the Women's Health Care Specialist programs comes primarily from author interviews of Carol Ervin, October 16, 2006, and Willow, December 30, 2009.

57. E-mail correspondence between Lorraine Carolan and the author, February 25, 2010; and between Susan Anderson and the author, February 26, 2010.

58. Anderson e-mail correspondence.

59. The Chico FWHC did apply through AB 1503 for a program to train lay women to provide abortions. It is unclear whether this program was denied or whether the women of FWHC withdrew their application.

60. Shauna Heckert and Diane (Dido) Hasper to Barbara Avid [*sic*], OFP, May 20, 1977, Chico FWHC files.

61. Willow interview.

62. Howard and Roederer, *Health Manpower Licensing,* 15, 22.

63. Lorraine Carolan to author, February 25, 2010, e-mail.

64. Richard M. Briggs, "The Use of Para-Medical Personnel in Obstetrics and Gynecology," *Northwest Medical Journal* 1 (1974): 9–11; and Evelyn T. Dravecky, "The Obstetric-Gynecologic Team: Focus on the Worker Member," *Clinical Obstetrics and Gynecology* 15 (1972): 319–32.

65. Susan Riesel, interview by the author, October 10, 2006; Susan Anderson, interview by the author, July 27, 2006; Lorraine Carolan, interview by the author, January 5, 2010; Willow interview.

CHAPTER SEVEN: "Safer Than Aspirin"

Research for this chapter was funded in part by National Institutes of Health Publication Grant LM009242-01A2 from the National Library of Medicine.

1. *Federal Register* 58 (January 21, 1993): 5400, 5401.

2. Michael McMahon, "Pondering a Non-prescription Pill," *USA Today,* March 16, 1992, 4D.

3. Elyse Tanouye and Rose Gutfeld, "Health: Talks Cancelled on Making 'Pill' Nonprescription," *Wall Street Journal,* January 28, 1993, B1.

4. Judy Norsigian, "Her Say: Don't Make the Pill Easier to Acquire," *Chicago Tribune,* April 11, 1993, sec. 8, p. 9.

5. Jeremy A. Greene, *Prescribing by Numbers: Drugs and the Definition of Disease* (Baltimore: Johns Hopkins University Press, 2007), 191–94.

6. Ibid., 194.

7. Gwen Kay, *Dying to Be Beautiful: The Fight for Safe Cosmetics* (Columbus: Ohio State University Press, 2005), 3–5.

8. Quoted in Elizabeth Siegel Watkins, *On the Pill: A Social History of Oral Contraceptives, 1950–1970* (Baltimore: Johns Hopkins University Press, 1998), 42. *Washington Post* investigative reporter Morton Mintz, who had written about the tragic side effects of the drug thalidomide in 1962, also accused the FDA of lax standards in approving the contraceptive pill in his book *The Therapeutic Nightmare* (Boston: Houghton Mifflin, 1965); Mintz, *The Pill: An Alarming Report* (Greenwich, CT: Fawcett, 1969). Mintz later exposed corporate malfeasance in the development of the Dalkon Shield intrauterine device in his book *At Any Cost: Corporate Greed, Women, and the Dalkon Shield* (New York: Pantheon Books, 1985).

9. The advisory committee included physicians and scientists from outside of the FDA, so that the agency could receive independent sources of input. The committee was given the task of determining the extent to which oral contraceptives increased the risk of blood clotting, as well as whether the pill caused cancer of the breast, cervix, or endometrium. The advisory committee addressed charges by consumer activists that the FDA was too easily dismissing women's complaints about the pill. Suzanne White Junod, "Women over 35 Who Smoke: A Case Study in Risk Management and Risk Communication," in *Medicating Modern America: Prescription Drugs in History,* ed. Andrea Tone and Elizabeth Siegel Watkins (New York: New York University Press, 2007), 100–101.

10. Barbara Seaman, *The Doctors' Case against the Pill* (New York: Peter H. Wyden, 1969).

11. At the end of the hearings, the FDA commissioner announced that the agency would require drug manufacturers to include a patient package insert in every package of birth control pills. The insert had to be written in lay language and include all known side effects and health risks associated with oral contraceptives. The FDA soon caved in to pressure from physicians, who claimed that the insert interfered with the doctor-patient relationship; and manufacturers, who argued that a prescription drug needed an insert only for the physician, not the patient. Feminist groups were angered by the resulting simplified patient insert, arguing that women had a right to full knowledge about the drugs they were taking. Elizabeth Siegel Watkins, "'Doctor, Are You Trying to Kill Me?' Ambivalence about the Patient Package Insert for Estrogen," *Bulletin of the History of Medicine* 76 (2002): 87–88; Suzanne White Junod and Lara Marks, "Women's Trials: The Approval of the First Contraceptive Pill

in the United States and Britain," *Journal of the History of Medicine and Allied Sciences* 57 (2002): 158–59.

12. A. L Herbst, H. Ulfelder, and D. C. Poskanzer, "Adenocarcinoma of the Vagina: Association of Maternal Stilbestrol Therapy with Tumor Appearance in Young Women," *New England Journal of Medicine* 284, no. 15 (1971): 878–81.

13. Lucile Kirtland Kuchera, "Postcoital Contraception with Diethlystilbestrol," *Journal of the American Medical Association* 218, no. 4 (October 25, 1971): 562.

14. Jane Brody, "Disturbing Hints of a Possible Link to Cancer," *New York Times*, October 31, 1971, E14.

15. "'Morning After' Pill Used at U," *Michigan Daily*, undated clipping, Scrapbook, box 4, University Health Service Records, Bentley History Library, University of Michigan, Ann Arbor.

16. Sandra Morgen, *Into Our Own Hands: The Women's Health Movement in the United States, 1969–1990* (New Brunswick, NJ: Rutgers University Press, 2002), 10.

17. Kay Weiss, "Cancer and the Morning-After Pill (Will You Be Mourning-After?)," *her-self* 1 (September 1972): 1.

18. "Fact Sheet: Diethylstilbestrol (DES): The Morning-After Pill," *her-self* 1, no. 7 (December–January 1972–73): 14. Other feminist health organizations also became involved in these efforts to inform women about the dangers of DES. The Boston Women's Health Book Collective disseminated information sheets on the drug. These leaflets included warnings about the morning-after pill, stating that the risks of further exposure to DES were uncertain and that the well-informed patient needed to weigh the risks and benefits of this treatment carefully. "Fact Sheet on Daughters of DES Mothers," n.d., box 99, FF 9, Boston Women's Health Book Collective Papers, Schlesinger Library, Radcliffe Institute for Advanced Study, Cambridge, MA; TO ALL WOMEN BORN BETWEEN 1945–1970: ARE YOU A 'D.E.S.' DAUGHTER? box 100, FF 16, ibid.

19. "Beginnings," *Network News*, October 1976, 1, box 6, National Women's Health Network Records, Sophia Smith Collection, Smith College Library, Northampton, MA (hereafter referred to as NWHN Records).

20. Wendy P. Kline, *Bodies of Knowledge: Sexuality, Reproduction, and Women's Health in the Second Wave* (Chicago: University of Chicago Press, 2010), 4–5.

21. Testimony of Judy Norsigian, presented on behalf of the National Women's Health Network, *Fertility and Contraception in America: Contraceptive Technology and Development: Hearings before the House Select Committee on Population*, March 8, 1978, 95th Cong., 2nd sess., 3:375–79.

22. Ibid.

23. William A. Nolen, "How Safe Is the 'Morning-After' Pill," *McCall's*, June 1973, 16, 116; "How Safe Is the New Morning-After Pill," *Good Housekeeping*, June 1973, 184. Television broadcasts: ABC, September 20, 1973, "FDA/Birth Control Pills"; CBS, September 20, 1973, "FDA/Birth Control Pills"; NBC, September 20, 1973, "FDA/Birth Control Pills"; NBC, February 27, 1975, "Diethylstilbestrol"; CBS, September 9, 1975, "DES Ban." These television news broadcasts are available from the Vanderbilt Media

Archives, Vanderbilt University, Nashville, TN. I am very grateful to Suzanne White Junod for telling me about this archive.

24. Mary Henry, "DES Dispute: Contradictory Points of View Swirl over Estrogen's Use as Rape Treatment," *Florida Times-Union*, September 21, 1982, A-1, A-3, box 7, NWHN Records.

25. Dianne Glover, Meghan Gerety, Shirley Bromberg, Susan Fullam, Peter Divasto, and Arthur Kaufman, "Diethylstilbestrol in the Treatment of Rape Victims," *Western Journal of Medicine* 125 (1975): 334.

26. "Expert on Cancer Wary of New Pill," *New York Times*, February 23, 1973, 7.

27. A. Albert Yuzpe, telephone interview by the author, September 24, 2009.

28. A. A. Yuzpe, H. J. Thurlow, I. Ramzy, and J. I. Leyshon, "Post Coital Contraception—A Pilot Study," *Journal of Reproductive Medicine* 13, no. 2 (1974): 53–58; Yuzpe and William J. Lancee, "Ethinylestradiol and dl-Norgestral as a Postcoital Contraceptive," *Fertility and Sterility* 28, no. 9 (1977): 932–36; Yuzpe, "Postcoital Hormonal Contraception: Uses, Risks, Abuses," *International Journal of Gynaecology and Obstetrics* 15, no. 2 (1977): 133–36; R. Percival Smith and Adrianne Ross, "Post-coital Contraception Using Dl-norgestral/Ethinyl Estradiol Combination," *Contraception* 17, no. 3 (1978): 247–52.

29. Lee H. Schilling, "An Alternative to the Use of High-Dose Estrogens for Postcoital Contraception," *Journal of the American College Health Association* 27 (1979): 247–49. See also Schilling, "Awareness of the Existence of Postcoital Contraception among Students Who Have Had a Therapeutic Abortion," *Journal of American College Health* 32 (1984): 244–46.

30. "Ovral Touted as Morning-After Pill," *Contraceptive Technology Update* 1, no. 1 (1980): 11–13.

31. Robert A. Hatcher, "Morning-After Pill Can Unlock Health Care Door," *Contraceptive Technology Update* 1, no. 4 (1980): 54.

32. Yuzpe interview.

33. "Postcoital Contraception: A Delicate Political Issue," *Contraceptive Technology Update* 5, no. 4 (April 1984): 41–43.

34. "Lack of Data on 'Morning After' Pill Confounds Clinicians," *Contraceptive Technology Update* 8, no. 11 (1987): 137–39.

35. "Lack of Awareness Reason Most Do Not Prescribe Postcoital OCs," *Contraceptive Technology Update* 7, no. 9 (1986): 106–7.

36. Eve W. Paul, director of legal services, to Mr. Jack Shettle, Brooks, Shettle & Garman, September 20, 1978; Paul to Winston Gaye, July 27, 1978, Planned Parenthood Federation of America Papers, Series II, box 68, folder 61, Sophia Smith Collection, Smith College Library, Northampton, MA.

37. L. L. Wynn, "US: Sexual Archetypes from the DIY to Post-dedicated Product Eras," in *Emergency Contraception: The Story of a Global Reproductive Health Technology*, ed. Angel M. Foster and L. L. Wynn (New York: Palgrave Macmillan, 2011).

38. Alison Piepmeier, *Girl Zines: Making Media, Doing Feminism* (New York: New York University Press, 2009), 190.

39. R. William Soller, "Evolution of Self-Care with Over-the-Counter Medications," *Clinical Therapeutics* 20, suppl. C (1998): C134–C140.

40. Pauline Vaillancourt Rosenau, "Rx-to-OTC Switch Movement," *Medical Care Research and Review* 51 (1994): 429–40.

41. Soller, "Evolution of Self-Care," C134–C140.

42. Davina C. Ling, Ernst R. Berndt, and Margaret K. Kyle, "Deregulating Direct-to-Consumer Marketing of Prescription Drugs: Effects on Prescription and Over-the-Counter Product Sales," *Journal of Law and Economics* 45 (2002): 691–92.

43. James Trussell, Felicia Stewart, Malcolm Potts, Felicia Guest, and Charlotte Ellertson, "Should Oral Contraceptives Be Available without Prescription?" *American Journal of Public Health* 83 (1993): 1094–99.

44. Norsigian, "Her Say," 9.

45. Elyse Tanouye and Rose Gutfeld, "Health: Talks Cancelled on Making 'Pill' Nonprescription," *Wall Street Journal,* January 28, 1993, B1.

46. Norsigian, "Her Say."

47. Sarah E. Samuels, Jeff Stryker, and Mark D. Smith, "Over-the-Counter Birth Control Pills: An Overview," in *The Pill from Prescription to Over the Counter,* ed. Sarah E. Samuels and Mark D. Smith (Menlo Park, CA: Henry J. Kaiser Family Foundation, 1994), 2.

48. Francine Coeytaux and Amy Allina, "The Pill without Prescription: The International Experience," in Samuels and Smith, *The Pill,* 41, 48–49.

49. James Trussell, Felicia Stewart, Charlotte Ellertson, Felicia Guest, and Malcolm Potts, "Efficacy Implications of Making the Pill Available over the Counter," in Samuels and Smith, *The Pill,* 121, 130.

50. Natalie Anger, "Future of the Pill May Lie Just over the Counter," *New York Times,* August 8, 1993, E5.

51. James Trussell, Felicia Stewart, Felicia Guest, and Robert A. Hatcher, "Emergency Contraceptive Pills: A Simple Proposal to Reduce Unintended Pregnancies," *Family Planning Perspectives* 24 (1992): 269–73.

52. Joyce Price, "Report Says Several Drugs Can Act as Abortion Pills," *Washington Times,* January 3, 1993, A5.

53. National Women's Health Network, "Position Paper: Postcoital Contraception: Time for Cautious Approval," 1993, box 7, NWHN Records. For more on the Network's continuing opposition to OTC status for oral contraceptives, see "Network's Position on Oral Contraceptive Availability without a Prescription," *Network News,* May–June, 1993, 5, 7–8.

54. Heather Munro Prescott, *The Morning After: A History of Emergency Contraception in the United States* (New Brunswick, NJ: Rutgers University Press, 2011).

55. Geoffrey Cowle, "Right off the Shelf," *Newsweek,* July 10, 2000, 50.

56. Sarah Lueck, "FDA Examines More Switches of Drugs to Be Nonprescription," *Wall Street Journal* (Eastern ed.), June 29, 2000, B26.

57. Kirsten Moore, Comments before Food and Drug Administration Public

Hearing on Over-the-Counter Drug Products, June 28, 2000, Docket 00N-1256, www.fda.gov/ohrms/dockets/DOCKETS/00n1256/ts00013.pdf, accessed June 22, 2010.

58. Jack Stover, ibid.

59. Tara Shochet, ibid.

60. Beverly Winikoff, ibid.

61. Amy Allina, ibid.

62. Ibid.

63. Ibid.

64. Elisa S. Wells, Jane Hutchings, Jacqueline S. Gardner, Jennifer L. Winkler, Timothy S. Fuller, Don Downing, and Rod Shafer, "Using Pharmacies in Washington State to Expand Access to Emergency Contraception," *Family Planning Perspectives* 30 (1998): 288–90.

65. Gina Kolata, "Without Fanfare, Morning-After Pill Gets a Closer Look," *New York Times*, October 8, 2000, 1, 22.

66. Randall W. Lutter to Bonnie Scott Jones and Simon Heller, June 9, 2006, Docket No. 2001P-0075/CP1, Regulations.gov, www.regulations.gov, accessed August 15, 2011.

67. Quoted in "FDA Advisory Panels Recommend EC Be Sold without Prescription," *Kaiser Daily Women's Health Policy*, December 17, 2003, www.kaisernetwork.org/daily_reports/rep_index.cfm?DR_ID=21382, accessed June 22, 2010.

68. Lutter to Scott Jones and Heller.

69. Quoted in L. L. Wynn, Joanna N. Erdman, Angel M. Foster, and James Trussell, "Harm Reduction or Women's Rights? Debating Access to Emergency Contraceptive Pills in Canada and the United States," *Studies in Family Planning* 38 (2007): 254.

70. Quoted ibid.

71. Center for Reproductive Rights, "Center for Reproductive Rights Questions Timing of Plan B Announcement," July 31, 2006, http://reproductiverights.org/en/press-room/center-for-reproductive-rights-questions-timing-of-fda-plan-b-announcement, accessed June 22, 2010.

72. "Morning-After Pill Conspiracy: Interview with Annie Tummino," http://reproductiverights.org/en/document/morning-after-pill-conspiracy-interview-with-annie-tummino, accessed June 22, 2010.

73. "Morning-After Pill Conspiracy—History," www.mapconspiracy.org/history.html, accessed June 22, 2010.

74. Jenny Brown, "One Million March for Women's Reproductive Rights," May 15, 2004, http://jfbrown.wordpress.com/2004/05/15/one-million-march-for-womens-reproductive-rights/, accessed June 23, 2010.

75. "Morning-After Pill Conspiracy—History."

76. "Morning-After Pill Conspiracy: Interview with Annie Tummino."

77. "Morning-After Pill Conspiracy—History."

78. Jenny Brown, "FDA Tries to Divide Women by Age to Deny Us Our Rights!" www.mapconspiracy.org/age.html, accessed June 22, 2010.

79. Gardner Harris, "Official Quits on Pill Delay at the F.D.A." *New York Times*, September 1, 2005, A12.

80. U.S. Government Accountability Office, "Decision Process to Deny Initial Application for Over-the-Counter Marketing of the Emergency Contraceptive Drug Plan B Was Unusual," GAO-06-109, November 14, 2005, www.gao.gov/new.items/d06109.pdf, accessed June 22, 2010.

81. Center for Reproductive Rights, "Former Head of FDA to Be Deposed on Tuesday in 'Morning-After Pill' Case as Agency Rejects Five-Year-Old Citizen's Petition," http://reproductiverights.org/en/press-room/former-head-of-fda-to-be-desposed-on-tuesday-in-morning-after-pill-case-as-agency-rejects, accessed June 23, 2010.

82. Center for Reproductive Rights, "Federal Court Rules FDA Must Reconsider Plan B Decision," http://reproductiverights.org/en/document/federal-court-rules-fda-must-reconsider-plan-b-decision-0, accessed June 23, 2010.

83. "Victory for Women! Ruling in Morning-After Pill Lawsuit," www.mapconspiracy.org/, accessed June 23, 2010.

84. Sandra Levy, "The New Gatekeepers: Changing Market Conditions are Catapulting Pharmacists into the Pivotal Role of Patrolling a Third Class of Drugs," *Drug Topics* 149, no. 14 (July 25, 2005): 24.

85. Oral Contraceptives Over-the-Counter Working Group, "Statement of Purpose," http://ocsotc.org/?page_id=5, accessed July 13, 2010. Working group participants include representatives from the National Women's Health Network and other former critics of the OTC switch.

86. "Behind the Counter Availability of Certain Drugs," Public Meeting Wednesday, November 14, 2007, transcript, Docket FDA-2007-N-00832, Regulations.gov, www.regulations.gov, accessed August 15, 2011.

CHAPTER EIGHT: The Prescription as Stigma

The funding support of the American Pain Society and the Milbank Memorial Fund is gratefully acknowledged. The author would also like to thank the interview participants for their openness and generosity in sharing their stories.

1. Raymond W. Houde, "Systematic Analgesics and Related Drugs; Narcotic Analgesics," in *International Symposium on Pain of Advanced Cancer*, ed. John J. Bonica and Vittorio Ventafridda, Advances in Pain Research and Therapy, vol. 2 (New York: Raven Press, 1979), 263–73, quote on 263.

2. E.g., see R. Chou, G. J. Fanciullo, P. G. Fine, J. A. Adler, J. C. Ballantyne, P. Davies, M. I. Donovan, et al., American Pain Society–American Academy of Pain Management Opioids Guidelines Panel, "Clinical Guidelines for the Use of Chronic Opioid Therapy in Chronic Noncancer Pain," *Journal of Pain* 10, no. 2 (February 2009): 113–30.

3. Norman Howard-Jones, "A Critical Study of the Origins and Early Development of Hypodermic Medication," *Journal of the History of Medicine and Allied Sciences* 2 (1947): 201–45.

4. George Wood, *A Treatise on Therapeutics and Pharmacology or Materia Medica,* 3rd ed. (Philadelphia: Lippincott, 1868), 711 ("exaltation"); Francis Anstie, "The Hypodermic Injection of Medicines," *Practitioner* 1 (1868): 32 ("of danger"); Virgil G. Eaton, "How the Opium Habit Is Acquired," *Popular Science Monthly* 33 (September 1888), reprinted in H. Wayne Morgan, *Yesterday's Addicts: American Society and Drug Abuse, 1865–1920* (Norman: University of Oklahoma Press, 1974).

5. Ernest S. Bishop, *The Narcotic Drug Problem* (New York: Macmillan, 1920), 70.

6. Harrison Act, Public Law 63-223, 63rd Cong., 3rd sess., December 17, 1914.

7. David F. Musto, *The American Disease: Origins of Narcotic Control,* 2nd expanded ed. (New York: Oxford University Press, 1987), 174–75.

8. For examples of these cases and their impact, see Musto, *American Disease;* and Steven R. Belenko, ed., *Drugs and Drug Policy in America: A Documentary History* (Westport, CT: Greenwood Press, 2000).

9. Samuel Hopkins Adams, "How People Become Drug Addicts," *Colliers,* March 1, 1924, 9.

10. Lawrence Kolb and A. G. Du Mez, "The Prevalence and Trend of Drug Addiction in the United States and Factors Influencing It," *Public Health Reports* 39 (May 23, 1924): 1179–1204.

11. Lawrence Kolb, "Types and Characteristics of Drug Addicts," *Mental Hygiene* 8 (1925): 300–313.

12. Warren H. Cole, introduction to *Management of Pain in Cancer,* ed. M. J. Schiffrin and E. J. Gross (Chicago: Year Book, 1956), 8.

13. Schiffrin and Gross, *Management of Pain in Cancer,* 14.

14. Marcia L. Meldrum, "The Property of Euphoria: Research and the Cancer Patient," in *Opioids and Pain Relief: A Historical Perspective,* ed. Marcia L. Meldrum (Seattle: IASP Press, 2003), 196–99.

15. Carol P. H. Germain, *The Cancer Unit: An Ethnography* (Wakefield, MA: Nursing Resources, 1979), 194. The pseudonyms in this account were assigned by the original author.

16. Richard M. Marks and Edward J. Sachar, "Undertreatment of Medical Inpatients with Narcotic Analgesics," *Annals of Internal Medicine* 78 (1973): 173–81, quote on 181; John P. Morgan and David L. Pleet, "Opiophobia in the United States: The Undertreatment of Severe Pain," in *Society and Medication: Conflicting Signals for Prescribers and Patients,* ed. John P. Morgan and Doreen V. Kagan (Lexington, MA: Lexington Books, 1983), 313–26. In comparison trials, 100 mg. of meperidine has been shown to be less effective than 650 mg. of aspirin.

17. Ada K. Jacox, *Pain: A Source Book for Nurses and Other Health Professionals* (Boston: Little, Brown, 1977), 384.

18. Nathan B. Eddy and Everette L. May, "The Search for a Better Analgesic," *Science* 181 (1973): 407–14; Kenner C. Rice, "Analgesic Research at the National Institutes of Health: State of the Art 1930s to the Present," in Meldrum, *Opioids and Pain Relief,* 57–83, Hunt quoted on 58. The Committee on Drug Addiction has had several

names and homes over the years; it continues its work today as an independent body, the College on Problems of Drug Dependence.

19. For further discussion, see Meldrum, "Property of Euphoria."

20. Raymond W. Houde, Oral History Interview by Marcia L. Meldrum, 1995, John C. Liebeskind History of Pain Collection, UCLA. Houde and Rogers's work was under the auspices of the NRC Committee, discussed in Eddy and May, "Search for a Better Analgesic," 407–14. Their collaborator Stanley Wallenstein analyzed the data.

21. Houde, "Systematic Analgesics and Related Drugs," 269.

22. Houde Oral History Interview.

23. Ada G. Rogers, Oral History Interview by Marcia L. Meldrum, 1995, John C. Liebeskind History of Pain Collection, UCLA.

24. Robert G. Twycross, "Choice of Strong Analgesic in Terminal Cancer: Diamorphine or Morphine?" *Pain* 3 (1977): 93–104.

25. Cicely Saunders, Oral History Interview by John C. Liebeskind, 1993, John C. Liebeskind History of Pain Collection, UCLA; Robert G. Twycross, "Overview of Analgesia," in Bonica and Ventafridda, *International Symposium on Pain*, 617–33, quote on 632.

26. World Health Organization, *Cancer Pain Relief* (Geneva: World Health Organization, 1986). There was initial hesitation by WHO in releasing these guidelines, which have since become widely accepted. See Marcia L. Meldrum, "The Ladder and the Clock: Cancer Pain and Public Policy at the End of the Twentieth Century," *Journal of Pain and Symptom Management* 29, no. 1 (January 2005): 41–54.

27. "Chronic Pain in America: Roadblocks to Relief," 1999, report for the American Pain Society, the American Academy of Pain Medicine, and Janssen Pharmaceutica, at www.ampainsoc.org/links/roadblocks/, accessed September 28, 2010.

28. Chou et al., "Guidelines for Opioid Therapy."

29. Richard M. Kanner and Kathleen Foley, "Patterns of Drug Use in a Cancer Pain Clinic," *Annals of the New York Academy of Sciences* 362 (1981): 161–72, "*dearth of clinical studies*" on 171; Russell K. Portenoy and Kathleen Foley, "Chronic Use of Opioid Analgesia in Non-malignant Pain: Report of 38 Cases," *Pain* 25 (1986): 171–86.

30. Steven D. Passik, Kenneth L. Kirsh, Margaret V. McDonald, Sam Ahn, Simcha M. Russak, Lisa Martin, Barry Rosenfeld, William S. Breitbart, and Russell K. Portenoy, "A Pilot Study of Aberrant Drug-Taking Attitudes and Behaviors in Samples of Cancer and AIDS Patients," *Journal of Pain and Symptom Management* 19 (2000): 274–86, quote on 278–80.

31. Interview by author, July 30 and August 1, 2001, transcript available in the John C. Liebeskind History of Pain Collection, UCLA Biomedical Library. "Mary" and "Tom" are pseudonyms. All subsequent quotations in this section are from this transcript.

32. Peter Whoriskey, "Rush Limbaugh Turns Himself In on Fraud Charge in Rx Drug Probe," *Washington Post*, April 29, 2006, www.washingtonpost.com/wp-dyn/content/article/2006/04/28/AR2006042801692.html, accessed February 12, 2010.

33. United States v. Hurwitz, U.S. Court of Appeals, 4th cir., No. 05-4474 (August 22, 2006).

34. Norman S. Miller and Andrea Greenfield. "Patient Characteristics and Risk Factors for Development of Dependence on Hydrocodone and Oxycodone," *American Journal of Therapeutics* 11 (2004): 26–32.

35. See Barry Meier, *Pain Killer: A "Wonder" Drug's Trail of Abuse and Death* (Emmaus, PA: Rodale Books, 2003).

36. DEA Information Bulletin on Oxycontin Diversion and Abuse, January 2001, www.justice.gov/ndic/pubs/651, accessed January 7, 2010.

37. Office of National Drug Control Policy, Executive Office of the President, *Teens and Prescription Drugs: An Analysis of Recent Trends on the Emerging Drug Threat,* February 2007, 6, www.theantidrug.com/pdfs/TEENS_AND_PRESCRIPTION_DRUGS.pdf, accessed January 7, 2010.

38. Theodore J. Cicero, James A. Inciardi, and Alvaro Muñoz, "Trends in Abuse of OxyContin and Other Opioid Analgesics in the United States, 2002–2004," *Journal of Pain* 6 (October 2005): 662–72.

39. Lloyd D. Johnston, Patrick M. O'Malley, Jerald G. Bachman, and John E. Schulenberg, *Monitoring the Future: National Results on Adolescent Drug Use: Overview of Key Findings, 2008,* NIH Pub. 09-7401 (Bethesda, MD: NIDA, 2009), 28–29.

40. Office of National Drug Control Policy, *Teens and Prescription Drugs,* 3. Rates were studied in children in grades eight, ten, and twelve.

41. Ibid., 4.

42. DEA, "Action Plan to Prevent the Diversion and Abuse of Oxycontin," www.deadiversion.usdoj.gov/drugs_concern/oxycodone/abuse_oxy.htm, accessed January 7, 2010.

43. DEA, "Drugs and Chemicals of Concern: Oxycodone," www.deadiversion.usdoj.gov/drugs_concern/oxycodone/summary.htm, accessed January 7, 2010. At least two common sources of illegal prescription drugs are overlooked in these statements: "borrowing" from a parent or other family member's legitimate supply and Internet sales.

44. DEA, "Drugs and Chemicals of Concern: Hydrocodone," www.deadiversion.usdoj.gov/drugs_concern/hydrocodone.pdf, accessed January 7, 2010.

45. "U.S. Drug Prevention, Treatment, Enforcement Agencies Take on 'Doctor Shoppers,' 'Pill Mills,'" March 1, 2004, Office of National Drug Control Policy Press Release, www.drugs.com, accessed January 7, 2010.

46. DEA, "Oxycontin Diversion and Abuse," www.justice.gov/ndic/pubs/651.

47. "The Myth of the 'Chilling Effect,'" www.justice.gov/dea/pubs/pressrel/pr103003p.html, accessed January 7, 2010. Sanctions include fines, letters of admonition, and revocation of registration.

48. As David Herzberg notes in chapter 9, prescription volume was often the telltale clue, sometimes the only one that led regulators to "pill mills" and "gray markets" dispensing drugs under medical auspices.

49. Michele M. Leonhart, deputy administrator, DEA Office of Diversion Con-

trol, "Dispensing of Controlled Substances for the Treatment of Pain," *Federal Register* 69, no. 220 (November 16, 2004): 67171.

50. David Joranson, director of the Pain and Policy Studies Group at the University of Wisconsin, to Michele Leonhart, November 24, 2004, www.painpolicy.wisc.edu/DEA/IPSresponse.pdf, accessed January 10, 2010.

51. David Joranson, Russell K. Portenoy, and Steven D. Passik to Karen Tandy, October 26, 2004, www.painpolicy.wisc.edu/DEA/letter%20to%20DEA.pdf, accessed January 10, 2010.

52. Leonhart, "Dispensing of Controlled Substances." The ten "recurring patterns" were based on an appellate court ruling of 1978, United States v. Rosen, 582 F.3d 1032 (5th cir). The seven other pointers to "condemned behavior" included prescribing without a physical examination, warning the patient to fill multiple prescriptions at different pharmacies, and prescribing drugs at intervals or for conditions inconsistent with "legitimate medical treatment." The final rule was published nearly two years later as 21 CFR Part 1306, *Federal Register* 71, no. 172 (September 6, 2006).

53. Scott S. Fishman, "Commentary: From Balanced Pain Care to Drug Trafficking: The Case of William Hurwitz and the DEA," *Pain Medicine* 5 (2005): 162–64. The Hassenbusch-Fishman letter is reproduced as part of this commentary.

54. John Tierney, "Juggling Figures, and Justice, in a Doctor's Trial," *New York Times*, July 3, 2007, www.nytimes.com, accessed January 3, 2010.

55. Ibid., 1 (on the Web site).

56. Fishman, "Commentary."

57. *United States v. Hurwitz.*

58. William Hurwitz, "The Challenge of Prescription Drug Misuse: A Review and Commentary," *Pain Medicine* 5 (2005): 152–61, quotes on 156. His description of adolescent users is supported by Office of National Drug Control Policy, *Teens and Prescription Drugs*. *Pain Medicine* is the journal of the American Academy of Pain Medicine.

59. John Tierney, "At Trial, Pain Has a Witness," *New York Times*, April 24, 2007, www.nytimes.com, accessed January 3, 2010. Lohrey later became Hurwitz's patient, and he treated her until his arrest.

60. Tierney, "Juggling Figures," 1 (on the Web site).

61. Tierney, "At Trial," 1 (on the Web site).

62. Jerry Markon, "Virginia Pain Doctor's Prison Term Is Cut to 57 Months," *Washington Post*, July 14, 2007, www.washingtonpost.com, accessed January 3, 2010.

63. Leonhart, "Dispensing of Controlled Substances"; 21 CFR Part 1306, *Federal Register* 71, no. 172 (September 6, 2006): 52720.

64. "Issuance of Multiple Prescriptions for Schedule II Controlled Substances," 21 CFR Part 1306, *Federal Register* 72, no. 222 (November 19, 2007): 64923.

65. "Prescription Medicine News Releases," www.justice.gov/dea/pubs/pressrel/prescription_index.html, accessed February 12, 2010.

66. Tina Rosenberg, "When Is a Pain Doctor a Drug Pusher?" *New York Times Magazine*, June 17, 2007, www.nytimes.com, accessed January 10, 2010.

67. Ibid., 10, at www.nytimes.com/2007/06/17/magazine/17pain-t.html, accessed January 10, 2010.

68. See, e.g., Howard S. Smith, Kenneth L. Kirsh, and Steven D. Passik, "Chronic Opioid Therapy Issues Associated with Opioid Abuse Potential," *Journal of Opioid Management* 5, no. 5 (September–October 2009): 287–300.

69. Chou et al., "Guidelines for Opioid Therapy," 117, 120.

70. Cf. David Herzberg's analysis in chapter 9.

71. Cf. Jeremy Greene's discussion in chapter 10 of the uses of the prescription for surveillance.

72. For lifetime misuse: Substance Abuse and Mental Health Services Administration, *Results from the 2009 National Survey on Drug Use and Health: Volume 2, Technical Appendices and Selected Prevalence Tables*, 2010, Office of Applied Studies, NSDUH Series H-38B, HHS Publication No. SMA 10-4586 Appendices, Rockville, MD, p. 98, available at www.oas.samhsa.gov/NSDUH/2k9NSDUH/2k9ResultsApps.pdf, accessed August 9, 2011. For admissions: Substance Abuse and Mental Health Services Administration, Office of Applied Studies, *The TEDS Report: Characteristics of Substance Abuse Treatment Admissions Reporting Primary Abuse of Prescription Pain Relievers: 1998 and 2008*, September 23, 2010, Rockville, MD, p. 1, available at www.oas.samhsa.gov/2k10/230b/230bPainRelvr2k10Web.pdf, accessed August 9, 2011.

73. Chronic pain patients very probably pursued such strategies before the 1970s as well and suffered severe stigmatization as a result.

CHAPTER NINE: Busted for Blockbusters

I would like to thank the volume editors and Nicolas Rasmussen for their close and intelligent attention to earlier drafts of this chapter. Where I was able to follow their suggestions, the chapter is much better for it.

1. Prosecutors' summation, United States v. Witt, court transcript from February 14, 1983, 15121, in box 10, Case File 82 Cr. 33-CSH, General Case Files, U.S. District Court, Southern District of New York, Records of District Courts of the United States, RG 021, National Archives and Records Administration, Central Plains Region, Lee Summit, MO. Hereafter cited as *Witt*.

2. Lawrence Glass testimony, October 12, 1982, 2305, box 7, *Witt*.

3. Data from Drug Abuse Warning Network and the DEA, cited in E. B. Staats (comptroller general), *Retail Diversion of Legal Drugs: A Major Problem with No Easy Solution*, March 10, 1978 (Washington, DC: General Accounting Office, 1978), 3–7. See also U.S. General Accounting Office, *Comprehensive Approach Needed to Help Control Prescription Drug Abuse*, October 29, 1982 (Washington, DC: General Accounting Office, 1982).

4. See, e.g., Elizabeth Siegel Watkins, *On the Pill: A Social History of Oral Contraceptives* (Baltimore: Johns Hopkins University Press, 2001); Jeremy Greene, *Prescribing by Numbers: Drugs and the Definition of Disease* (Baltimore: Johns Hopkins University Press, 2008); Nicolas Rasmussen, *On Speed: The Many Lives of Amphetamine* (New

York: New York University Press, 2008); David Herzberg, *Happy Pills in America: From Miltown to Prozac* (Baltimore: Johns Hopkins University Press, 2009); Andrea Tone, *The Age of Anxiety: A History of America's Turbulent Affair with Tranquilizers* (New York: Basic Books, 2009); Robert Bud, *Penicillin: Triumph and Tragedy* (Oxford: Oxford University Press, 2009).

5. See Rasmussen, *On Speed*, for the many illicit markets for amphetamine, including the story of celebrity "scrip doctor" Max Jacobson (168–71).

6. Rasmussen, *On Speed*, is a notable exception. But see David Musto, *The American Disease: Origins of Narcotics Control*, 3rd ed. (Oxford: Oxford University Press, 1999); Nancy Campbell, *Using Women: Gender, Policy, and Social Justice* (New York: Routledge, 2000); David Courtwright, *Dark Paradise: A History of Opiate Addiction in America* (Cambridge, MA: Harvard University Press, 2001); Joseph Spillane, *Cocaine: From Medical Marvel to Modern Menace in the United States, 1884 to 1920* (Baltimore: Johns Hopkins University Press, 2002); Curtis Marez, *Drug Wars: The Political Economy of Narcotics* (Minneapolis: University of Minnesota Press, 2004); Caroline Jean Acker, *Creating the American Junkie: Addiction Research in the Classic Era of Narcotic Control* (Baltimore: Johns Hopkins University Press, 2002); Eric Schneider, *Smack: Heroin and the Postwar City* (Philadelphia: University of Pennsylvania Press, 2008).

7. David Rothman, *Strangers at the Bedside: A History of How Law and Bioethics Transformed Medical Decision Making* (New York: Basic Books, 1991); Paul Starr, *The Social Transformation of American Medicine: The Rise of a Sovereign Profession and the Making of a Vast Industry* (New York: Basic Books, 1984).

8. See, e.g., Dominique Tobbell, "Allied against Reform: Pharmaceutical Industry–Academic Physician Relations in the United States, 1945–1970," *Bulletin of the History of Medicine* 82 (2008): 878–912; and chapter 2, by Podolsky. For rare counterexamples, see chapter 1, by Rasmussen; Rasmussen, *On Speed*, chap. 7, "Amphetamine's Decline."

9. John Swann, "The FDA and the Practice of Pharmacy: Prescription Drug Regulation before 1968," in *Federal Drug Control*, ed. Jonathan Erlen and Joseph Spillane (Binghamton, NY: Pharmaceutical Products Press, 2004), 145–74.

10. Herzberg, *Happy Pills*, chap. 3; Rasmussen, *On Speed;* chapter 1, this volume.

11. Drug Abuse Control Amendments of 1965, Public Law 89-74, 79 Stat. 226; David Musto and Pamela Korsmeyer, *The Quest for Drug Control: Politics and Federal Policy in a Period of Increasing Substance Abuse, 1963–1981* (New Haven, CT: Yale University Press, 2002), 38–106; Herzberg, *Happy Pills*, chap. 3.

12. Comprehensive Drug Abuse Prevention and Control Act of 1970, Public Law 91-513, 84 Stat. 1236 (October 27, 1970). See Joseph F. Spillane, "Debating the Controlled Substances Act," *Drug and Alcohol Dependence* 76 (2004): 17–29; David Courtwright, "The Controlled Substances Act: How a "Big Tent" Reform Became a Punitive Drug Law," *Drug and Alcohol Dependence* 76 (2004): 9–15; Musto and Korsmeyer, *Quest for Drug Control*, 38–106.

13. U.S. Senate, Subcommittee to Investigate Juvenile Delinquency of the Committee on the Judiciary, *Methaqualone (Quaalude, Sopor) Traffic, Abuse, and Regulation,*

93rd Cong., 1st sess. (Washington, DC: Government Printing Office, 1973), 88; Staats, *Retail Diversion of Legal Drugs*, 12–24; Spillane, "Debating the Controlled Substances Act"; DEA, Office of Diversion Control, *Drug Diversion—A Historical Perspective* (Washington, DC: Drug Enforcement Administration, Office of Diversion Control, 1993), 10.

14. Quote from "A Summary of the NIDA Technical Review," in James R. Cooper, ed., *Impact of Prescription Drug Diversion Control Systems on Medical Practice and Patient Care*, NIDA Research Monograph 131 (Rockville, MD: National Institute on Drug Abuse, 1993), 9; Staats, *Retail Diversion of Legal Drugs*, 14–15.

15. Numbers tabulated from the *Federal Register*, where DEA decisions were published.

16. Herzberg, *Happy Pills*.

17. Rasmussen, *On Speed*, 91–99, 171, 178–79; Swann, "FDA and the Practice of Pharmacy"; data drawn from IMS America Ltd., *Drug Abuse Warning Network: Phase V Report, May 1976–April 1977*, U.S. Department of Justice, Drug Enforcement Administration, 1978.

18. IMS America Ltd., *DAWN Annual Report, 1979*, U.S. Department of Justice, Drug Enforcement Administration, 1979, p. 22; see also Staats, *Retail Diversion of Legal Drugs*, 5–7.

19. Sherwin Gardiner testimony, in U.S. Senate, *Methaqualone*, 160–61.

20. M. Falco, *Methaqualone: A Study of Drug Control* (Washington, DC: Drug Abuse Council, 1975), 12–34, quote on 24.

21. Quaalude advertisement, *Archives of General Psychiatry* 25, no. 5 (1971): 30–32.

22. George Gay, MD (Haight-Ashbury Free Clinic), in U.S. Senate, *Methaqualone*, 191.

23. Falco, in U.S. Senate, *Methaqualone*, 12–34.

24. Dr. Richard Kunnes, 146; Sen. Birch Bayh, 151; Dr. George Gay, 191, all in U.S. Senate, *Methaqualone;* Julia Cameron, "Pretty Poison: A Cold Look at a Hot Drug," *Potomac (Washington Post)*, February 11, 1973; John Russell, "Methaqualones—Heroin for Lovers," *Journal of the Mississippi State Medical Association* 14, no. 11 (1973): 496; Peter Ognibene, "There's Gold in Them There Pills," *New Republic*, April 21, 1973, 14.

25. Quaalude advertisement, from U.S. Senate, *Methaqualone*, 296–97.

26. E. Pascarelli, "Methaqualone Abuse: The Quiet Epidemic," *Journal of the American Medical Association* 224, no. 11 (1973): 1512–14; S. Ager, "Luding Out," *New England Journal of Medicine* 281, no. 1 (1972): 51; D. Zwerdling, "Methaqualone: The "Safe" Drug That Isn't Very," *Washington Post*, November 12, 1972, in U.S. Senate, *Methaqualone*, 360–64. See also David Herzberg, "Blockbusters and Controlled Substances: Miltown, Quaalude, and Consumer Demand for Drugs in Postwar America," *Studies in the History and Philosophy of Biological and Biomedical Sciences*, in press, available online June 17, 2011. For moral panics, see Erich Goode and Nachman Ben-Yehuda, "Moral Panics: Culture, Politics, and Social Construction," *Annual Review of Sociology* 20 (1994): 149–71.

27. U.S. Senate, *Methaqualone*, 94.

28. Parke-Davis testimony, 246–47; Rorer testimony, 265, both ibid.

29. *Federal Register* 37 (October 4, 1973c): 27517–19. See also Herzberg, "Blockbusters and Controlled Substances." Quotas were tracked in the *Federal Register;* impact on medical use from D. L. Crosby, L. B. Burke, and J. S. Kennedy, "Drug Scheduling—What Effects?" *Proceedings of the Annual Meeting of the American Pharmaceutical Association* (1978), cited in Spillane, "Debating the Controlled Substances Act," 26.

30. Schneider, *Smack;* Rasmussen, *On Speed.*

31. U.S. Senate, *Methaqualone,* 4, 91, 131, 363.

32. See, e.g., Julie Baumgold, "Down at the Juice Bars," *New York Magazine,* December 4, 1972.

33. Rasmussen *On Speed,* 168–71.

34. Martin Siegel testimony, September 23, 1982, 214–19, box 3; Ronald Asherson testimony, November 2, 1982, 4320, box 12; Lawrence Glass testimony, October 12, 1982, 2220–21, box 7, all in *Witt.*

35. "Prescription Information from Manhattan Center," n.d., box 3, *Witt.*

36. Asherson testimony, November 2, 1982, 4411, box 12, *Witt.*

37. Prosecutor summation, February 14–17, 1983, 15077, box 10, *Witt.*

38. Dr. Peter Sarosi testimony, December 3, 1982, 7559–80, box 13, *Witt.*

39. "Statement given by Joseph DeBeneditto at the New York District Office of the DEA on August 12, 1980," box 7, *Witt.* One patient also told the jury that Ungar had overridden a physician's refusal to prescribe for him. See Steven Greenberg testimony, October 5, 1982, 1429–30, box 4, *Witt.*

40. Shawn DePietto testimony, October 29, 1982, 4845, box 12, *Witt.*

41. See, e.g., Gordon Schauer testimony, October 12, 1982, 2110, box 7, *Witt.*

42. Benjamin Rose testimony, December 8, 1982, 8043–47, box 4; Jaimie Kahn testimony, October 4, 1982, 1102–3, box 8; DePietto testimony, October 29, 1982, 3843, box 12, all in *Witt.*

43. Ben Rose Ex Parte Affidavit, 3–4, box 7, *Witt.*

44. Asherson testimony, November 2, 1982, 4325, box 12; Dr. Cary English testimony, December 7, 1982, 7844, box 4, both in *Witt.*

45. Richard Bedenkop testimony, January 28, 1983, 13078, 13136, box 5, *Witt.*

46. Kahn testimony, October 4, 1982, 1125–27, box 8, *Witt.*

47. Greenberg testimony, October 5, 1982, 1384–85, box 4, *Witt.*

48. Sarosi testimony, December 3, 1982, 7542–57, box 13, *Witt.*

49. Siegel testimony, September 23, 1982, 214–15, box 3; Asherson testimony, November 2, 1982, 4320, box 12; Rose testimony, December 8, 1982, 8627, box 4; Kahn testimony, October 4, 1982, 1112, box 8, all in *Witt.*

50. DePietto testimony, October 29, 1982, 3943–44, box 12, *Witt.*

51. Herzberg, *Happy Pills;* Tone, *Age of Anxiety;* Rasmussen, *On Speed.*

52. Kahn testimony, October 4, 1982, 1234–50, box 8, *Witt.*

53. Greenberg testimony, October 5, 1982, 1500, 1549–1603, box 8, *Witt.*

54. Gregory Drezga testimony, January 14, 1983, 11288–89, box 4, *Witt.*

55. Defense opening arguments, September 22, 1982, 90, box 3, *Witt.*

56. Ibid., 130.

57. Prosecutors' summation, February 14, 1983, 15080, box 10, *Witt.*

58. Martin Siegel, Grand Jury testimony, 6, box 3; Siegel testimony, September 30, 1982, 761–62, box 3, both in *Witt.*

59. Siegel testimony, September 30, 1982, 316–17, box 3, *Witt.*

60. Ibid.

61. Glass testimony, October 12, 1982, 2220–43, box 7, *Witt.* Interestingly, Glass and Siegel were both DOs (Doctors of Osteopathy) rather than MDs. Osteopathy began in the nineteenth century as one of many medical "sects" opposed to the pharmaceuticals of the era, but as osteopaths fought to survive the professional winnowing of early-twentieth-century medicine, they increasingly abandoned that opposition in return for inclusion in new medical licensing laws—and, ironically, the legal right to prescribe medicines. By the 1950s most DOs prescribed pharmaceuticals, but there is no evidence that Glass and Siegel's osteopathic backgrounds reflected any larger patterns linking osteopathy to illicit prescribing. See Norman Gevitz, *The DOs: Osteopathic Medicine in America* (1982; reprint, Baltimore: Johns Hopkins University Press, 2004), 101–15.

62. Glass testimony, October 12, 1982, 2216–18, box 7, *Witt.*

63. Ibid., 2234–36.

64. Ibid., 2262–71.

65. Ibid., 2274–2337.

66. Trial documents do not specify which Swiss companies. Asherson testimony, November 2, 1982, 4134–36, 6308, box 12, *Witt.*

67. Ibid., 4134–52, 4496.

68. Ibid., 4157.

69. Ibid., 4411; DePietto testimony, October 29, 1982, 3847, box 12, *Witt.*

70. Asherson testimony, November 2, 1982, 4411, box 12, *Witt.*

71. Ibid., 4519–24.

72. Glass testimony, October 13, 1982, 2456–57, box 7, *Witt.*

73. Asherson testimony, November 2, 1982, 4309–10, box 12, *Witt.*

74. Sarosi testimony, December 3, 1982, 7526, box 13, *Witt.*

75. Dr. Emily Cole testimony, December 3, 1982, 7490, box 13, *Witt.*

76. Drezga testimony, January 14, 1983, 11294, box 4, *Witt.*

77. English testimony, December 7, 1982, 7838–58, box 4, *Witt.*

78. Information about Manhattan Center, n.d., box 3, *Witt.*

79. Defense opening arguments, September 22, 1982, 86, box 3, *Witt.*

80. Defense summation, February 14, 1983, 15324, box 10, *Witt.*

81. Ibid., 15323.

82. Defense opening arguments, September 22, 1982, 91, box 3, *Witt.*

83. Cole testimony, December 3, 1982, 7502–3, box 13, *Witt.*

84. Drezga testimony, January 14, 1983, 11353, 11403, quote from 11504, box 4, *Witt.*

85. Defense opening arguments, September 22, 1982, 87, box 3, *Witt.*

86. See State of New Jersey v. Michael C. Barry, M.D., "Judgment of Conviction

and Order for Commitment," Indictment S-301-81-02, March 17, 1981; State of New Jersey v. Alfred C. Gaymon, M.D., "Judgment of Conviction and Order for Commitment," Indictment S-301-81-06, March 17, 1981.

87. U.S. Senate, Subcommittee on Security and Terrorism of the Committee on the Judiciary, *Drug Enforcement Administration Oversight and Budget Authorization and S. 2320*, 97th Cong., 2nd sess., April 23, 1982 (Washington, DC: Government Printing Office, 1982), 88; Spillane, "Debating the Controlled Substances Act." Numbers tabulated from the *Federal Register*, where DEA decisions were published.

88. Cooper, *Impact of Prescription Drug Diversion Control Systems*, 1–18.

89. Jeremy A. Greene, "Pharmaceutical Marketing Research and the Prescribing Physician," *Annals of Internal Medicine* 146, no. 10 (2007): 742–48.

CHAPTER TEN: The Afterlife of the Prescription

The author wishes to acknowledge the assistance of Andrea Bainbridge at the American Medical Association Archives, Greg Higby and Elaine Stroud at the American Institute for the History of Pharmacy, and the History of Medicine Division of the National Library of Medicine for access to archival materials. Jerry Avorn, Arthur Daemmrich, Scott Podolsky, Nicolas Rasmussen, and Elizabeth Watkins commented on earlier drafts of this chapter.

1. O. L. Wade, "The Computer and Drug Prescribing," in *Computers in the Service of Medicine*, ed. G. McLachlan and R. A. Shegog (Oxford: Oxford University Press, 1969), 160–61.

2. This chapter employs new documentary materials to extend a series of themes developed elsewhere by the author in Jeremy A. Greene, "Pharmaceutical Marketing Research and the Prescribing Physician," *Annals of Internal Medicine* 146, no. 10 (2007): 742–48; Scott H. Podolsky and Jeremy A. Greene, "A Historical Perspective of Pharmaceutical Promotion and Physician Education," *Journal of the American Medical Association* 300 (2008): 831–33; Jeremy A. Greene and Scott H. Podolsky, "Keeping Modern in Medicine: Pharmaceutical Promotion and Physician Education in Postwar America," *Bulletin of the History of Medicine* 83 (2009): 331–77; Arthur Daemmrich and Jeremy A. Greene, "From Visible Harm to Relative Risk: Overcoming Fragmented Pharmacovigilance," in *The Fragmentation of U.S. Health Care: Causes and Solutions*, ed. Einer Elhauge (Oxford: Oxford University Press, 2010).

3. On surveillance, statistics, and statecraft, see Amy Fairchild, Ronald Bayer, and James Colgrove, *Searching Eyes: Privacy, the State, and Disease Surveillance in America* (Berkeley: University of California Press, 2007); Nikolas Rose, *The Politics of Life Itself: Biomedicine, Power, and Subjectivity in the Twenty-First Century* (Princeton, NJ: Princeton University Press, 2006); Theodore Porter, *Trust in Numbers: The Pursuit of Objectivity in Scientific and Public Life* (Princeton, NJ: Princeton University Press, 1996).

4. E.g., Gerald Oppenheimer, "Profiling Risk: The Emergence of Coronary Heart Disease Epidemiology in the U.S., 1947–1970," *International Journal of Epidemiology* 35

(2006): 515–19; George Weisz, "The US National Survey on Chronic Disease," paper presented at the conference "Drugs, Standards, and Chronic Diseases," Manchester University, UK, November 28, 2010.

5. Paul de Haen, *Development Schedule of New Drug Products* (New York: Romaine Pierson, 1949), 28.

6. Rebecca Lemov, "Towards a Data Base of Dreams: Assembling an Archive of Elusive Materials, c. 1947–61," *History Workshop Journal* 67, no. 1 (2009): 44–68; Sarah Igo, *The Averaged American: Surveys, Citizens, and the Making of a Mass Public* (Cambridge, MA: Harvard University Press, 2007); Peter Simonson, "Politics, Social Networks, and the History of Mass Communications Research: Re-Reading Personal Influence," *Annals of the American Academy of Political and Social Science* 608 (November 2006), special issue, 233–50.

7. E.g., Nancy Tomes, "Merchants of Health: Medicine and Consumer Culture in the United States, 1900–1940," *Journal of American History* 88, no. 2 (2001): 519–47; Rima D. Apple, *Vitamania: Vitamins in American Culture* (New Brunswick, NJ: Rutgers University Press, 1996).

8. Harry M. Marks, "Revisiting 'The Origins of Compulsory Drug Prescriptions,'" *American Journal of Public Health* (1995): 109–15.

9. Jeremy A. Greene, "Attention to Details: Etiquette and the Pharmaceutical Salesman in Postwar America," *Social Studies of Science* 34, no. 2 (2006): 271–92.

10. W. D. McAdams, "Three Major Marketing Problems on the Desks of Pharmaceutical Management Today," *Proceedings of the American Pharmaceutical Manufacturer's Association Midyear Meeting* (December 17, 1947): 272–80.

11. U.S. Bureau of the Census, U.S. Census of Manufacturers, 1954, Bulletin MC-28C, Drugs and Medicines, cited in *Facts about Pharmacy and Pharmaceuticals* (New York: Health News Institute, 1958), 51.

12. Richard J. Hull, "Marketing Concepts," in *Workings and Philosophies of the Pharmaceutical Industry*, ed. Karl Raiser (New York: National Pharmaceutical Council, 1959), 58–59.

13. Greene, "Attention to Details."

14. R. L. McQuillan, *Is the Doctor In? The Story of a Drug Detail Man's Fifty Years of Public Relations with Doctors and Druggists* (New York: Exposition Press, 1963).

15. De Haen, *Development Schedule of New Drug Products*, 56; Arthur F. Peterson, *Pharmaceutical Selling, 'Detailing,' and Sales Training* (New York: McGraw Hill, 1949).

16. Hull, "Marketing Concepts," 59.

17. Igo, *Averaged American.*

18. Raymond Gosselin, "Massachusetts Prescription Survey," master's thesis, Massachusetts College of Pharmacy, 1950.

19. Hull, "Marketing Concepts," 59.

20. Ibid.

21. George Morris Piersol, medical director of the National Disease and Therapeutic Index, n.d., Harry F. Dowling Archives (HFD), "Physicians—Drug Lists," box 11, MSC 372, National Library of Medicine (NLM), Bethesda, MD.

22. Lea Associates Inc., n.d., HFD, "Physicians—Drug Lists," box 11, MSC 372, NLM.

23. Within a few years, Lea Associates was producing regular reports as well as drug- and disease-specific analyses of its data for thirty-seven major pharmaceutical firms, including Merck and Upjohn. *Lea Associates, Inc.*, pamphlet, n.d., HFD, "Physicians—Drug Lists," box 11, MSC 372, NLM.

24. Specifically, these included the NDTI, the NPA, the National Journal Audit, the National Mail Audit, the National Detailing Audit, the National Detailing Aid Survey, the National Hospital Audit, the Hospital Record Study, the Semi-Annual Audit of Hospital Laboratory Tests, the Dental Study, the Audit of Pathology Cultures, the Podiatry Study, the Small Animal Veterinary Study, the Physician Omnibus, the Specialty Omnibus, the Pharmacy Omnibus, and the Syndicated Journal Ad Test. A year's subscription to the NDTI, the NPA, and the four major source audits in 1968 would cost fifty thousand dollars per client. Medical Data Services Inc., *Medical Data Services, Inc.*, 1968, Kremers Reference Files (KRF), C18(d), American Institute of the History of Pharmacy (AIHP), University of Wisconsin, Madison.

25. J. White, "Medical Data Processing in Pharmaceutical Research, Sales, and Advertising," in *A Working Partnership: Pharmaceutical Marketing Research and Medical Data Processing* (Clifton, NJ: Fisher-Stevens, 1962), KRF, C18(d), AIHP.

26. Hull, "Marketing Concepts," 62.

27. "Miss Fourteen O. One Is Quite a Girl: Computer Dabbles in Song and Dance," *Upjohn Intercom,* December 1963, 5. Upjohn Collection, Kalamazoo Public Library, Kalamazoo, MI.

28. Greene, "Pharmaceutical Marketing Research."

29. "The AMA Keeps Records on All Physicians," promotional flier, BD 991, American Medical Association Archives (AMAA), Chicago.

30. R. Steinbrook, "For Sale: Physicians' Prescribing Data," *New England Journal of Medicine* 354, no. 26 (2006): 2745–47.

31. "As more and more physicians become specialists it is increasingly important that we pinpoint our promotion through special mailing lists, by advertising in specialty journals, and by having our professional service representatives use discrimination in the selection of products to be detailed." Hull, "Marketing Concepts," 57.

32. Ibid., 62.

33. Theodore Caplow, "Market Attitudes: A Research Report from the Medical Field," *Harvard Business Review* 30 (1952): 105–12. Caplow's work was preceded by other studies in trade journals that indirectly or partially addressed the issue of physicians' source of information about new drugs, e.g., Charles C. Rabe, "The Doctor Measures the Detailman," *Medical Marketing* 11 (1952): 19–25.

34. Greene and Podolsky, "Keeping Modern in Medicine."

35. Robert Ferber and Hugh Wales, "The Effectiveness of Pharmaceutical Advertising: A Case Study," *Journal of Marketing* 22 (1958): 398–407; see also Ferber and Wales, *The Effectiveness of Pharmaceutical Promotion* (Urbana: University of Illinois, 1958); and Harvard Business School, "Wolff Drug Company [Case Study, 1959]," HFD, "Physicians—Information They Have," box 11, NLM.

36. Institute for Motivational Research, *Research Study on Pharmaceutical Advertising* (Croton-on-Hudson: Pharmaceutical Advertising Club, 1955), 7, 9.

37. Ibid., 39.

38. E. Katz and G. Menzel, "Social Relations and Innovation in the Medical Profession: The Epidemiology of a New Drug," *Public Opinion Quarterly* (Winter 1955): 337–72. See also J. Coleman, Menzel, and Katz, "Social Processes in Physicians' Adoption of a New Drug," *Journal of Chronic Disease* 9, no. 1 (1959): 1–19; and Coleman, Katz, and Menzel, *Medical Innovation: A Diffusion Study* (New York: Bobbs-Merrill, 1966).

39. Subsequent observational researchers argued for the importance of detail men in providing a context-rich individuated approach to the study of the prescription. "The detail man's unique interaction with both mass and interpersonal communication channels suggests that factors influencing medical opinion formation cannot be effectively evaluated by statistical sampling of respondents treated as isolated individuals. The considerable hiatus between the American medical stereotype and the individual physician's human cognitive limitations, as well as the indications cited above, indicate the need for gaining additional medical in-group identification in future research." Robert R. Rehder, "Communication and Opinion Formation in a Medical Community: The Significance of the Detail Man," *Journal of the Academy of Management* 8 (1965): 282–91.

40. Norman G. Hawkins, "The Detailman and Preference Behavior," *Southwestern Social Science Quarterly* 40 (1959): 213–24; Christophe Van den Bulte and Gary L. Lilien, "Medical Innovation Revisited: Social Contagion versus Marketing Effort," *American Journal of Sociology* 106 (2001): 1409–35.

41. An earlier AMA-Gaffin study had involved focus groups with key figures in pharmaceutical marketing regarding the relationship between medical education and pharmaceutical promotion. *Drug Industry Antitrust Act: Hearings pursuant to S. Res. 52 on S. 1552, before the Subcommittee on Antitrust and Monopoly of the Committee on the Judiciary*, 87th Cong., 1st sess., part 1, 1961, 129. The AMA distributed subanalyses of the Gaffin study to all prominent firms in the pharmaceutical industry through a series of twenty color mailings that detailed the influence of pharmaceutical marketing on the practicing physician. Reports 1–20, 33-14, BD 991, AMAA.

42. See "The Fond du Lac Study," in *Drug Industry Antitrust Act*, part 1, 698–806.

43. Ibid., 700–701.

44. Statement of Dr. Hugh Hussey, chairman of the American Medical Association, *Drug Industry Antitrust Act*, 126, 1961.

45. Morton Mintz, *The Therapeutic Nightmare* (Boston: Houghton Mifflin, 1965).

46. Daemmrich and Greene, "From Visible Harm to Relative Risk."

47. W. Lenz, "Thalidomide and Congenital Abnormalities," *Lancet* 279, no. 7219 (June 6, 1962): 45. For more on thalidomide and the origins of adverse event reporting, see Daemmrich and Greene, "From Visible Harm to Relative Risk"; and Trent Stephens and Rock Brynner, *Dark Remedy: The Impact of Thalidomide and Its Revival as a Vital Medicine* (New York: Basic Books, 2001).

48. W. Kunz, H. Keller, and H. Mückter, "N-Phthalyl-glutaminsäure-imid: Experimentelle Untersuchungen an einem neuen synthetischen Produkt mit sedativem Eigenschaften," *Arzneimittelforschung* 6 (1956): 426–30; H. Jung, "Klinische Erfahrungen mit einem neuen Sedativum," *Arzneimittelforschung* 6 (1956): 430–34.

49. Helen Taussig, "A Study of the German Outbreak of Phocomelia," *Journal of the American Medical Association* 180 (June 30, 1962): 1106–14. On the differential regulatory reception of the thalidomide crisis in Germany and the United States, see Arthur Daemmrich, "A Tale of Two Experts: Thalidomide and Political Engagement in the United States and West Germany," *Social History of Medicine* 15 (2002): 137–58.

50. The initiation of federal drug regulation with the Biologicals Act of 1902 (predating the formation of the FDA by four years) came in response to public outcry over the deaths of children receiving tainted diphtheria immunizations. Similarly, the passage of the 1938 Food, Drug, and Cosmetic Act, which dramatically expanded the powers of the FDA, was achieved largely in response to the Elixir Sulfanilamide disaster, in which more than one hundred people, many of them children, died after consuming a sulfa drug prepared with ethylene glycol. Daniel P. Carpenter, *Reputation and Power: Organizational Image and Pharmaceutical Regulation at the FDA* (Princeton, NJ: Princeton University Press, 2010); Charles O. Jackson, *Food and Drug Legislation in the New Deal* (Princeton, NJ: Princeton University Press, 1970); Harry M. Marks, *The Progress of Experiment: Science and Therapeutic Reform in the United States, 1900–1990* (Cambridge: Cambridge University Press, 2000).

51. See Daemmrich and Greene, "From Visible Harm to Relative Risk"; Thomas Maeder, *Adverse Reactions* (New York: William Morrow, 1994); American Medical Association Council on Drugs, "Blood Dyscrasias Associated with Chloramphenicol (Chloromycetin) Therapy," *Journal of the American Medical Association* 172 (April 30, 1960): 2044–45; Charles M. Huguley Jr., Allan J. Erslev, and Daniel E. Bergsagel, "Drug-Related Blood Dyscrasias," *Journal of the American Medical Association* 177 (July 8, 1961): 23–26.

52. American Medical Association–Boston Collaborative Drug Surveillance Program Pilot Drug Surveillance Study, "Progress Note Number 1," John Adriani Papers (JAP), box 28, folder 2, NLM, 1.

53. Kefauver Harris Amendments, Public Law 87-781 (October 10, 1962), 76 Stat. 783.

54. See "FDA Seminar on Adverse Reactions," September 6–9, 1966, JAP, box 27, NLM; and *Federal Register* 32 (June 6, 1967): 8080–89. On the subsequent evolution of the FDA Drug Experience Form, see *Federal Register* 33 (May 11, 1968): 7077–78; Stanley A. Edlavitch, "Adverse Drug Event Reporting: Improving the Low US Reporting Rates," *Archives of Internal Medicine* 148, no. 7 (1988): 1499–1503; Joseph E. Johnson, "Leighton E. Cluff (1923–2004)," *Transactions of the American Clinical Climatology Association* 116 (2005): xlv–l.

55. As Cluff noted in 1964, "Surveillance and epidemiological investigation of untoward drug reactions should include methods for calculating drug usage and defining the population at risk in addition to methods for detecting adverse reactions.

Without an accurate denominator there is no means of satisfactorily determining the incidence of adverse reactions to any medication." Leighton E. Cluff, George F. Thornton, and Larry G. Seidl, "Studies on the Epidemiology of Adverse Drug Reactions. I. Methods of Surveillance," *Journal of the American Medical Association* 188 (1964): 977.

56. Ibid., 978.

57. Ibid., 982.

58. D. J. Finney, "The Design and Logic of a Monitor of Drug Use," *Journal of Chronic Disease* 18 (1965): 77–98.

59. On mechanization: "The development of computers with large memories and quick access to stored information is making possible systems of medical record linkage, whereby complete medical records for a segment of a population can be maintained centrally." Ibid., 82.

60. Ibid., 88. "Instead of relying solely on individual medical practitioners, the computer will analyze the situation further . . . The number of possible complexities of adverse drug reactions far exceeds the availability of men and resources for intensive study of each. A well programmed computer is admirably suited to the task of screening a great number of records in respect of a wide range of possibilities, and so of conserving medical skill and resources for medical tasks" (93–94).

61. Dennis Slone, Hershel Jick, and Ivan Borda, "Drug Surveillance Utilizing Nurse Monitors: An Epidemiological Approach," *Lancet* 2, no. 7469 (October 22, 1966): 901–3.

62. Dennis Slone, Leonard F. Gaetano, Leslie Lipworth, Samuel Shapiro, George Parker Lewis, and Hershel Jick, "Computer Analysis of Epidemiologic Data on Effect of Drugs on Hospital Patients," *Public Health Reports* 84, no. 1 (January 1969): 40. The quotation continues, "The onus for recording the data is removed from the physician himself and is made the direct responsibility of another member of the ward team."

63. Ibid., 41.

64. "A basic vital-statistics chart was completed for each new patient admitted to the ward, and as each drug was prescribed a drug-starting sheet . . . was filled in. Whenever possible the data were collected as the doctor wrote the prescription; otherwise the nurse interviewed the doctor later the same day (or within 48 hours on weekends). A complementary sheet . . . was filled in when the drug was discontinued. If a drug was stopped because of a suspected adverse reaction, a special notification sheet was sent by the nurse monitor to the division of clinical pharmacology." Slone, Jick, and Borda, "Drug Surveillance," 901–2.

65. Hershel Jick, Olli S. Miettinen, Samuel Shapiro, George P. Lewis, Victor Siskind, and Dennis Slone, "Comprehensive Drug Surveillance," *Journal of the American Medical Association* 213, no. 9 (1970): 1455–60.

66. Ivan T. Borda, Dennis Slone, and Hershel Jick, "Assessment of Adverse Reactions within a Drug Surveillance Program," *Journal of the American Medical Association* 205, no. 9 (1968): 645–47.

67. Ivan T. Borda, Hershel Jick, Dennis Slone, Barbara Ginan, Barry Gilman, and

Thomas C. Chalmers, "Studies of Drug Usage in Five Boston Hospitals," *Journal of the American Medical Association* 202, no. 6 (1967): 170–74.

68. Slone et al., "Computer Analysis of Epidemiologic Data."

69. Samuel Shapiro, Dennis Slone, George P. Lewis, and Hershel Jick, "The Epidemiology of Digoxin: A Study in Three Boston Hospitals," *Journal of Chronic Diseases* 22, no. 5 (1969): 361–71; Hershel Jick, Dennis Slone, Ivan T. Borda, and Samuel Shapiro, "Efficacy and Toxicity of Heparin in Relation to Age and Sex," *New England Journal of Medicine* 279, no. 6 (August 8, 1968): 284. Subsequent papers by the group demonstrated a much lower incidence of drug reactions across a larger subset of drugs and patients. Hershel Jick, "Drugs—Remarkably Nontoxic," *New England Journal of Medicine* 291, no. 16 (October 17, 1974): 824–28.

70. A. L. Herbst, H. Ulfelder, and D. C. Poskanzer, "Adenocarcinoma of the Vagina: Association of Maternal Stilbestrol Therapy with Tumor Appearance in Young Women," *New England Journal of Medicine* 284, no. 15 (1971): 878–81.

71. Hershel Jick, Alexander M. Walker, and Claude Spriet-Pourra, "Postmarketing Follow-up," *Journal of the American Medical Association* 242, no. 21 (1979): 2310–14.

72. Edmund D. Pellegrino, "Meddlesome Medicine and Rational Therapeutics," *Drug Intelligence and Clinical Pharmacy* 9, no. 9 (1975): 482.

73. Ibid., 483–84.

74. "From the very beginning of the Baltimore City Medical Care Program, the costs of drugs supplied to patients by prescriptions have constituted a significant segment of the total budget of the program . . . Since there were few restrictions on the participating physicians with regard to prescription writing, examination of the drug problem was considered necessary for proper administration." Frank F. Furstenberg, Matthew Taback, Harry Goldberg, and J. Wilfrid Davis, "Prescribing, an Index to the Quality of Medical Care: A Study of the Baltimore City Medical Care Program," *American Journal of Public Health* 43 (1953): 1299.

75. Ibid., 1306–8. As a further aside, another comparison looked at use of drugs as specified by the "cost status of equivalent official preparation."

76. Charlotte Muller, "Medical Review of Prescribing," *Journal of Chronic Diseases* 18 (1965): 689–96. See also Muller, "Outpatient Drug Prescribing Related to Clinic Utilization in Four New York City Hospitals," *Health Services Research* 3, no. 2 (Summer 1968): 142–54.

77. Charles F. Federspiel, Wayne A. Ray, and William Schaffner, "Medicaid Records as a Valid Data Source: The Tennessee Experience," *Medical Care* 14, no. 2 (1976): 166–72.

78. The few infections included typhoid fever, *Haemophilus influenzae* meningitis, pneumococcal and meningococcal meningitis in patients unable to take penicillin, Gram-negative bacteremias, and certain rickettsial infections. See Milton Silverman, *The Drugging of the Americas* (Berkeley: University of California Press, 1976). Also see William Kitto, "Chloramphenicol Today—Reply," *Journal of the American Medical Association* 234, no. 10 (1975): 1016; Milton Silverman and Philip R. Lee, *Pills, Profits, and Politics* (Berkeley: University of California Press, 1974), 283–88.

79. T. W. Meade, "Prescribing of Chloramphenicol in General Practice," *British Medical Journal* 1 (1967): 671–73; O. L. Wade, "Prescribing of Chloramphenicol and Aplastic Anemia," *Journal of the College of General Practitioners* 12 (1966): 277–86; P. D. Stolley and L. Lasagna, "Prescribing Patterns of Physicians," *Journal of Chronic Diseases* 22 (1969): 395–405. Stolley, M. H. Becker, and J. D. McEvilla, "Drug Prescribing and Use in an American Community," *Annals of Internal Medicine* 76 (1972): 537–40. Although several studies had documented the frequent prescribing of chloramphenicol in the United States and the United Kingdom, none had been able to comprehensively evaluate appropriate and inappropriate prescriptions on such a large scale as Wayne Ray's 1976 Medicaid study.

80. Wayne Ray, "Prescribing of Chloramphenicol in Ambulatory Practice—An Epidemiological Study among Tennessee Medicaid Recipients," *Annals of Internal Medicine* 84, no. 3 (1976): 266, 270. Out of nearly 4.2 million people living in Tennessee at that time, slightly over 300,000 were eligible for Medicaid benefits, and claims were paid for just over 100,000 individuals to 3,409 Tennessee-based MDs participating in Medicaid. Out of this population, 1,761 prescriptions for chloramphenicol were written for 992 outpatient Medicaid patients, from a total of 205 physicians. Full linkage of diagnosis, provider, and patient was possible for 593 prescriptions for chloramphenicol.

81. Joint Commission on Prescription Drug Use (JCPDU), *Final Report of the Joint Commission on Prescription Drug Use* (Washington, DC: Government Printing Office, 1980), 20: "Data bases (e.g., those created by IMS America, PAID, Medicaid, etc.) can be used to provide information on drug use. The FDA currently purchases some of these data and is experimenting with new ways to interpret them. Such data may provide information about drug use that is necessary to establish risk/benefit assessments and that is needed by researchers and policy makers to explore how our society uses medicines."

82. "A consensus exists that . . . drugs are not optimally used . . . We simply don't know how different kinds of doctors use different categories of drugs; we don't know the true incidence of adverse effects nor do we appreciate the very real benefits of appropriate drug usage." Cited in Hershel Jick, Alexander M. Walker, and Claude Spriet-Pourra, "Postmarketing Follow-up," *Journal of the American Medical Association* 242, no. 21 (1979): 2311.

83. JCPDU, *Final Report*, i. See also "Minutes of Meeting, November 30, 1976," ibid.

84. JCPDU, *Final Report*, 73. Early on, the commission had "achieved consensus that a 'National Center for Drug Experience' should be an independent epidemiologic unit to deal with problems of prescription drug use." This was later replaced by the less volatile Program on Medical Surveillance. Minutes of Dr. Robbins's Subcommittee, Joint Commission on Prescription Drug Use, Atlanta, Georgia, December 1, 1977, in JCPDU, *Final Report*, 6.

85. Drug utilization review typically refers to retrospective studies summarizing who prescribes a drug, which drug is prescribed in a given circumstance, and the time

course of drug prescription. Drug use evaluation denotes more prospective study of the same facts that also focuses on the logic behind the prescription and may also measure associated clinical outcomes. See, e.g., Darrel C. Bjornson, Joaquima Serradell, and Abraham G. Hartzema, "Drug Utilization—Measurement, Classification, and Methods," in *Pharmacoepidemiology: An Introduction*, 3rd ed., ed. Abraham G. Hartzema, Miquel Porta, and Hugh Hanna Tilson (Cincinnati: Harvey Whitney Books, 1998), 131–60, 132.

86. World Health Organization, *Consumption of Drugs: Report on a Symposium* (EURO 3102) (Copenhagen: Regional Office for Europe, 1970); Ulf Bergman, "The History of the Drug Utilization Research Group in Europe," *Pharmacoepidemiology and Drug Safety* 15, no. 2 (2006): 95–98. The early studies in the county of Jämtland, in Sweden, had utilized pharmaceutical marketing data obtained from a private company. In 1978, however, Sweden developed the nationalized Diagnosis and Therapy Survey, in which "copies of prescriptions, containing not only drugs, but also diagnosis and/or symptoms, are collected from a stratified, randomized sample of physicians. This survey connects the diagnosis/symptoms for which the drugs are prescribed and allows for the important diagnosis therapy link." Bjornson et al., 143. The reference cited here is A. Ekedahl et al., "Benzodiazepine Prescribing Patterns in a High-Prescribing Scandinavian Community," *European Journal of Clinical Pharmacology* 44 (1993): 141–46. See also Ulf Bergman and F. Sjoqvist, "Measurement of Drug Utilization in Sweden: Methodological and Clinical Implications," *Acta Medica Scandinavia*, suppl. 683 (1984): 15–22.

87. "Minutes of Meeting, November 30, 1976," 4, in JCPDU, *Final Report.*

88. JCPDU, *Final Report*, 65.

89. Ibid., 66.

90. "Minutes of Meeting, November 30, 1976," 4, ibid.

91. "The importance of this approach lies not so much in any particular HMO's organizational structure as in the underlying concept of a prepaid health plan that can provide most or all health care for its members. At the end of 1978, 203 HMOs provided basic medical services, including prescription drugs, to a total of approximately 7.5 million people. If the current five largest HMOs (each with roughly one million members) were included in [the surveillance program], approximately 75 cases per year of an illness with an incidence rate of 1/50,000 would be expected to live in this base. From these five, a cohort of 10,000 exposed patients could have been assembled in the first year of marketing of 23% of the new chemical entities introduced in 1973–77. If all currently operating HMOs were included (7.5 million members), the figures rise to 150 cases per year and 40% of the drugs . . . Patient identity would remain within the files of the HMO." JCPDU, *Final Report*, 61.

92. Hershel Jick, "The Commission on Professional and Hospital Activities—Professional Activity Study: A National Resource for the Study of Rare Illnesses," *American Journal of Epidemiology* 109, no. 6 (1979): 625–27; "Minutes of Third Meeting, Stanford Court Hotel, July 26–27, 1977," in JCPDU, *Final Report.*

93. JCPDU, *Final Report*, 67.

94. "[IMS] will attempt to survey existing informational systems of postmarketing surveillance for detecting both adverse and beneficial effects of prescription drugs and will, if necessary, attempt to identify and recommend new methods or technology in this area." Minutes of Dr. Azarnoff's Subcommittee "Strategy and Tasks of Surveillance," JCPDU, February 5, 1978, Sonesta Beach Hotel, Key Biscayne, Florida, in JCPDU, *Final Report*, 1.

95. Minutes of Meeting, JCPDU, January 27–28, 1977, Washington, DC, in JCPDU, *Final Report*. By 1979 another staffer in Kennedy's office suggested that the concept of a fully private marketing surveillance organization would be acceptable, as long as it produced relevant data. Minutes of Twelfth Meeting, JCPDU, November 12–13, 1979, Watergate Hotel, Washington, DC, in JCPDU, *Final Report*.

96. Ed F. Lindner, "Medical Advertising: Its Role, Ethics, and Concepts," *New York State Journal of Medicine* 64 (August 1, 1964): 1990.

Contributors

Julie A. Fairman, professor in the School of Nursing at the University of Pennsylvania, is the author of *Critical Care Nursing: A History* (1998) and *Making Room in the Clinic: Nurse Practitioners and the Evolution of Modern Health Care* (2006). She is the director of the Barbara Bates Center for the Study of the History of Nursing and is currently working on the politics of primary care.

Jeremy A. Greene is assistant professor in the Department of the History of Science, Harvard University, instructor in medicine in the Brigham and Women's Hospital Division of Pharmacoepidemiology and Pharmacoeconomics, and on the teaching faculty of the Department of Global Health and Social Medicine, Harvard Medical School. His first book, *Prescribing by Numbers: Drugs and the Definition of Disease* (2007), was awarded the 2009 Rachel Carson Prize from the Society for the Social Studies of Science. He is currently writing a book on the history of generic drugs and conducting research on the evolving role of pharmaceuticals in global public health.

David Herzberg, associate professor in the Department of History at the University of Buffalo (SUNY), is the author of *Happy Pills in America: From Miltown to Prozac* (2009). He is currently researching a book on the history of prescription drug abuse and addiction.

Judith A. Houck, associate professor in the departments of Medical History and Bioethics, Gender and Women's Studies, and History of Science at the University of Wisconsin, Madison, is the author of *Hot and Bothered: Women, Medicine, and Menopause in Modern America* (2006). She is currently working on a history of feminist health clinics in California.

Marcia L. Meldrum, associate researcher at the UCLA Center for Health Sciences and Society, has written extensively on the history of pain and pain management, opioid use, and randomized clinical trials; her work is published in peer-reviewed journals in both the historical and the medical literature. She is currently working on the history of public mental health services in Los Angeles County.

Scott H. Podolsky is assistant professor of global health and social medicine at Harvard Medical School and director of the Center for the History of Medicine at the Countway Library of Medicine. He is the author of *Pneumonia before Antibiotics: Therapeutic Evolution and Evaluation in Twentieth-Century America* (2006) and coauthor of *Generation of Diversity: Clonal Selection Theory and the Rise of Molecular Immunology* (1997) and *Oliver Wendell Holmes: Physician and Man of Letters* (2009). He is presently working on a monograph on the history of antibiotics.

Heather Munro Prescott is professor of history at Central Connecticut State University. She is the author of *A Doctor of Their Own: The History of Adolescent Medicine* (1998) and *Student Bodies: The Influence of Student Health Services on American Society* (2007). Her latest book is *The Morning After: A History of Emergency Contraception in the United States* (2011).

Nicolas Rasmussen, professor in the School of History and Philosophy at the University of New South Wales (Sydney, Australia), is author of *On Speed: The Many Lives of Amphetamine* (2008) and *Picture Control: The Electron Microscope and the Transformation of Biology in America, 1940–1960* (1997). He is currently working on the history of relationships between academic biomedical science and the drug industry, the development of the first-generation recombinant DNA pharmaceuticals, and popular and medical conceptions of obesity in early cold war America.

Dominique A. Tobbell, assistant professor in the Program in the History of Science, Technology, and Medicine at the University of Minnesota, is the author of *Pills, Power, and Policy: The Struggle for Drug Reform in Cold War America and Its Consequences* (2012). She is oral historian for the University of Minnesota's Academic Health Center History Project and is currently working on a political history of academic health centers in the United States.

Elizabeth Siegel Watkins, professor in the Department of Anthropology, History, and Social Medicine at the University of California, San Francisco, is the author of *On the Pill: A Social History of Oral Contraceptives, 1950–1970* (1998) and *The Estrogen Elixir: A History of Hormone Replacement Therapy in America* (2007) and coeditor of *Medicating Modern America: Prescription Drugs in History* (2007). She has recently published articles on the history of Norplant and the history of male menopause and testosterone replacement.

Index